P9-CQR-129

GETTING TO KNOW Web GIS

SECOND EDITION

Pinde Fu

Esri Press
REDLANDS | CALIFORNIA

Esri Press, 380 New York Street, Redlands, California 92373-8100
 First edition 2015; second edition 2016.

Printed in the United States of America
20 19 18 17 16 1 2 3 4 5 6 7 8 9 10

Ask for Esri Press titles at your local bookstore or order by calling 800-447-9778, or shop online at esri.com/esripress. Outside the United States, contact your local Esri distributor or shop online at eurospanbookstore.com/esri.

Esri Press titles are distributed to the trade by the following:

In North America:

Ingram Publisher Services
Toll-free telephone: 800-648-3104
Toll-free fax: 800-838-1149
E-mail: customerservice@ingrampublisherservices.com

In the United Kingdom, Europe, Middle East and Africa, Asia, and Australia:

Eurospan Group
3 Henrietta Street
London WC2E 8LU
United Kingdom
Telephone: 44(0) 1767 604972
Fax: 44(0) 1767 601640
E-mail: eurospan@turpin-distribution.com

Contents

Preface

By any measure, today is an exciting time to be engaged in the GIS field. Web GIS is emerging and extending the reach of GIS far beyond what anyone could have imagined.

ArcGIS has grown to become a web GIS platform that allows users to deliver authoritative maps, analytics, and geographic information to a wider audience, using lightweight clients and custom apps on web and smart devices. Virtually anyone can use web GIS to find, use, create, and share maps; perform GIS analysis; collaborate with others in communities; and easily deploy configurable apps to many users.

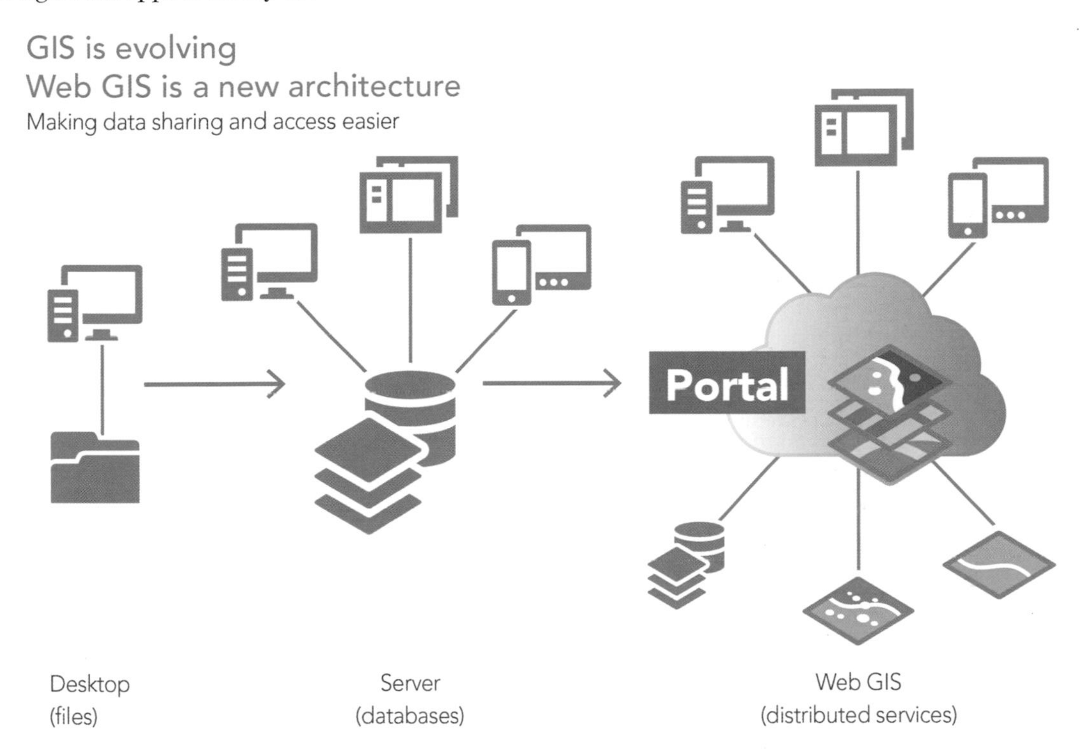
GIS is evolving
Web GIS is a new architecture
Making data sharing and access easier
Portal
Desktop
(files)
Server
(databases)
Web GIS
(distributed services)

A living atlas of basemaps, imagery, and geographic information is built into this platform. The Living Atlas of the World is available for anyone to use, along with ArcGIS Open Data, which includes thousands of datasets and web services that have been shared and registered in ArcGIS by users from around the world.

The new ArcGIS provides an online infrastructure for making maps, apps, and geographic information available throughout an organization, across a community, and openly on the web. This new vision for web GIS fully complements, integrates, and extends existing professional GIS workflows.

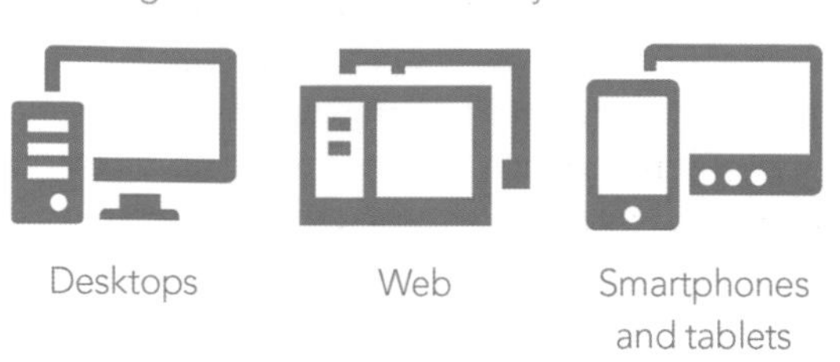
The app revolution
Is making web GIS available everywhere
Desktops
Web
Smartphones and tablets

Getting to Know Web GIS will help readers begin their journey to learn about, understand, and apply web GIS. This workbook teaches users how to share resources online and build web GIS apps easily and quickly.

The second edition explores the latest web GIS advances, such as smart mapping, Story Maps, ArcGIS Pro, 3D web scenes, App Studio for ArcGIS, Operations Dashboard, and ArcGIS API for JavaScript 4.0. The book is a practical manual for classroom lab work and on-the-job training for GIS students, instructors, GIS analysts, managers, web developers, and a broad range of GIS professionals. The second edition also covers the Esri suite of web GIS technologies, including ArcGIS Online, Portal for ArcGIS, ArcGIS for Server, configurable web apps, Web AppBuilder for ArcGIS, and Collector for ArcGIS mobile app.

Web GIS is a promising field with great applicability to e-government, e-business, e-science, and daily life. The societal need for both the web and GIS has generated a strong and increasing demand for good web GIS professionals. Professors need a lab book to teach web GIS, and on-the-job professionals need a guide to teach themselves. This book fits the need. In writing this book, I have tried to keep the following principles in mind:

- **Easy to apply:** You do not have to be a developer to build web apps. This book facilitates immediate productivity and teaches how to build engaging web apps without a single line of programming. Even the JavaScript programming chapter is relatively easy to follow, teaching readers how to adapt sample code.
- **Current:** Web GIS technologies advance rapidly. This book teaches state-of-the-art technical skills needed for building applications and managing projects.
- **Holistic:** Unlike books that focus on individual products, this book teaches web GIS technologies as a holistic platform, from the server side to the browser, mobile, and desktop client side.

Each of the 10 chapters includes the following sections:

- A conceptual discussion that gives readers the big picture and the underlying principles
- System requirements that help instructors set up the lab
- A detailed tutorial with screen captures that confirm progress along the way
- Common questions and answers
- Assignments that allow readers to practice what they have learned
- Online resources with links to videos

The sample data for this book is available on the Esri Press book resource page at **esripress .esri.com/bookresources**. The Esri website also provides a 60-day trial of ArcGIS for Desktop.

The book is the result of the author's extensive working experience at Esri and teaching experience at Harvard University Extension, University of Redlands, and Henan University. This course and the set of labs have been well received by these universities and their students. Professors can use this book as the lab book for their web GIS courses, and professionals at work can use this book for on-the-job training.

I welcome your feedback at **esripress@esri.com** and hope this book sparks your imagination and encourages creative uses of web GIS.

Acknowledgments

I would like to thank everyone who supported this book project at Esri. Special thanks go to Kathleen Morgan for inviting me to write this book; Claudia Naber for managing everything from acquisitions and scheduling to data credit lines; Mark Henry for his excellent editing that greatly improved the quality of the book; and Molly Zurn, Catherine Ortiz, Mike Livingston, Julia Nelson, Sandi Newman, Monica McGregor, Riley Peake, Brandy Perkins, and Derick Mendoza for their valuable contributions.

I am also grateful for the support of many other Esri colleagues and other contributors. I sincerely thank Mourad Larif, Brian Cross, and Bill Derrenbacher for giving me the flexibility to work on the book; Clint Brown for providing guidance on the book contents; Derek Law and Geri Miller for reviewing the first edition; and Professor Marci Meixler at Rutgers University for carefully testing and reviewing the second edition of this book. Thanks to Jeremy Bartley, Urban MacGillivray, Ismael Chivite, Moxie Zhang, Jianxia Song, Jay Chen, Tif Pun, Michael Gould, Li Lin, Wei Zhao, Allen Carroll, Nathan Shephard, Jean Gea, David Asbury, and Ty Fitzpatrick for sharing enlightened discussions on web GIS. This book would not be possible without the inspiration and support of Esri President Jack Dangermond.

This book was developed on the basis of my work experience at Esri and my teaching and lecturing experience at Harvard University Extension, Henan University, the University of Redlands, the University of California, California State University, and many other universities. I especially want to thank Professors Yaochen Qin and Yu Chen at Henan University and all my students for providing feedback that has improved the content and structure of this book.

Finally, and most importantly, I would like to thank my family for their love and support.

1 2 3 4 5 6 7 8 9 10

Chapter 1
Start with the cloud: Build web apps using ArcGIS Online

This chapter introduces the concept of web GIS with the Esri web GIS platform. It begins with an overview of web GIS, cloud GIS, and ArcGIS Online and then demonstrates a quick and easy way to build web GIS apps using the Story Map Tour configurable app. This chapter familiarizes you with ArcGIS Online basic operations and workflows and introduces flexible ways to build web GIS apps that you will learn in other chapters.

Learning objectives

- *Grasp the basics of the new generation web GIS platform.*
- *Understand the different approaches for building web GIS apps.*
- *Learn the ArcGIS Online workflow for creating web apps.*
- *Work with GIS data in comma-separated value (CSV) files.*
- *Create and share web maps and web apps.*
- *Use the Story Map Tour configurable app.*

What is web GIS?

Web GIS is the combination of the web and GIS. The web removes the constraint of distance in cyberspace and thus allows people the freedom to interact with GIS apps globally and access information almost instantly. Web GIS uses web technologies, including but not limited to Hypertext Transfer Protocol (HTTP), Hypertext Markup Language (HTML), uniform resource locator (URL), JavaScript, and WebSocket.

The first operational GIS was developed in the 1960s by Roger Tomlinson. Since then, GIS has continuously evolved from a local file-based single computer system to a central database-based client/server system, often with multiple servers and many more client computers. The invention of the Internet in the late 1960s and the World Wide Web in the early 1990s laid the foundation for an evolutionary leap toward web GIS. In 1993, the Xerox Corporation Palo Alto Research Center (PARC) developed a mapping web page, which marked the origin of web GIS. In the 2000s, web GIS evolved into a new generation—a system of distributed web services you can access in the cloud, as represented by the Esri ArcGIS platform.

Inheriting the power of the Internet and the web, web GIS offers many advantages:

- **Global reach:** You can share your geographic information easily within your organization and with people all over the world.
- **Large number of users:** You can share your app with dozens, or even millions, of users supported by the scalable cloud technology.
- **Low cost per user:** The cost of building one web GIS app is often less than the cost of building a standalone desktop solution and installing it for every user.
- **Better cross-platform capabilities:** Web apps, especially those built with JavaScript, can run on desktop and mobile browsers running a wide range of operating systems, from Windows, Mac OS, and Linux to iOS, Android, and Windows Phone.
- **Easy to use:** Web GIS apps typically incorporate simplicity, intuition, and convenience into their design. Therefore, public users can use these apps without having prior knowledge.
- **Easy to maintain:** Web clients can benefit from the latest program and data updates each time they access a web app. The web administrator does not have to update all the clients separately.

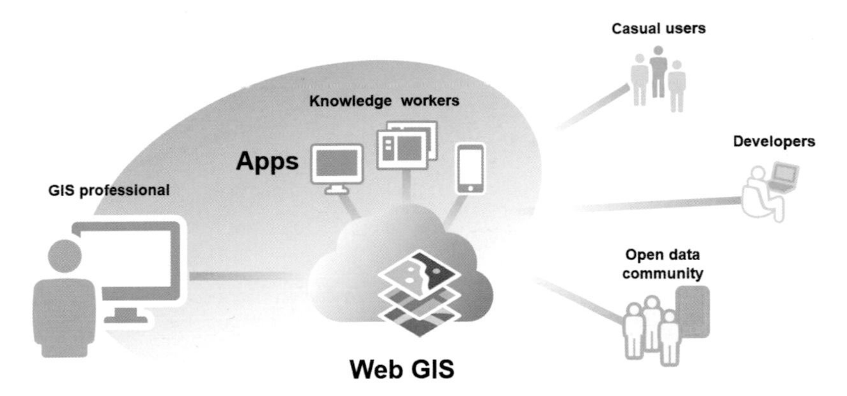

Web GIS presents a pattern for delivering GIS capabilities. Web GIS enables all members of an organization to easily access and use geographic information within a collaborative environment. GIS professionals working on the desktop create and share information to the web GIS and extend geospatial intelligence to broad users across organizations and throughout communities.

Web GIS unlocks and delivers the power of geospatial intelligence to offices and homes and puts technology in the hands of billions of people. Web GIS demonstrates immense value to government, business, science, and daily life. In recent years, the concept and importance of spatial location have become more mainstream, and web GIS awareness is increasingly prominent in many organizations.

- **For government,** web GIS offers an ideal channel for sharing public information services and delivering open data, an engaging medium for encouraging public participation, and a powerful framework for supporting decision making.
- **For business,** web GIS helps create novel business models and reshape existing ones. It enhances the power of location-based advertising, business analysis, and volunteered geographic information, generating tremendous revenue both directly and indirectly.
- **For science**, web GIS creates new research areas and renews existing avenues of research.
- **In daily life,** web GIS helps people decide where to eat, stay, and shop and how to get from here to there.

ArcGIS is a web GIS platform

Web GIS is central to the strategic direction of Esri in implementing GIS as a platform. ArcGIS represents a cutting-edge and complete web GIS platform that enables every employee and contractor to easily discover, use, make, and share maps from any device, anywhere, anytime.

ArcGIS is a new-generation web GIS platform that provides mapping, analysis, data management, and collaboration.

- At the center of this web GIS pattern is a portal, either ArcGIS Online or Portal for ArcGIS, which represents a gateway for accessing all spatial products in an organization. The portal helps organize, secure, and facilitate access to geographic information products.
- Client applications on desktops, web apps, tablets, and smartphones interact with the portal to search, discover, and access maps and other spatial content.
- In the back-office infrastructure, the portal is powered by two components: GIS servers and ready-to-use content.

Web GIS deployment models

The ArcGIS web GIS platform offers three deployment models:

- **ArcGIS Online**, the cloud-based offering, in which all components are hosted in the cloud. There is no hardware infrastructure for an organization to maintain because Esri manages and maintains ArcGIS Online.
- **Portal for ArcGIS**, the on-premises model, in which an organization manages the hardware and software infrastructure to operate the ArcGIS platform.
- **Hybrid deployment**, which combines parts of the cloud-based model with parts of the on-premises model. Hybrid deployment is by far the most common web GIS deployment pattern. Details of such a model depend on an organization's business workflows and security requirements.

Paths to building web GIS applications

Getting to Know Web GIS, second edition, teaches you how to build web GIS apps. The ArcGIS suite of web GIS products offers many paths to this goal.

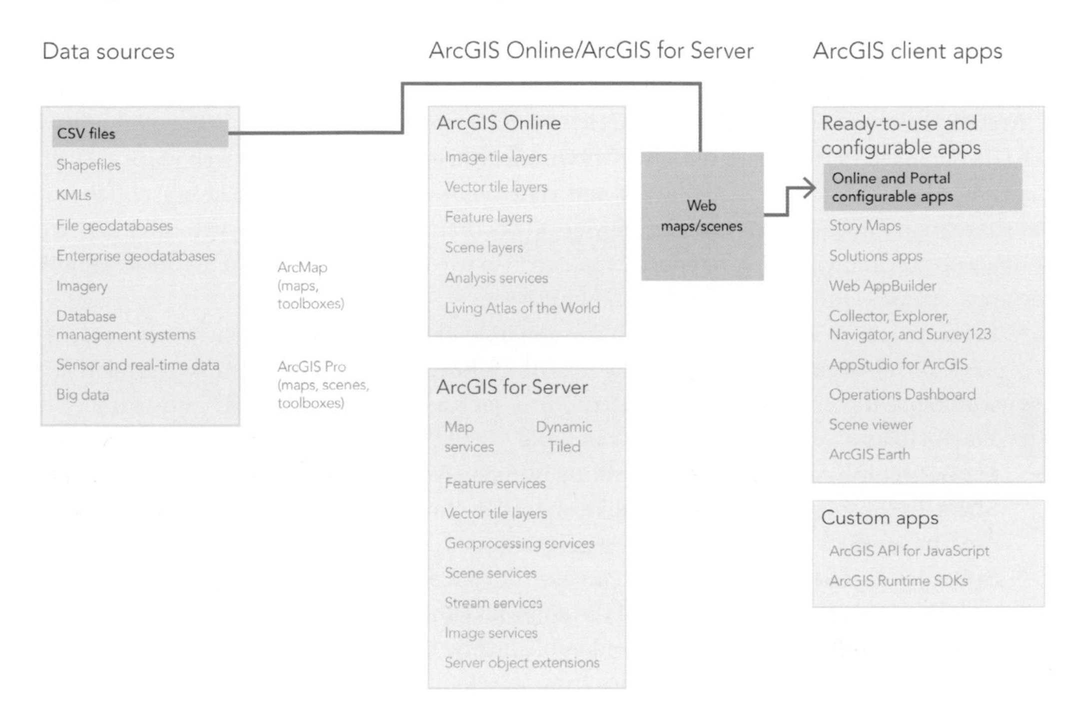

ArcGIS offers many ways to build web apps. The green line in the figure highlights the technology taught in this chapter.

- The data tier (on the left side of the figure) contains formats that range from simple CSV files managed with Microsoft Excel to sophisticated geodatabases managed with enterprise databases. These formats enable you to create a map, toolbox, and 3D scene in ArcGIS for Desktop software, including ArcGIS Pro.
- Using the tools in the middle tier of the figure, you can publish desktop resources to ArcGIS Online as various types of web services. You can then add those web services to ArcGIS Online to create web maps. Organizations that do not want to place their resources in the public cloud can use ArcGIS for Server and Portal for ArcGIS, a form of ArcGIS Online (with some differences) used in a private cloud.
- Options for the presentation (or client) tier on the right side of the figure range from ready-to-use apps that are configured without programming to custom apps that use various web application programming interfaces (APIs) or software development kits (SDKs) to meet special requirements.

Start with ArcGIS Online

Cloud computing is an important research area and technology trend. Cloud computing is based on the idea that many of the computing tasks that individual computers handle locally could operate more efficiently using multiple computer centers connected through the Internet. Cloud GIS uses cloud computing technology to enhance GIS capabilities that help users lower costs, reduce complexity, and quicken scalability.

ArcGIS Online (**www.arcgis.com**) is a cloud GIS. The technology is an online, collaborative web GIS deployed in the cloud. With ArcGIS Online, you can use and create web maps (2D) and scenes; access ready-to-use maps, layers, and analytics; publish data as web layers; collaborate and share maps; access maps from any device; make maps from your data; customize the ArcGIS Online website; and view status reports. You can also use ArcGIS Online as a platform to build web apps. ArcGIS Online provides the following services:

- **Infrastructure as a Service (IaaS):** You can upload your data and publish web services to ArcGIS Online and host them on the ArcGIS Online infrastructure. In this perspective, you would use the ArcGIS Online infrastructure, such as storage, CPU, and bandwidth.
- **Platform as a Service (PaaS):** You can build web GIS apps without programming by using configurable apps or with programming by using ArcGIS web APIs and ArcGIS Runtime SDKs. In this perspective, you would use ArcGIS Online as a development platform for creating apps.
- **Software as a Service (SaaS):** You can use the rich collection of basemaps, the analytical capabilities of thematic layers, and the countless and ever-increasing number of apps that are hosted in ArcGIS Online and published by Esri and its user communities. These capabilities are provided as a service from the cloud.

Adoption of ArcGIS Online and its quality of service

Before organizations add cloud GIS to their enterprise architecture, they first must assess the quality of services (QoS) of the cloud GIS. The following main factors represent QoS:

- **Performance:** How efficiently the system responds to user requests, usually measured in response time
- **Scalability:** The ability to support a growing number of users without dramatically reducing performance
- **Availability:** A measure of how often a system is accessible to end users, often measured in the percentage of time—for example, 99.99 percent
- **Security:** The ability to provide confidentiality and secure access by authenticating the parties involved, encrypting messages, and providing access control

ArcGIS Online provides reliable and trustworthy services in these four areas of QoS. Enabled by a large number of servers in the cloud and the use of high-performance computing technologies, ArcGIS Online hosts millions of map layers and serves hundred millions of visitors. It responds to tens of thousands of requests per second with fast performance, high scalability and availability. You can monitor ArcGIS Online availability in its health dashboard (**status.arcgis.com**). ArcGIS Online follows a robust and effective framework to enforce security and protect user privacy. ArcGIS Online is certified as compliant with many federal and international security and privacy standards (see more information at **http://doc.arcgis.com/en/trust/compliance/compliance-tab-intro.htm**). Partly because ArcGIS Online provides a high QoS level, the technology has been quickly adopted by numerous government and commercial organizations around the world, from local to national governments as well as oil and gas corporations, educational institutions, healthcare organizations, and law enforcement agencies.

ArcGIS Online and Portal for ArcGIS information-sharing model

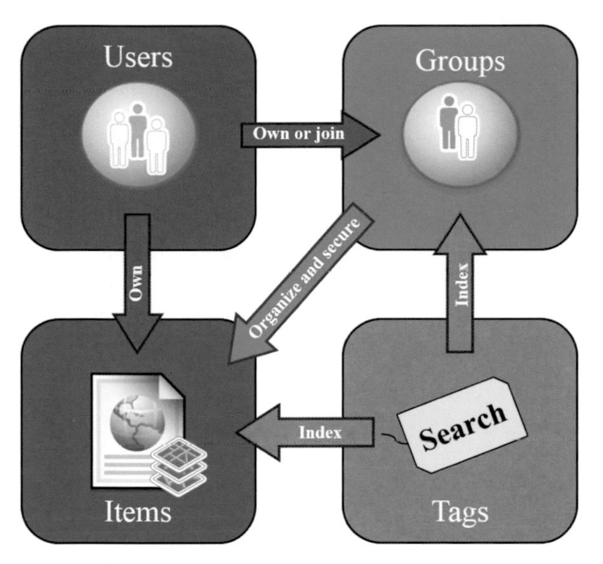

ArcGIS Online and Portal for ArcGIS information-sharing model.

The elements of the ArcGIS Online and Portal for ArcGIS information-sharing model include users, groups, content, and tags.

- Users can create and join groups.
- Users sign in to create and share content items, which can be a large variety of data, layers, and web maps and apps.
- Content items have tags, which are indexed so that users can search and discover items more efficiently.
- Users can keep information to themselves, share with certain groups (not with individual users), share to their organizations, or share with everyone—the public. Sharing allows other users to see and access the shared items. ArcGIS supports a variety of sharing levels.

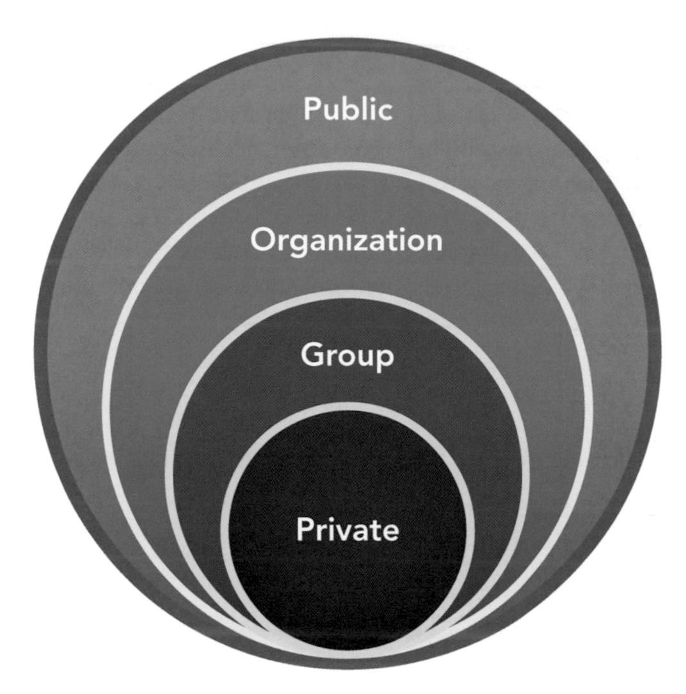

ArcGIS Online sharing levels. You can share your items with certain groups, your organization, or everyone.

Types of ArcGIS Online user accounts

Anonymous users can access the contents and apps shared with the public in ArcGIS Online as long as an organization has enabled anonymous access. However, you must have an account with ArcGIS Online to save your work and create web apps. Two main types of user accounts are provided:

- **Public accounts:** An ArcGIS Online public account is a personal account with limited usage and capabilities. You can create a public account at ArcGIS Online. A public user can add simple data; create web maps and web apps; and access public data, services, maps, and apps shared by others. However, users with public accounts cannot publish hosted services or access many ArcGIS analytical functions. Public accounts are for personal use only.

- **Organizational accounts:** To become a member of an ArcGIS Online organization, you or your organization's administrator must create an ArcGIS Online for Organizations account. An organizational user can assume one of the following roles:
  - **User:** In addition to the functions available to public users, organizational users can access the data, services, maps, and apps shared within the organization.
  - **Publisher:** In addition to user-level functions, publishers can publish hosted geospatial web services to ArcGIS Online and perform spatial analyses.
  - **Administrator:** In addition to publishing functions, administrators can configure their organization's ArcGIS Online website (such as its featured content gallery) and manage its users and groups.
  - **Custom:** The ArcGIS Online administrator can define a custom role with specialized permissions (for example, view-only).

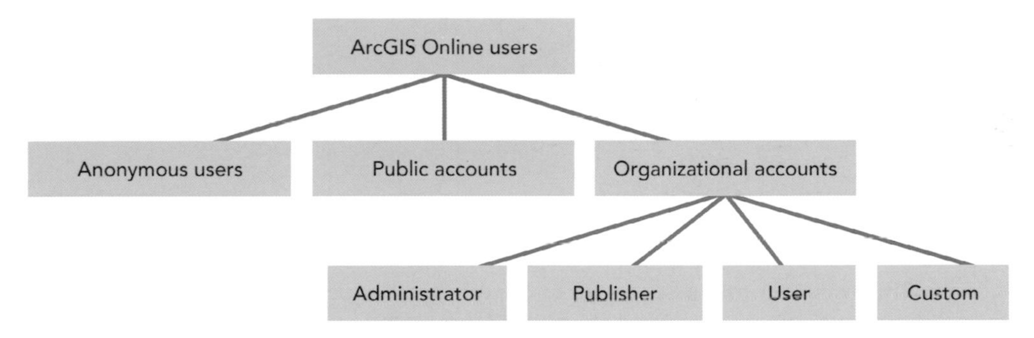

ArcGIS Online user types.

Main types of content items in ArcGIS Online and Portal for ArcGIS

Five main types of content in ArcGIS Online—data, layers, web maps, tools, and web apps—relate closely to this book's main goal: learning how to build web GIS apps.

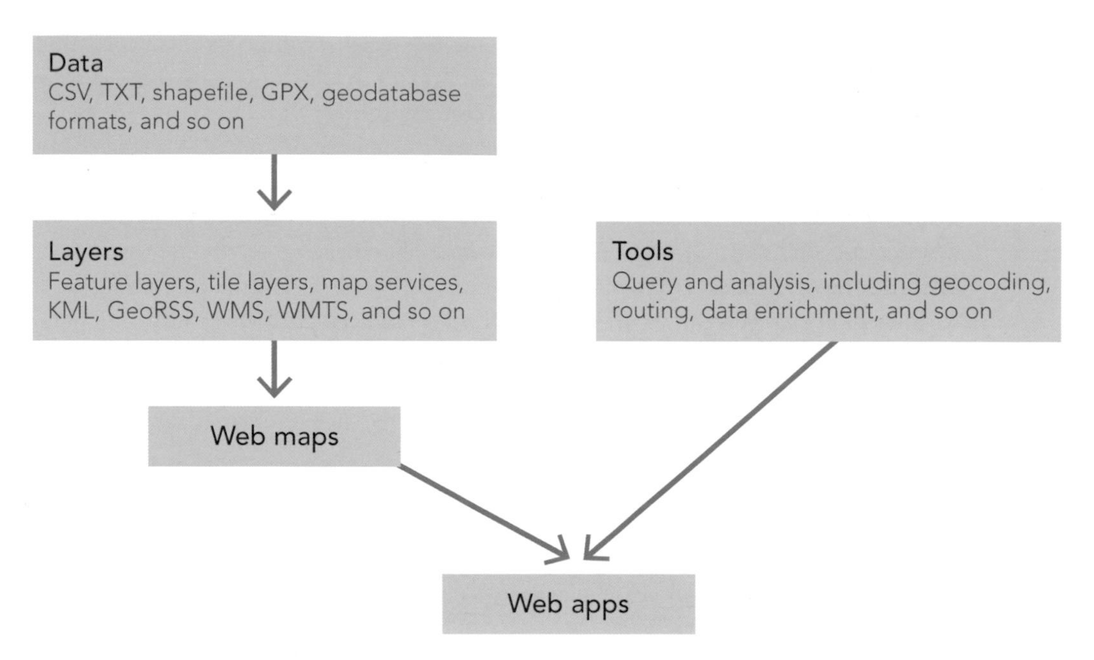

The main types of content items in ArcGIS Online and Portal for ArcGIS.

Typically, a web app comprises one or more web maps, which in turn include or reference one or more layers. A layer can take the form of a CSV file, a shapefile, or a web service.

- **Data:** ArcGIS Online supports data in a variety of formats, including CSV, TXT, shapefile, GPS Exchange Format (GPX), and geodatabase.
- **Layers:** ArcGIS Online can host layers, including the data, and can reference layers, including GeoRSS, map services, feature services, image services, Keyhole Markup Language (KML), and the Web Map Service (WMS) standard defined by the Open Geospatial Consortium (OGC).
- **Web maps and scenes:** These maps interactively display geographic information that you can use to answer questions. A web map or scene (the 3D counterpart to a web map) comprises or references multiple layers.
- **Tools:** App tools perform analytical functions, such as geocoding, routing, generating PDFs, summarizing data, finding hotspots, and analyzing proximity.
- **Web apps:** These apps are created for a targeted audience and purpose. Developers can program with ArcGIS web APIs to build web apps. However, you do not have to be a developer to create a web app. ArcGIS Online provides many configurable apps that you can use to create impressive web apps without any programming. In ArcGIS Online, a web app sometimes comprises a single web map and sometimes multiple web maps, such as web apps created based on comparison analysis (side by side) and swipe configurable app templates.

Steps to creating web GIS apps

Here is the typical workflow used to create web apps using ArcGIS Online:

1. Define the objectives of your information product.
2. Search for data layers in ArcGIS Online, and/or add your own data to ArcGIS Online.
   - For simple forms and small sizes, add data directly through the map viewer.
   - Otherwise, publish data, maps, and toolboxes as web services or web layers, and add them to your web map.
3. Create and share your web map using the ArcGIS Online map viewer.
   - Add your service or layer (or other available types of layers) to your web map.
   - Symbolize your layer (for some types of layers only) and configure pop-ups.
   - Save and share your web map.
4. Create and share your web app.

Browse the configurable apps to find one that best suits your needs, and then use it to transform your web map into a web app. If no configurable app meets your requirements, use ArcGIS web APIs or Runtime SDKs to create your own app. After the app is created, your app is private. You need to share it for others to search, discover, and use. ArcGIS Online has different sharing levels.

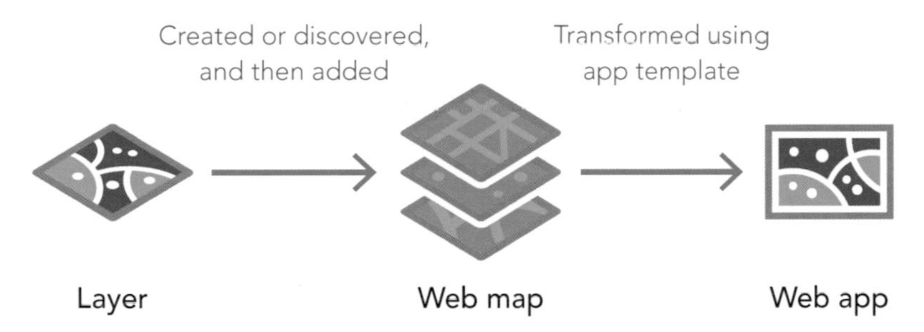

ArcGIS Online allows users to easily create web maps by assembling various formats of layers and to create web apps from web maps by applying configurable app templates.

This tutorial

In this tutorial, you will create a web GIS app that introduces the main points of interest (POIs) in the City of Redlands, California.

Data: A CSV file contains data for the main POIs in Redlands, including longitude, latitude, names, descriptions, photo or video URLs, and thumbnail URLs.

The sample data for this entire book is available at **http://esripress.esri.com/bookresources**. To get the data, navigate to the webpage, find the title, *Getting to Know Web GIS*, second edition, download the sample data, and extract the files to **C:\EsriPress** on your computer, or follow your instructor's instructions to download the data.

Requirements:

- Your web app should display a basemap (a street map or satellite imagery) of the city and POI locations along with their descriptions and photos or videos.
- The web app should be engaging and easy to use.
- The web app should work on desktops, tablets, and smartphones.

Solution: To build this web app, select **Story Map Tour**, one of the most popular configurable apps in ArcGIS Online. See two screen captures of the app in the figure and a live sample at **http://storymaps.esri.com/stories/maptour-palmsprings**.

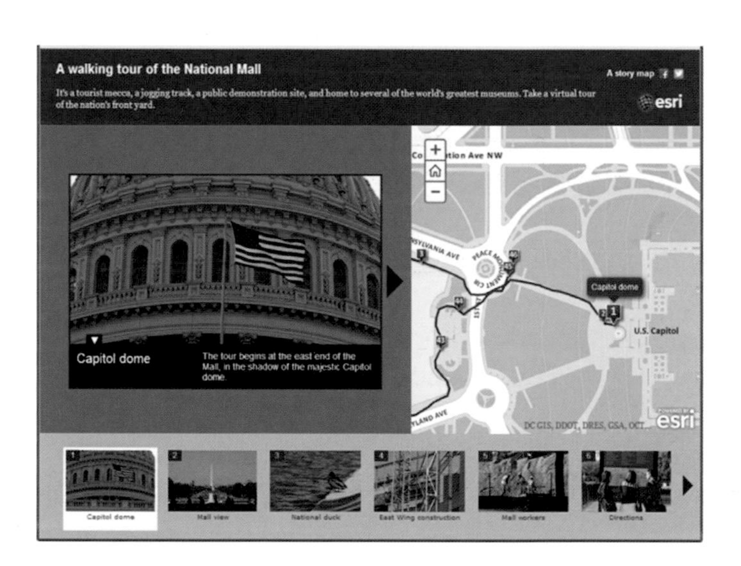

The Story Map Tour configurable app working in a desktop browser (left) and on a smartphone.

The Story Map Tour configurable app produces attractive, easy-to-use web apps that help you present geographic information with compelling photographic and video story elements. The template layout automatically rearranges itself to adapt to various screen sizes and can display a set of places on a map in a numbered sequence made for browsing. The template is designed for use in web browsers on desktops, smartphones, and tablets.

This template benefits the following scenarios:

- Show the world the work your government department, organization, or agency is doing or has done.
- Showcase key attractions of a city or region.
- Introduce a park and its features.

- Provide a tour of a campus, an outdoor art collection, or a historical district.
- Educate people about areas of scientific or geographic interest.
- Direct public attention to places you want to improve or protect.
- Create online photo or video journals of a trip or event.

System requirements:

- Microsoft Excel or a text editor to create and edit your CSV data
  - CSV format easily represents points, though not complex geometric forms such as lines and polygons.
  - Excel automatically maintains correct CSV format (for example, adding correct quotation marks).
- A web browser
- ArcGIS Online (or Portal for ArcGIS)
  - For the work you will do in this chapter, a user-level account will suffice; however, you will need a publisher-level account later in the book, so get a publisher account now.
  - If you do not have access to an organizational account, create a 60-day free trial account.

Note to instructors: You can create a group for your students in which they can share their work with other members.

1.1 Create an ArcGIS Online trial account

Skip this section if you already have an account for ArcGIS Online or Portal for ArcGIS.

If your organization has ArcGIS Online for Organizations or Portal for ArcGIS, please ask your administrator or instructor to create an account for you.

1. **Open your web browser, navigate to ArcGIS Online (www.arcgis.com), and then click Sign In in the upper-right corner of the page.**

2. **Click Try ArcGIS.**

ArcGIS Features Plans Gallery Map Scene Help

Sign In

Don't have an ArcGIS account?

Sign up for a 60-day trial.

TRY ARCGIS

Sign In esri

Username

3. Fill out the **Sign Up for the ArcGIS Trial** form:

   - **Input your name, email, and other requested information.**
   - **Click Start Trial to submit the form. You will know the form has been submitted correctly when a new page comes up that says, "Confirmation email sent!"**

Esri will send you a confirmation email for you to activate your account.

4. Check your email, and click the activation URL link in the **Activate Your Free ArcGIS Trial** email.

5. On the activation page, fill in the fields, accept the terms and conditions, and click **Create My Account.**

Having created an ArcGIS Online for Organizations trial account, you are made the administrator for your organization. You will be directed to the **Set up Your Organization** page.

6. On the **Set up Your Organization** page, fill in the fields. Then click **Save and Continue.** (Do not select **Allow access to the organization through HTTPS only.**)

You have now created your trial organization account. If you are prompted to download ArcMap and other software, click **Continue with ArcGIS Online**. You will need ArcMap and ArcGIS Pro later in this book, but not now.

1.2 Prepare your data

Configurable apps require certain kinds of data content. The Story Map Tour app, for example, requires a list of points (a point layer) and the locations, captions, descriptions, photos or videos, and thumbnails associated with them. You can organize your data in a CSV or point shapefile, feature service, map service, or other formats.

This chapter provides a sample CSV dataset with coordinates for the main POIs in the City of Redlands. Examine the sample data to familiarize yourself with the required fields.

1. **If you have not already done so, navigate to esripress.esri.com/bookresources, or follow your instructor's directions to download the sample data for the second edition of this book. Extract the files to C:\EsriPress.**

2. **In Microsoft Excel, navigate to C:\EsriPress\GTKWebGIS\Chapter1\Locations .csv, and study its data format.**

Name	Caption	Icon_color	Long	Lat	URL	Thumb_URL
Welcome to the City of Redlands	Located about 60 miles east of Los Angeles, replete with cultural, artistic and historical sites. Redlands, emerging as a regional leader, boasts small-town charm, and features a world center of geospatial information technologies. (<i><a href="http://en.wikipedia.org/wiki/Redlands,_California" target="_blank">More info</a></i>)	R	-117.182421	34.055448	https://googledrive.com/host/0BwLwen7VgLIFdW5uemtLeDJ0S3c/Cover.JPG	https://googledrive.com/host/0BwLwen7VgLIFMUk4UUZ0Q2d6YWM/cover.png
Esri	Headquartered in Redlands, Esri is a world leader in GIS software. Founded by Jack and Laura Dangermond in 1969, Esri now has 10 regional offices in the U.S. and a network of 80 international distributors, with about a million users in 200 countries. (<i><a href="http://www.esri.com" target="_blank">Website</a></i>)	R	-117.195688	34.056932	https://googledrive.com/host/0BwLwen7VgLIFdW5uemtLeDJ0S3c/Esri.JPG	https://googledrive.com/host/0BwLwen7VgLIFMUk4UUZ0Q2d6YWM/esri.png

The first row of your spreadsheet provides the header. Below that, each row contains one tour point. For each point, the Story Map Tour app expects the following fields:

- **Name:** A short name identifying the point.
- **Caption:** A description of the point. Keep it short (less than 350 characters is recommended). The caption can include HTML tags to format the text or provide hyperlinks.
- **Icon_color (optional):** The color of each point. The valid values—R, G, B, and P—indicate red, green, blue, and purple, respectively.
- **Geographic Location:** You can describe geographic location by specifying longitude and latitude as **Long** and **Lat** (in decimal degrees), a single **Address** field containing a complete street address, or multiple fields (such as **Address**, **City**, **State**, and **ZIP**). This tutorial uses Long and Lat.
- **URL:** The full web address for the full-size image or video, starting with http://, https://, or //. The recommended image size is 1000 × 750 pixels, but other sizes will also work.
  - For videos: The app does not include a generic video player. Instead, use the URL that a video hosting service, such as YouTube, provides for embedding videos via a link. Make sure to append **#isVideo** to the end of the URL (for example, **http://www.youtube.com/**

embed/RM0eMdrPhEA#isVideo). For YouTube videos, right-click the video being played, click **Copy Embed Code**, paste the code into Notepad, find the URL in the code, and append **#isVideo** to the end of the URL.
- To use photos or videos on your computer, you must first upload them to some form of online storage, such as Flickr, Picasa, Facebook, Google Drive, Microsoft SkyDrive, YouTube, or your own web server.
- If you have not yet collected your own images and videos, you can search for media through search engines and then copy their URLs.
- For images: Right-click an image. Select **Copy Image Location** in Firefox or **Copy Image URL** in Chrome. For Internet Explorer, select **Properties** and then copy the image address URL.

- **Thumb_URL:** The full web address of the thumbnail image (starting with http://, https://, or //). Images can fit to scale, but the recommended image size is 200 × 133 pixels. You will often need to find a point's latitude and longitude. For example, the last POI in the CSV dataset, Market Night, is missing both longitude and latitude. You will find these coordinates using the ArcGIS Online map viewer.

3. **Open a web browser, navigate to ArcGIS Online (www.arcgis.com) or your Portal for ArcGIS, and sign in.**

Familiarize yourself with the links at the top of the page:
- **Home** returns to the homepage.
- **Gallery** leads to featured maps and apps.
- **Map** goes to the map viewer.
- **Scene** goes to the 3D web scene viewer.
- **Groups** leads to the My Groups page where you can create and join groups.
- **My Content** links to the My Content page where users can view, add, and delete content items.
- **My Organization** leads you to a page about your organization. If you are an administrator of your organization, the page includes administrative tools for managing your organization.
- In the upper-right corner of the page, the **Search** box and button allow you to search for content in the ArcGIS Online catalog.

4. **Click Map to open the map viewer.**

Home Gallery Map Scene Groups My Content My Organization Pinde

If you know where this missing POI is, navigate there directly on the map. Here, you will use geocoding. Redlands Market Night takes place downtown at the intersection of Orange and East State Streets.

5. **Type Orange St & E State St, Redlands, CA in the Search text box, and then press Enter or click the Search button. After the address is found and the map is centered to the location, click the Zoom In button ⊞ until you can no longer zoom in. Remember the approximate location of the address on the map.**

You will get the coordinates of the location by measuring the longitude and latitude in the next step.

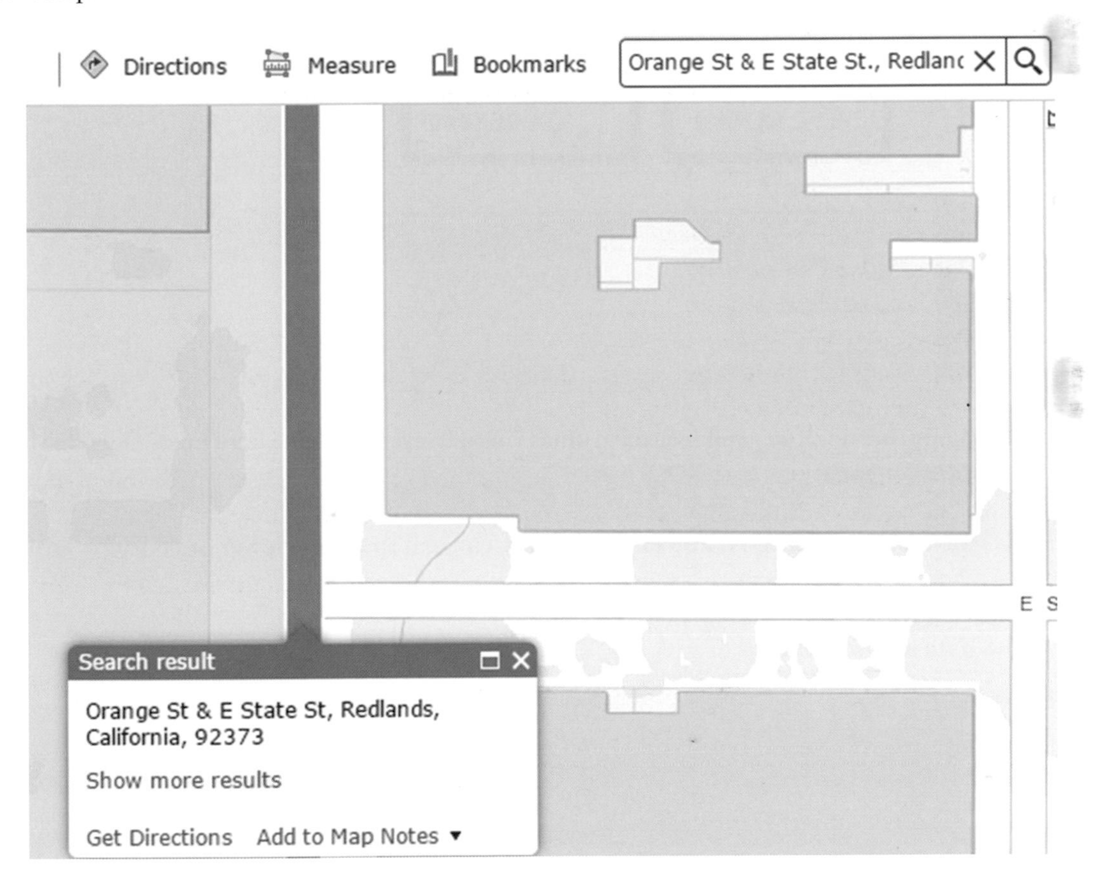

6. **At the top of the page, click Measure. Next, in the window, click the Location button, and then click the map near the pointer of the callout box. If your callout box has disappeared, simply click Search again.**

The location's longitude and latitude display under **Measurement Result**.

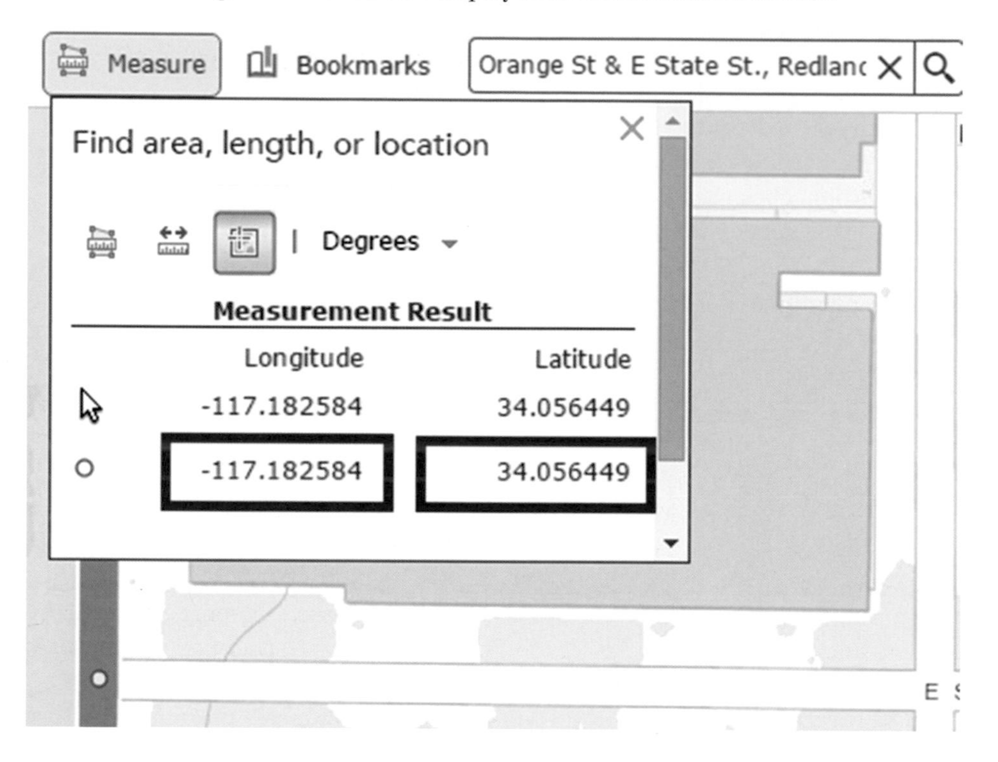

7. **Copy the longitude and latitude values you retrieved in step 6, and paste them to the Market Night row in the CSV file.**

8. **In Excel, save the CSV, and exit Excel. You can exit the map viewer or continue to step 2 of the next section.**

Your data is now complete.

1.3 Create a web map

You will want to make sure you are signed in before continuing with the remaining steps. Otherwise, you will not be able to save your web map, and you may lose your work.

1. **In a web browser, navigate to ArcGIS Online or the home page of your Portal for ArcGIS, sign in, and click Map to go to the map viewer.**

2. **Familiarize yourself with the map viewer menu bar.**

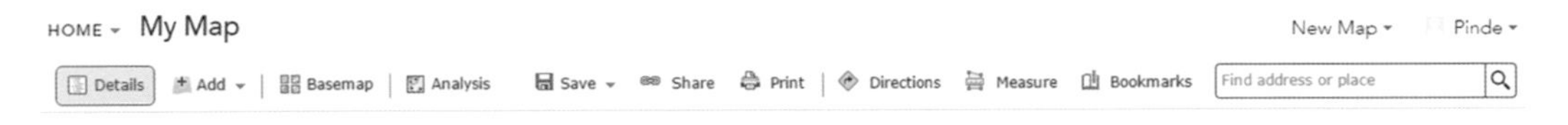

The ArcGIS Online map viewer helps users create, customize, and view web maps. On the menu bar, you will see the following buttons:

- The **Details** button toggles the panel on the left side of the map canvas. This panel can display a map's metadata, table of contents (TOC), or legend.
- The **Add** button is used to add a variety of layers into the map.
- The **Basemap** button displays a gallery of underlying imagery that you can choose from.
- The **Analysis** button leads to a rich set of analysis functions.
- The **Save** button allows you to save your web map.
- The **Share** button lets you select the people who will have access to your web map and choose how you will share it, either by embedding the map in a webpage or by creating a web app from a configurable app template.
- The **Print button** creates a hard copy of the current map view.
- The **Directions** button can calculate the best route from a starting location to the destinations you specify.
- The **Measure** button helps determine areas, distances, and a location's longitude and latitude.
- The **Bookmarks** button allows you to save a list of map areas so that you can quickly select one and zoom to that map area.
- In the **Find address or place** text box, you can specify an address or place and find its location on the map viewer.

3. Add the CSV file to the map viewer.

If you are using a web browser that supports the drag-and-drop operation (such as Chrome, Firefox, or Internet Explorer 10+), you can simply drag the CSV file to the map canvas.

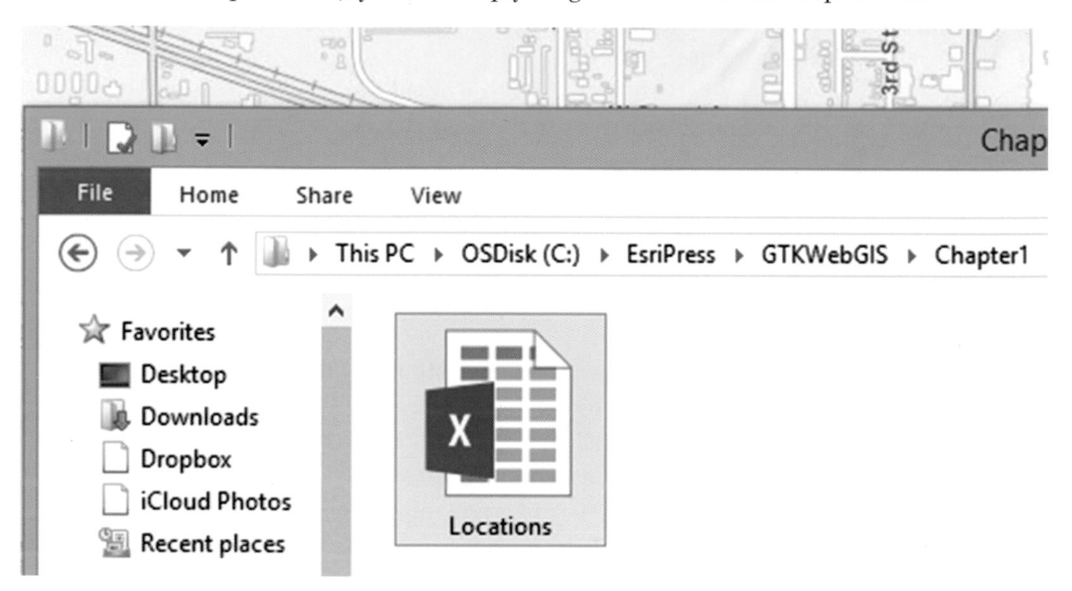

If your browser doesn't support drag and drop, click **Add**, select **Add Layer from File**, locate the CSV file on your computer, and click **Import Layer**.

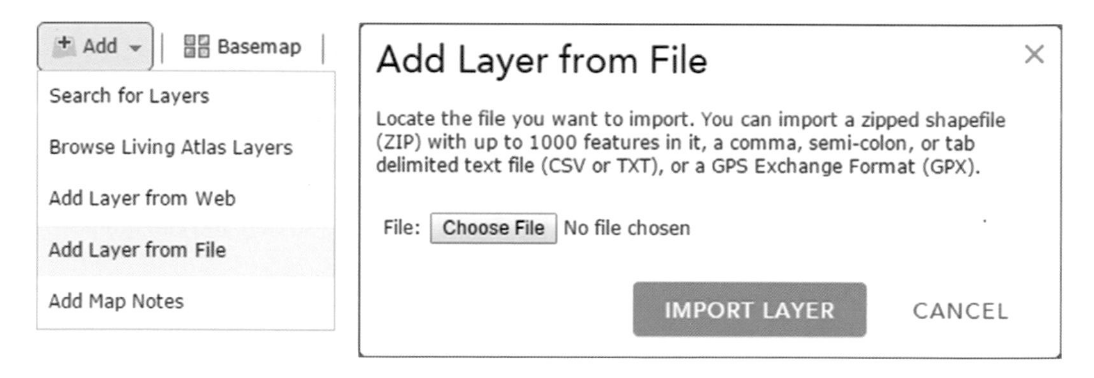

The map viewer displays your CSV data automatically.

4. Zoom the map to an extent that includes all the points.

This extent provides users with a view of all POI locations, and the extent can be used as the initial extent of your web app once you save your map.

5. **On the menu bar, click the Save button and choose Save.**

6. **In the Save Map window, enter the title, tags, and summary of your web map. Then click Save Map. Leave the web map open.**

Tip:

- For your homework, include your name in the title so that you and your instructor can easily find your web map.

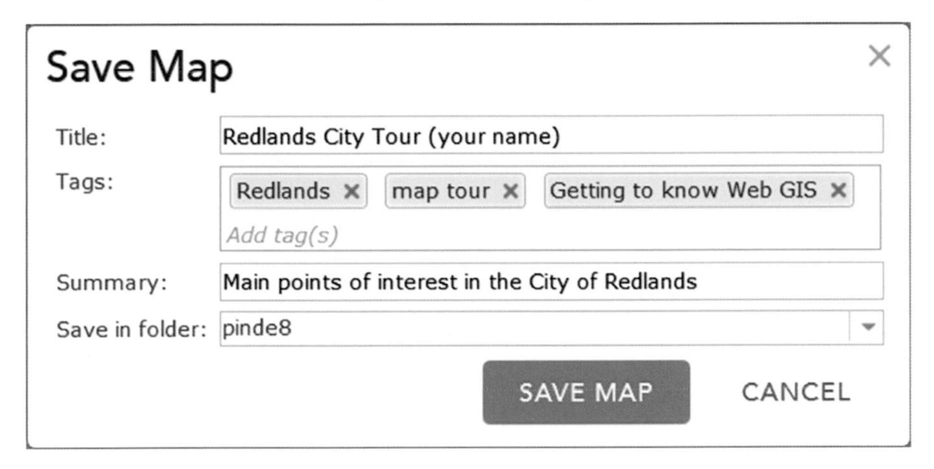

Congratulations! You have created a simple web map.

Typically, users need to configure pop-up windows and sometimes change symbols on map layers. You will learn these skills later in the book. For this tutorial, the Story Map Tour app automatically handles the style of your layer, so you do not need to change its style here.

1.4 Create a web app using a configurable app template

Completing this section will transform your web map into a web app using a Story Map Tour configurable app template.

1. **Continue from the last section, or sign in to ArcGIS Online or Portal for ArcGIS, and open the web map you just created. In the map viewer, click the Share button on the menu bar, which opens the Share window.**

2. **In the Share window, select the check box indicating Everyone (public) or the check box(es) indicating the organization and/or groups with which you would like to share your web map.**

Note: Unless you share your web map with everyone, a prompt will ask users to sign in whenever they open your web map and any web app that uses this map.

3. **Click Create a Web App.**

The **Create a Web App** window opens, presenting a gallery of the configurable apps. If your organization has configured custom galleries, you may not see the same configurable apps as those shown in the figure.

The apps are grouped, and the groups are listed in alphabetical order. You can use the scroll bar to review the full gallery, or you can click a group name on the left to see the apps in this group.

4. **Click Build a Story Map group on the left, find and click the Story Map Tour app, and then click Create App.**

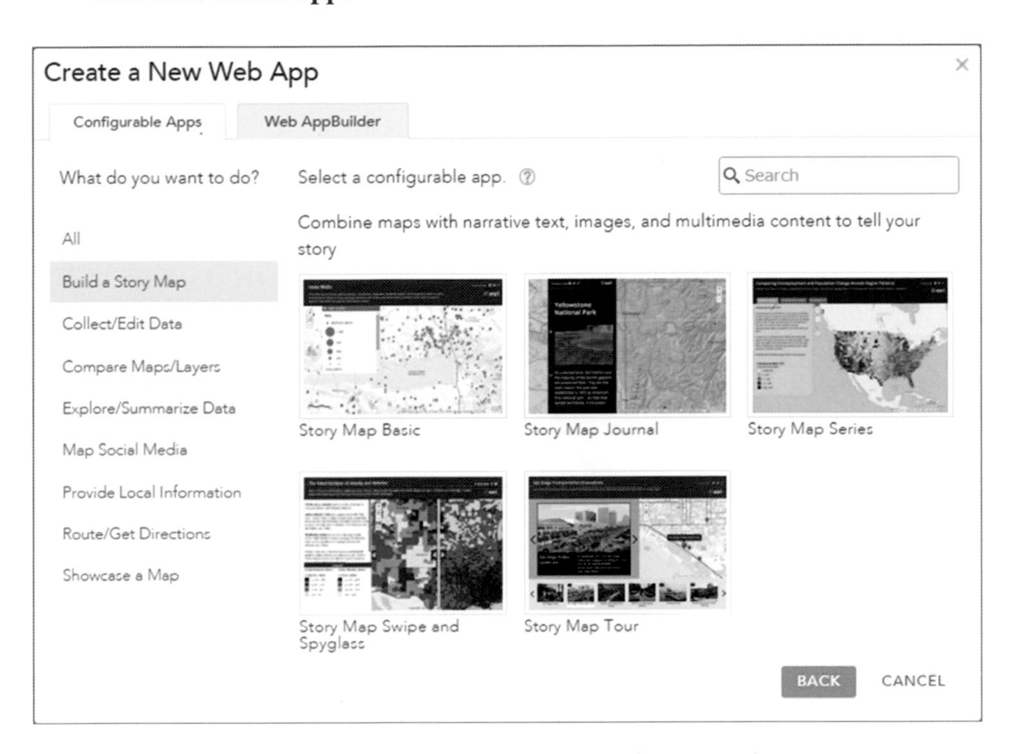

5. **Provide the appropriate title, tags, and summary information, and then click Done.**

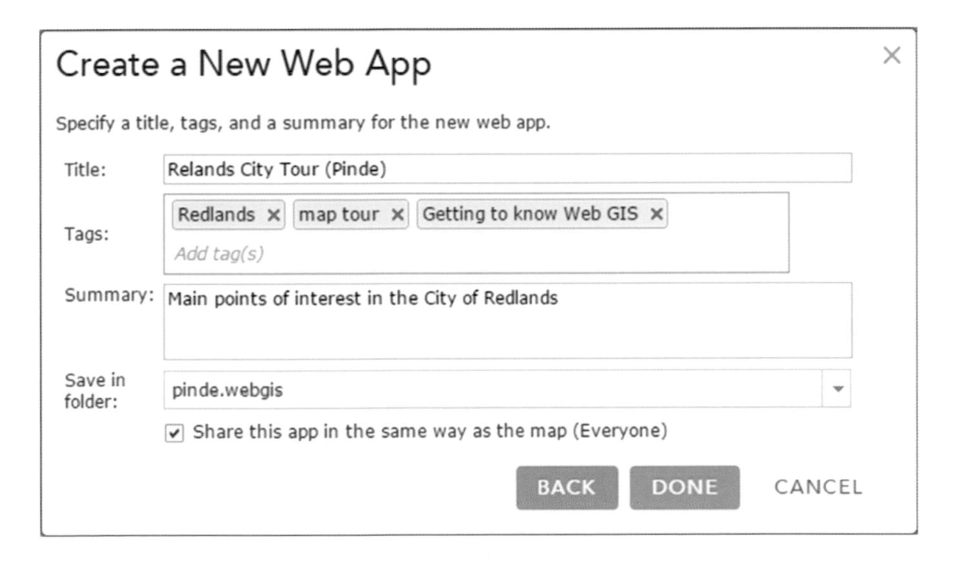

Note: The check box next to **Share this app in the same way as the map (Everyone)** is selected by default.

You have created your own informative and easy-to-use web app!

6. **Spend a few minutes exploring your new web app.**

You can navigate through the app's tour points by clicking the thumbnails, the arrows next to the photos, and the numeric icons on the map. If you click the thumbnail for the University of Redlands, for example, a video introducing the university appears.

Your web app is already created and saved. You will further configure it in the next section.

1.5 Configure your web app

Once you have determined that your app's tour points and their order, captions, and descriptions are correct, your app is complete. Optionally, you can enhance your application's features by using the app's builder mode (application configuration). In this mode, you can add or import new tour points; update and delete existing images; set or update locations and descriptions; update the app title, subtitle, and logo; and change the app layout.

1. **If you are continuing from the previous section, go to step 3; otherwise, sign in to ArcGIS Online or your Portal for ArcGIS.**

2. **In the My Content list, find and click the web app you just created to go to its item details page, and then click Configure App.**

3. **Familiarize yourself with the builder mode.**

- The **pencil** icon means that you can update nearby text, such as titles, subtitles, image captions, and descriptions.
- The **Change media** and **Change thumbnail** buttons can be used to change the URLs of media and thumbnail locations.
- The **Add**, **Organize**, and **Import** buttons allow you to interactively add more locations, change the order of points, and import media from Flickr, Picasa, Facebook, YouTube, or a CSV file.

Now, you will change the Esri photo into a video.

4. **Click the Esri thumbnail image. Click Change media, and then click Video. Remove the current URL, enter http://www.youtube.com/embed/RM0eMdrPhEA, and click Apply.**

The video loads into the picture frame.

Next, you will change the thumbnail for Esri to a new one that indicates a video.

5. **Click Change thumbnail. Replace the current URL with http://bit.ly/1nvc2PU (short URL equivalent to http://esrimapbook.esri.com/GTKwebgis/chapter1/thumbnails/esri_v.png), and then click Apply.**

6. **Click Save to save your changes.**

In the following steps, you should save your work regularly to prevent losing your changes.

7. **Above the thumbnail carousel, click Organize.**

The **Organize the tour** window allows you to delete tour points and drag pictures to change their order.

8. **Select the check box for Use the first point as introduction (does not appear in carousel). Click Apply to close the Organize the tour window.**

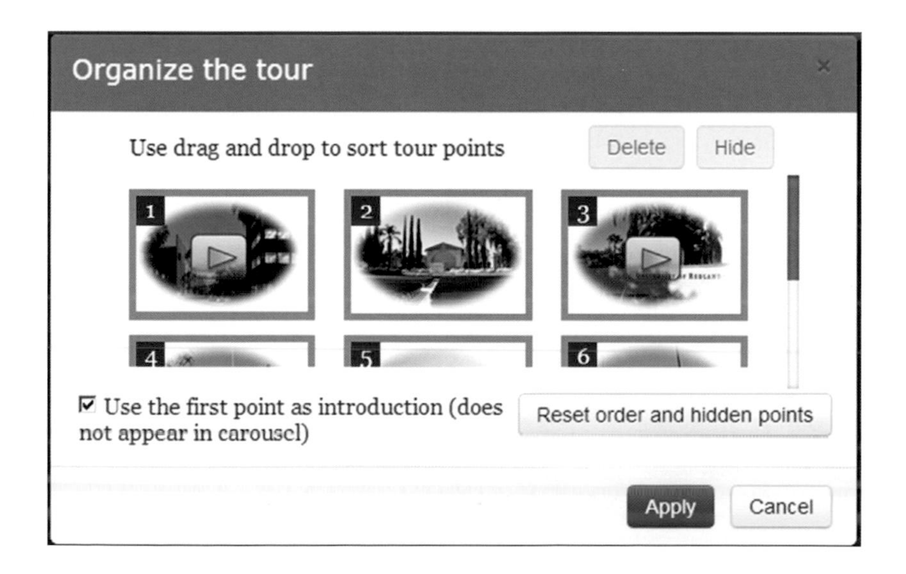

Selecting the check box sets the first record in your CSV as the introductory image to your app, allowing you to start your tour by showing a compelling image and an introductory caption to set the scene. The location of this record will not be shown on the map as a numbered point in your tour.

Optionally, you can import tour points and media from Flickr, Facebook, Picasa, YouTube, or an additional CSV file. To import a photo from Flickr, you would perform the following tasks:

- Click **Import** next to the **Organize** button.
- Click the **Flickr** icon (on the far left) in the **Import** window.

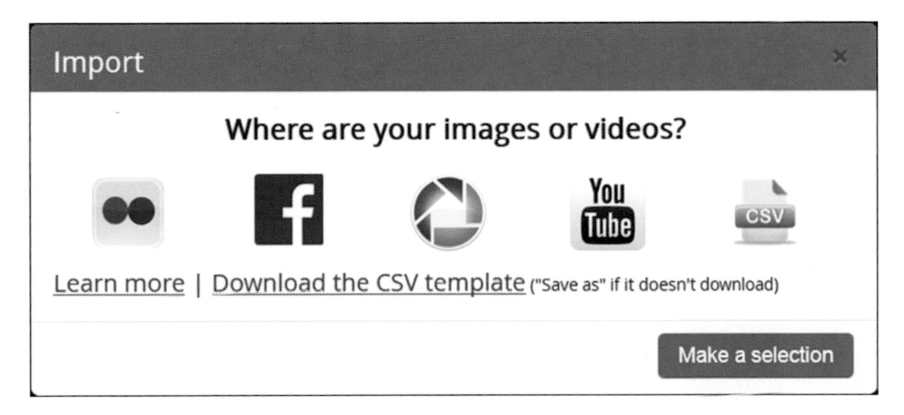

- Type **web GIS** as the Flickr user name, click **Look up**, select **MapTour(1)** from the **Select a Photo Set** box, and then click **Import**.

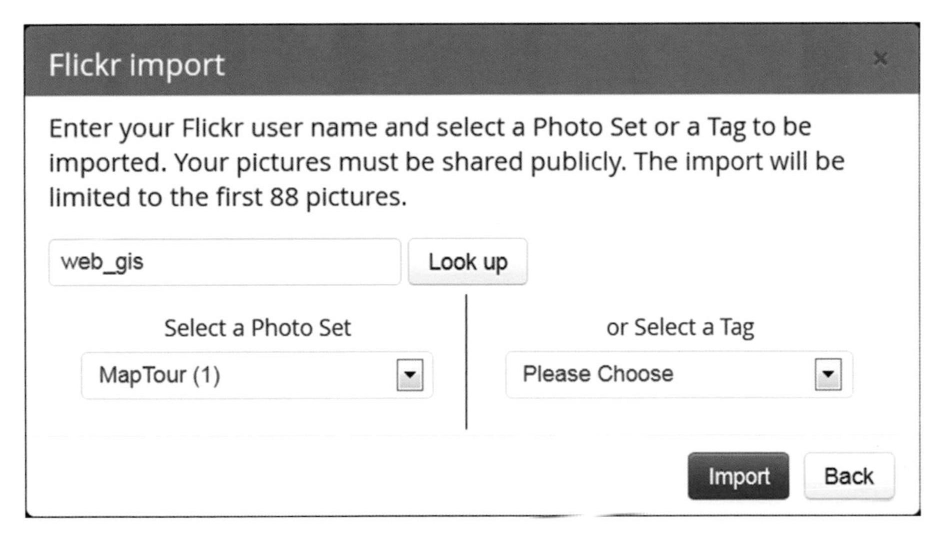

- Click **Located.**

The set contains one photo, which is already geotagged in Flickr. Flickr and similar websites can extract location information from the EXIF (exchangeable image file format) metadata in photos taken by GPS-enabled cameras, such as smartphones. These websites also allow users to specify photo locations manually, using maps.

- Click **Import** to import this photo.

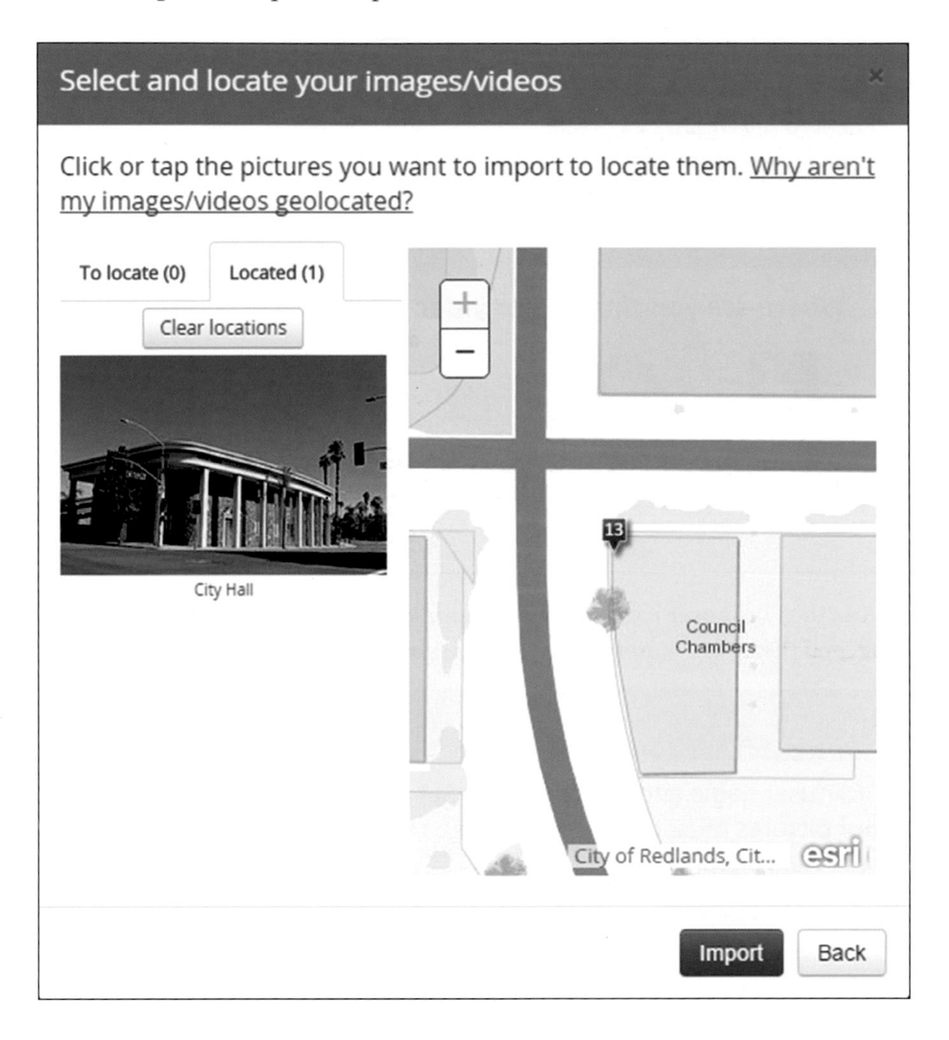

If you chose this option, your web map would include City Hall photo and the photo name and caption.

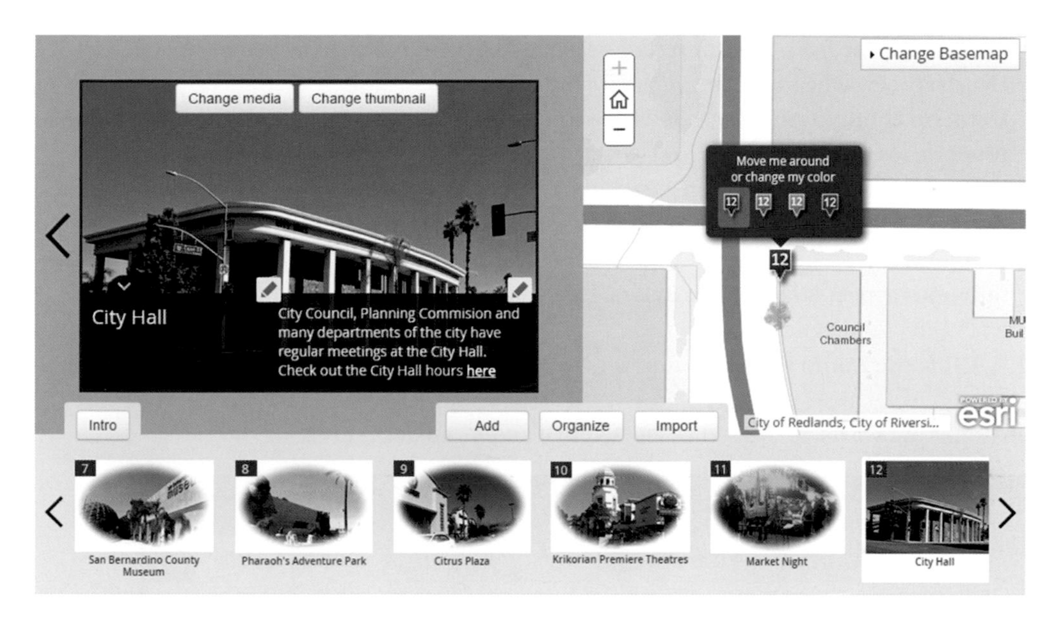

Optionally, you can add additional tour points manually by clicking the **Add** button next to the **Organize** button and filling in the media, name, and location information.

9. **In the page header, click Settings.**

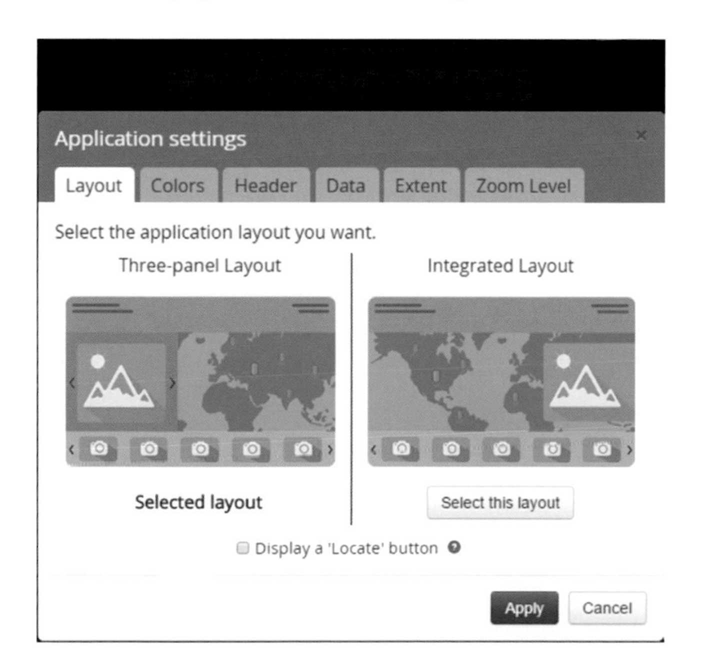

Clicking **Settings** opens the **Application settings** window, and you have the following tab options:

- **Layout:** Choose between **Three-panel Layout** and **Integrated Layout.**
- **Colors:** Choose from predefined color themes, or define your own theme.
- **Header:** Set the header logo, and share links.
- **Data:** No configuration is needed here. The sample CSV you use has all the fields named properly for the Story Map Tour to use.
- **Extent:** Define the initial map extent that users will see when the app first opens.
- **Zoom Level:** Specify a scale to which the map will automatically zoom whenever the app user goes from one tour point to another (but if users manually zoom in or out, the map tour app respects their choice and no longer applies your auto zoom level).

10. **Click the Zoom Level tab, and set the Scale/level to 1:5K (level 17) as illustrated.**

This scale allows users to see the selected POI and its adjacent area.

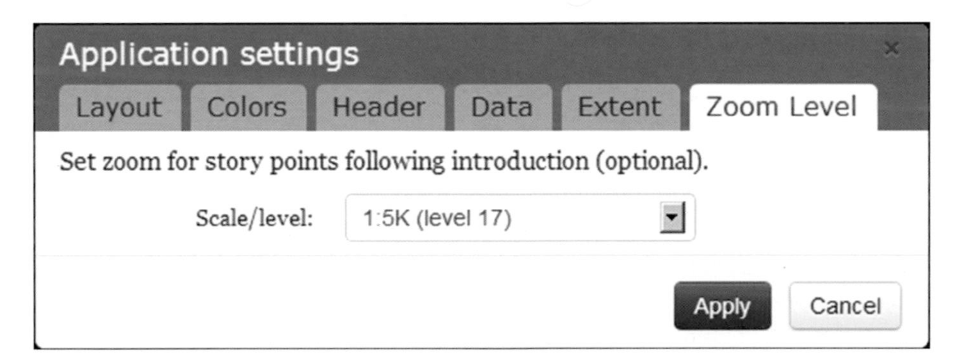

11. **Click the Header tab, change the logo and text if needed, and then click Apply.**

For example, you can add your name to the header so that your instructor can easily tell who created your application. Optionally, you can also exchange the logo for your organization's logo.

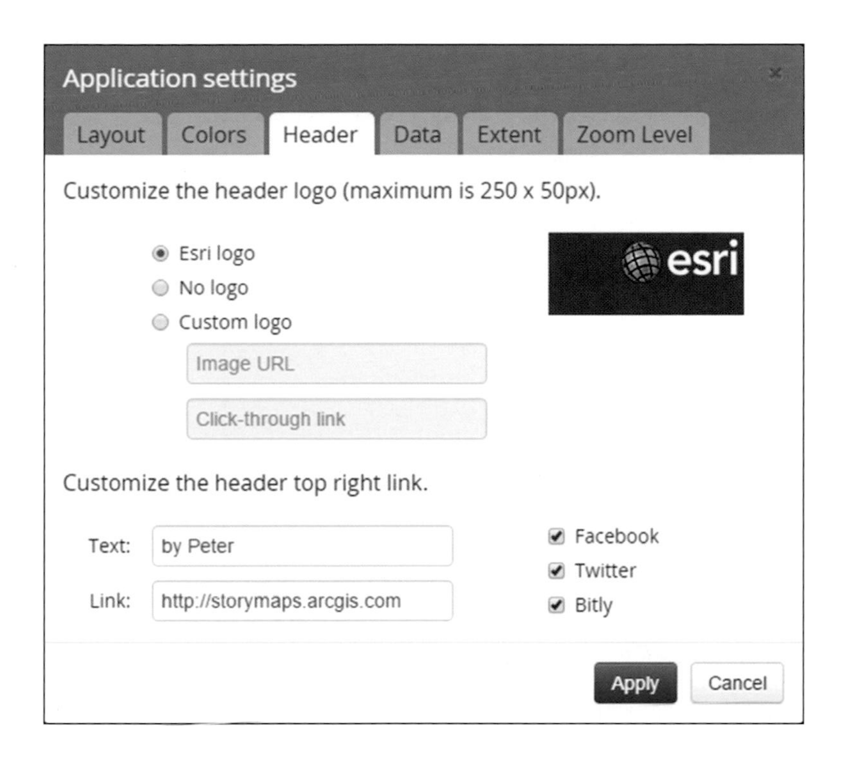

Examine the application to see if there is anything else you would like to configure. If so, you can make and apply further changes.

12. **In the page header, click Save to save your work.**

1.6 Share your web app

You have created and shared a web app with the same people with whom you shared your web map (see step 4 of section 1.4). Now you will share the URL of this web app with your audience so that they can see your web app.

1. **Click Share in the page header. If you see a message saying your tour is not shared, share your tour publicly.**

2. **In the Share your Tour window, click Open to preview your web app.**

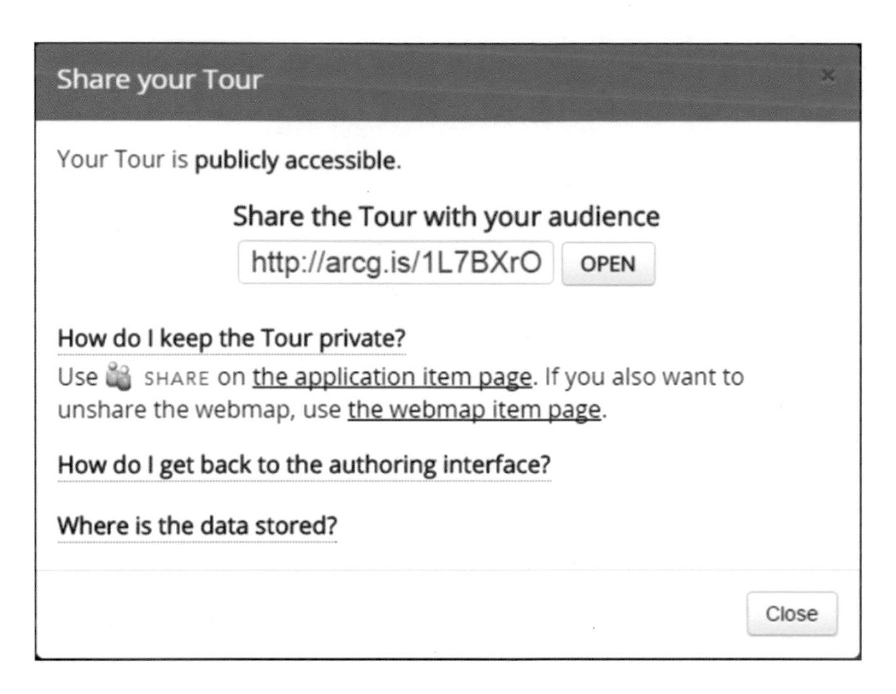

3. **Share the tour URL with your audience (for example, by copying and sending the URL by email or by displaying the URL link on your organization's homepage).**

4. **Test your web app on smart mobile devices.**

Open your app in the browser of your smart devices. An easy way to open the app is to send the URL to yourself via email, open that email on your smart device, and then click the URL.

Configurable apps use responsive web design technologies and can change their layouts to best fit various devices with different screen sizes. You will find they work well on iOS, Android, and Windows Phone tablets and phones.

In this tutorial, you have created a user-friendly, informative, and cross-platform web app. The app meets all the requirements listed early in this chapter—it displays a basemap and POI locations, their descriptions, and associated photos or videos; the app is engaging and easy to use; and it works on desktops, tablets, and smartphones, using the ArcGIS Online cloud platform. Additionally, your web app did not require a single line of programming.

You can create a Story Map Tour app in other ways. In addition to pictures and videos, you can display webpages and 3D web scenes. See the resources section.

QUESTIONS AND ANSWERS

1. **After uploading my CSV to the ArcGIS Online map viewer, I updated my CSV. Will the changes to my CSV automatically update in my web map and web app?**

   Answer: No.

   Once your CSV data has been added to the map viewer and saved with your web map, it is uploaded to the cloud. Your web map and web app will use this copy of the data rather than your local data.

   To use your new CSV data, remove the previous CSV data layer from your web map and replace it with the new CSV. Then save your web map.

   Consider the following alternatives:

   - You can first upload your CSV to a web folder, and then in your web map reference the CSV using its URL. In ArcGIS Online map viewer, you can configure the refresh rate of the layer to be one (1) minute, for instance. This way, when the CSV updates in the web folder, the updates will appear in your web map and web app automatically.

   - You can use a feature service layer (discussed in more detail later in the book) instead of a CSV. When someone edits the data (for example, collects a new point and adds new photos using the Collector mobile app), the updates will appear in your Story Map Tour app if you reload the app in your browser.

2. **In my map tour app, I would like to add a line layer to show the path of my tour. How do I add a line layer?**

   Answer: You can add a line layer in several ways.

   - If you hold the data in a shapefile, add it to your web map and configure its symbol using the map viewer.

   - If the data is in a geodatabase, create a map document file (MXD), publish it as a feature service or map service, and then add the service to your web map.

- If you do not have the tour path data, simply create it using a map notes layer. Open your web map in ArcGIS Online or Portal for ArcGIS map viewer, click **Add**, click **Add Map Notes**, give your layer a name—such as **Tour Path**—choose a template, and click **Create**. Choose a line symbol you like from the template on the left, and then apply your cursor on the map to draw your tour path.

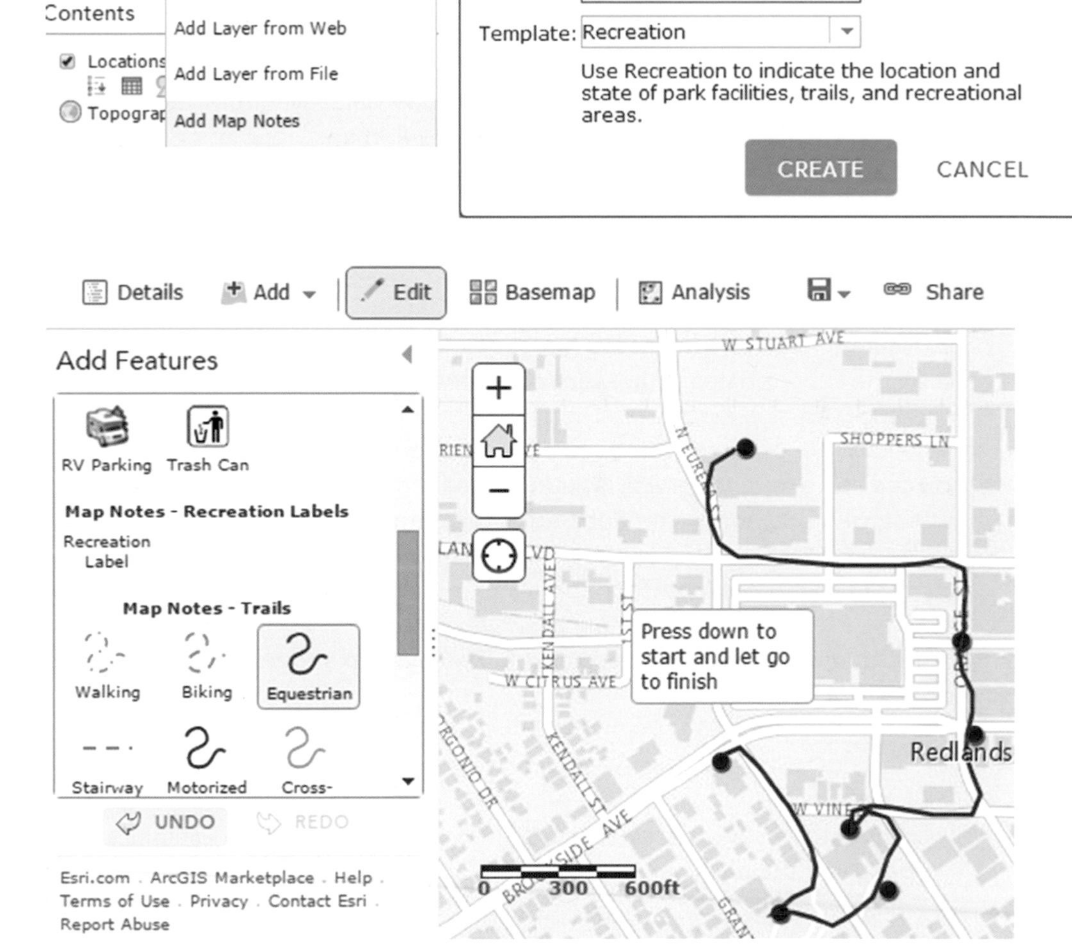

3. **What is the maximum number of tour points I can have?**

   Answer: Ninety-nine points for the hosted version.

   Most map tours will contain far fewer than the maximum 99 points per tour. However, you can download the configurable app source code, change the configuration to override this limit, and host the web app on your own web server.

4. **I found it slow work to locate longitudes and latitudes manually, one by one. Is there a more efficient way to define the locations of my points?**

   Answer: Use addresses, feature classes, or geotagged media if you have them.

   In your CSV, specify the addresses of your points in one or multiple address fields (such as **Address**, **City**, **State**, and **ZIP**). When you add this information to the ArcGIS Online map viewer, ArcGIS Online will geocode these addresses and find their locations automatically, as discussed in more detail later in the book.

   If you have your map points in a shapefile or a feature service, you do not need to create a dataset in CSV format. Add the shapefile or service to your web map directly.

   If you have geotagged media, such as photos taken using your smartphone with location enabled for your camera, you can create an empty web map and then import these photos in the configuration mode.

ASSIGNMENTS

Assignment 1: Choose from the following topics, and create an app using the Story Map Tour to showcase your topic.

- Your personal story (where you were born, where you moved, where you went to school or worked, and so on)
- Your city's key attractions
- The landmarks, buildings, and departments on your campus
- Places you have visited in the past or during a recent vacation
- Branches of a bank or supermarket in your city or region
- Projects that your organization has accomplished or is working on
- Locations of key environmental interest (for example, largest/oldest trees) or historic interest (for example, oldest houses)
- Other interests

What to submit: Email your web app URL to your instructor with the subject line **Web GIS Assignment 1: Your name**.

Resources

ArcGIS Online Help and tutorials

The ArcGIS Book, chapter 1, http://learn.arcgis.com/en/arcgis-book/chapter1.

ArcGIS Online help document site, http://doc.arcgis.com/en/arcgis-online.

"Get Started with Story Maps," http://learn.arcgis.com/en/projects/get-started-with-story-maps.

"Make a GeoPortfolio," http://learn.arcgis.com/en/projects/make-a-geoportfolio.

Esri blogs

"Add Layers to Your Story Map Tour," by Owen Evans, http://blogs.esri.com/esri/arcgis/2015/07/28/add-layers-to-your-map-tour.

"Add YouTube Videos to Your Story Map Tour," by Bern Szukalski, http://blogs.esri.com/esri/arcgis/2015/08/12/add-youtube-videos-story-map-tour.

"Adding Links to Captions in Your Story Map Tour," by Bern Szukalski, https://blogs.esri.com/esri/arcgis/2015/09/20/add-links-map-tour.

"Create a Story Map Tour from a Google Sheet," by Owen Evans, http://blogs.esri.com/esri/arcgis/2015/08/12/create-a-map-tour-from-a-google-sheet.

ArcNews and ArcWatch

"Intelligent Web Maps and ArcGIS Online," www.esri.com/news/arcnews/summer11articles/intelligent-web-maps-and-arcgis-online.html.

"Make a Map Tour Story Map," by Rupert Essinger, www.esri.com/esri-news/arcwatch/0513/make-a-map-tour-story-map.

Esri videos

"ArcGIS Online: Sharing your Content," by Ben Ramseth and John Thieling, http://video.esri.com/watch/4710/arcgis-online-sharing-your-content.

Chapter 2
More on web GIS layers, maps, and apps

This chapter introduces the best practices of web GIS app design and applies them to the creation of a web app. You will learn how to add data from comma-separated value (CSV) files using geocoding when you don't have latitudes and longitudes for your locations. You also will learn how to find layers from the ArcGIS Online catalog and the Living Atlas of the World, edit data in the map viewer, design layer styles using smart mapping, and configure layer pop-ups with multimedia. Additionally, you will work with a versatile configurable app.

Learning objectives

- *Learn basic components of a web GIS app.*
- *Map a comma-separated value (CSV) file using geocoding.*
- *Edit data in the map viewer.*
- *Design maps using smart mapping.*

1
2
3
4
5
6
7
8
9
10

- *Configure pop-ups with multimedia.*
- *Explore Living Atlas of the World contents.*
- *Use ArcGIS basic map viewer configurable app.*

Basic components of a web GIS app

Today's web GIS best practices recommend that a web GIS app should have basemaps, operational layers, and tools. ArcGIS Online supports this practice and makes it easy for you to create web GIS apps.

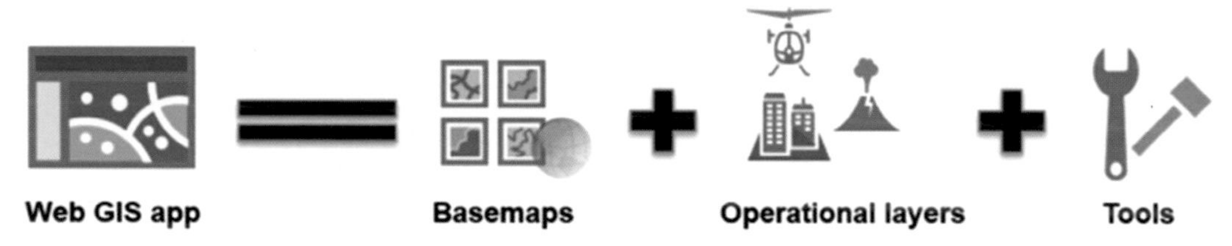

The basic components of a web GIS app.

Basemap layers

Basemaps provide a reference or context for your web GIS app. ArcGIS provides a collection of fast-responding basemaps. Most of the time, you can use them without worrying about creating them.

- ArcGIS provides a gallery of basemaps. These maps all have multiple scales with global coverage.
- In addition, you can use your own map services as basemaps (open the map viewer, click **Add** > **Add Layer from Web**, and check the **Use as Basemap** box).
- ArcGIS provides a global elevation service on which basemaps can be draped. This service supports scenes (the 3D counterparts of web maps).

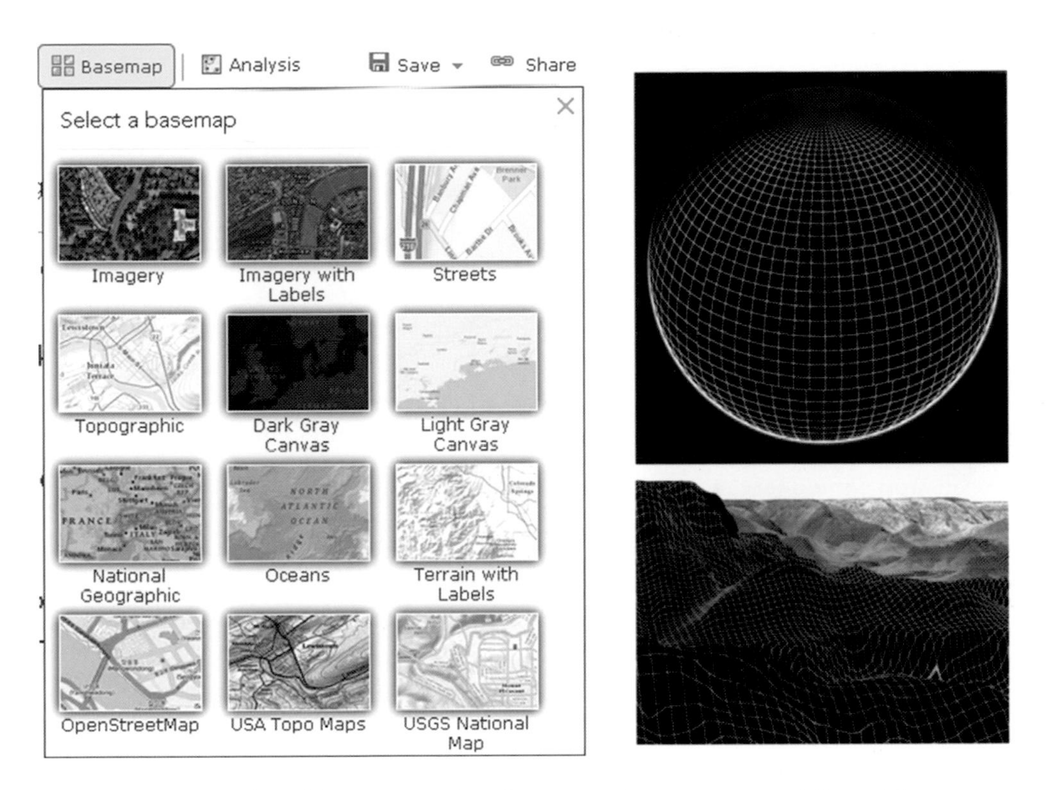

ArcGIS provides a gallery of basemaps and an elevation service to support both 2D and 3D web maps and apps.

Operational layers

Operational layers are theme layers that you and your end users work with most often.

- These layers are typically from your own data. For example, the tour points layer in the Story Map Tour app created earlier in the book is an operational layer. ArcGIS Online and Portal for ArcGIS support operational layers from a variety of formats, including CSV, TXT, GPS exchange (GPX) files, shapefiles, feature services, map services, image services, GeoRSS, Keyhole Markup Language (KML), Web Mapping Service (WMS), Web Map Tile Service (WMTS), and tile layers. You can also use map notes layers to create layers directly.
- You can also use other data as your operational layers. ArcGIS Online has a large number of layers, including thousands of layers in the Living Atlas of the World, created by Esri and the user community, and ArcGIS Open Data (**opendata.arcgis.com**). These layers span a range of subjects and can support maps and apps of almost every subject. You can search and discover authoritative layers that fit your needs.

Tools

Tools enable you to perform tasks beyond mapping, including common tasks, such as displaying pop-ups, querying, geocoding, routing, and generating heat maps, and more specialized tasks, such as performing analyses and workflows that implement specific logic for an enterprise.

- ArcGIS Online provides flexible ways for you to configure layer pop-ups.
- ArcGIS Online also provides rich and extensive spatial analysis functions for you to ask questions and solve spatial problems.
- ArcGIS for Server allows you to publish your own tools as geoprocessing services to support customized online spatial analysis.

Table 2.1 **Web GIS app components and their support in ArcGIS Online**

Web GIS app components	Support in the ArcGIS platform
Basemap layers	Basemap gallery ArcGIS map services, vector tile services, WMS, WMTS, tile layer as basemaps Elevation service to support 3D web scenes
Operational layers	Layers from files (CSV, shapefile, and so on) Layers from the Web (ArcGIS map/feature services, WMS, WMTS, KML, GeoRSS, and so on) Map notes layers
Tools	Pop-ups in map viewer Filter and query, geocoding, routing, and analysis in portal map viewer Geoprocessing services that can be published based on the models and scripts of users

Living Atlas of the World

Traditionally, you had to collect or prepare all or most of the data for your application and analysis. Today, you can find rich content from the ArcGIS Online catalog. The ArcGIS Living Atlas of the World collection is a selected subset of the ArcGIS Online contents. You can use Living Atlas content as your operational and basemaps layers.

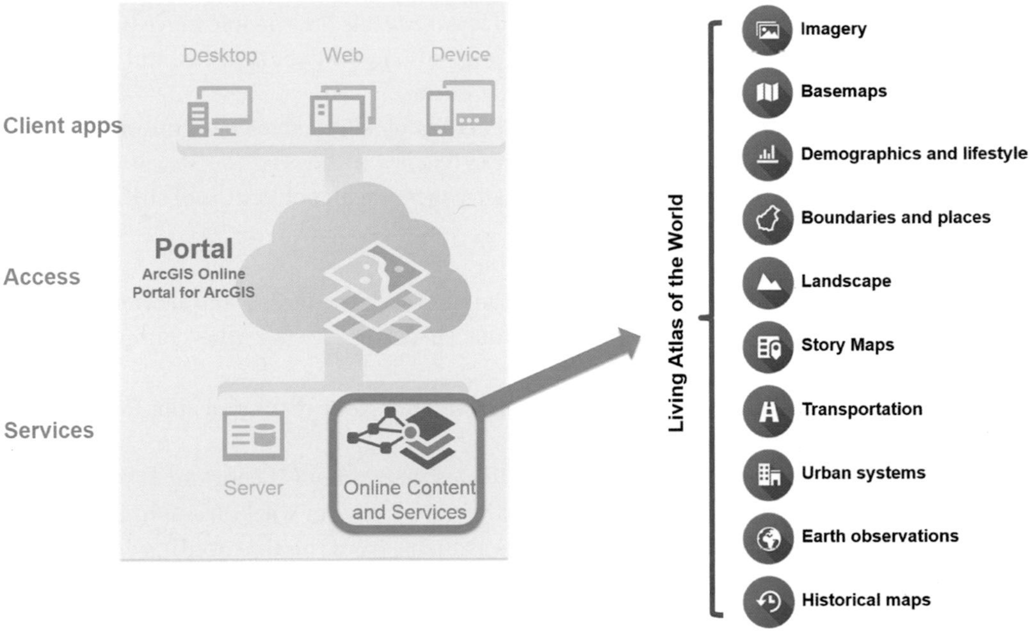

The Living Atlas of the World is a curated subset of ArcGIS Online information contributed to and maintained by Esri and the ArcGIS user community. The Living Atlas has thousands of layers covering many topics.

The Living Atlas is a dynamic collection of thousands of maps, data, imagery, tools, and apps produced by ArcGIS users worldwide and by Esri and its partners. This atlas is the foremost collection of authoritative, ready-to-use global geographic information ever assembled. You can combine content from this repository with your own data to create powerful new maps and apps. You can use these maps and apps when you perform diverse analyses in ArcGIS Online without having to collect the data yourself.

The Living Atlas provides the following content categories:

- **Imagery:** Detailed event, basemap, multispectral, and temporal imagery captures the historic and present state of our planet.
- **Basemaps:** Beautiful and authoritative maps provide reference for our world and context for your work.
- **Demographics and lifestyles:** Maps and data for the United States and more than 120 other countries reveal insights about populations and their behaviors.
- **Boundaries and places:** Boundaries help define where people live and work, and these layers span a variety of scales, from neighborhood to continental extents.
- **Landscape:** Data reflect both the natural environment and man-made influences to support land-use planning and management.
- **Story maps:** The map collection includes narrative text, images, and multimedia content to engage and inspire your audience.

- **Transportation:** The collection of maps and layers reveals how people move between places.
- **Urban systems:** More than half the world's population now lives in cities, and these layers allow analysis of how population impacts the world.
- **Earth observations:** These observations collect our planet's extreme events and conditions, from severe weather to earthquakes and fires.
- **Historical maps:** These maps reflect the changing physical, political, and cultural aspects of our world over time.

You can use and/or contribute to the Living Atlas.

- **As a user:** In this tutorial, you will use the Living Atlas by browsing and searching its content in the ArcGIS Online map viewer. You can also use Living Atlas content in ArcGIS Scene Viewer.
- **As a contributor:** You can contribute data or nominate your maps and apps for inclusion in the Living Atlas content collection.

The word "living" in the name Living Atlas indicates that content is continuously updated in minutes or hours (live traffic and real-time earthquakes), days or weeks (remote sensing imageries), and regularly as new content becomes available from the contributing community. For more information on the Living Atlas, go to **http://doc.arcgis.com/en/living-atlas/about**.

Smart mapping

Your layers must be displayed in meaningful styles or symbols for you and your end users to discover the hidden patterns and to deliver intended messages. If your layers do not come with styles or you do not like their existing styles, you can change styles using ArcGIS smart mapping capabilities (not all layers allow style changing).

Workflows analyze your data and suggest the best way to represent it.

Smart defaults take the guesswork out of setting up many of the map properties.

Based on the basemap you select, we can automatically suggest and coordinate colors and other map styling.

Suggested visible ranges allow you to see your data at sensible scales.

You can preview your styling choices on your screen.

Smart mapping aims to provide a strong, new "cartographic artificial intelligence" that enables all users to visually analyze, create, and share professional-quality maps easily and quickly with minimal mapping knowledge or software skills.

Smart mapping provides new and easy ways to symbolize your data and suggest the "smart" defaults. Smart mapping delivers continuous color ramps and proportional symbols, improved categorical mapping, heat maps, and new ways to use transparency effects to show additional details about your data via a streamlined and updated user interface. Unlike traditional software defaults that are static (the same every time), smart mapping analyzes your data quickly in many ways, suggesting the right defaults when you add layers and change symbolizing fields. The nature of your data, the map you want to create, and the story you want to tell all drive these smart choices.

Smart mapping does not oversimplify the map-authoring experience or take control away from you. You can still specify parameters manually to extend default capabilities. For more information on smart mapping, see **http://www.esri.com/landing-pages/arcgis-online/smart-mapping.**

User experience design principles

User experience is crucial in considering web GIS app design and implementation. A good web GIS app should deliver informative content and enhance necessary functionality for a fast, easy, and fun user experience.

- **Fast:** "Don't make me wait," say today's users. Web GIS apps should use caching, database tuning, appropriate client/server task partitioning, and load balancing to achieve optimal performance, scalability, and availability. For example, ArcGIS Online does most of these tasks for the web maps and apps that you create using the Esri technology.
- **Easy:** Today's users also say, "Don't make me think about which button to click," and "If I don't know how to use your site, it's your problem, and I will leave the site quickly!" Web GIS apps should focus on a specific purpose. Do not overwhelm users with unnecessary buttons and data layers. Make the user interface intuitive. The interface should provide feedback, such as visual cues, that lead users through a well-defined workflow and assure them that they are on the right track.
- **Fun:** Integrate photos, charts, videos, and animation into your web apps. Used properly, these media engage users, convey your key information, and improve user satisfaction.

Pop-ups

Pop-ups are windows that show geographic information and deliver geographic insight. They are a common tool that your end users rely on to interact with your operational layers. Today's users click or tap a location or feature on the map and expect to see a pop-up showing more information.

The default pop-up appearance for a layer is a plain list of attributes and values. You can configure the pop-ups to show custom-formatted text, attachments, images, and charts and to link to external web pages. Pop-ups enhance the attributes associated with each feature layer in the map and present the information in intuitive, interactive, and meaningful ways.

Create and edit data online

Typically, you will use desktop products such as ArcMap and ArcGIS Pro to create and edit GIS data. You can also create and edit lightweight data using the built-in ArcGIS Online map viewer.

In map viewer, you can create point, line, and polygon layers using map notes. Only the web map owner can modify your map notes layers (unless another copy is saved). You can also use the map viewer to edit feature layers, such as CSV, TXT, and GPX files and shapefiles. The ability to edit in map viewer helps if you lack access to ArcGIS for Desktop or have already created part of a web map configuration and do not want to start over again. You and your web app users also can edit feature layers from feature services.

This tutorial

In this tutorial, you will learn how to create a web GIS app that presents the spatial patterns of US population growth, and you will explore the reasons behind the patterns.

Data: For the operational layers, you are provided with a CSV file, **Top_50_US_Cities.csv**, which contains the 2010 to 2014 population growth of the 49 most populated cities in the United States.

- The CSV file contains no latitude and longitude fields.
- After adding data into the map viewer, you will need to fix the data there.
- Other operational layers are not provided. You will find them in the Living Atlas.

Requirements:

- Your app should display the population change patterns in major US cities and at various administrative unit levels.
- Your map symbols should be easy to understand.
- If a city or a region is clicked, your app should display associated details in intuitive ways.

System requirements:

- ArcGIS Online organizational account or Portal for ArcGIS account (user level is fine). You can use the trial account you created in chapter 1.
- A web browser.
- Microsoft Excel or a text editor.

2.1 Map CSV data using geocoding

In this section, you will map your first operational layer, a CSV file.

1. **In Microsoft Excel, navigate to C:\EsriPress\GTKWebGIS\Chapter2\Top_50_US_Cities.csv, and study the data fields.**

Rank_2014	City	State	Census_April_2010	Estimate_July_2010	Estimate_July_2011	Estimate_July_2012	Estimate_July_2013	Estimate_July_2014	Wikipedia_URL	Picture_URL
1	New York	New York	8,175,133	8,191,853	8,287,238	8,365,903	8,438,379	8,491,079	http://en.wikipedia.org/\	http://upl
2	Los Angeles	California	3,792,621	3,796,290	3,826,423	3,861,678	3,897,940		http://en.wikipedia.org/\	http://upl
3	Chicago	Illinois	2,695,598	2,697,319	2,705,627	2,715,415	2,722,307	2,722,389	http://en.wikipedia.org/\	http://upl
4	Houston	Texas	2,099,451	2,102,421	2,129,784	2,164,834	2,203,806	2,239,558	http://en.wikipedia.org/\	http://upl
5	Philadelphia	Pennsylvania	1,526,006	1,528,544	1,539,313	1,550,396	1,556,052	1,560,297	http://en.wikipedia.org/\	http://upl
6	Phoenix	Arizona	1,445,632	1,449,583	1,465,114	1,489,531	1,512,442	1,537,058	http://en.wikipedia.org/\	http://upl

Note: The CSV file contains no latitude and longitude fields. Instead, the file contains the following fields:

- **Rank_2014:** A city's 2014 rank by population
- **City:** City name
- **State:** Name of the state in which the city resides
- **Census_April_2010:** City population as of April 1, 2010
- **Estimate_July_2010:** Estimated city population as of July 1, 2010
- **Estimate_July_2011:** Estimated city population as of July 1, 2011
- **Estimate_July_2012:** Estimated city population as of July 1, 2012
- **Estimate_July_2013:** Estimated city population as of July 1, 2013
- **Estimate_July_2014:** Estimated city population as of July 1, 2014
- **Wikipedia_URL:** URL to the city's Wikipedia page
- **Picture_URL:** URL to the image of the city's seal or flag

Note: The map viewer has a limit of one thousand points per CSV file. If you are using a public user account and your CSV file needs geocoding, the limit is 250 points per CSV file. See the **Questions and Answers** section at the end of this chapter for more information.

2. **Open a web browser, navigate to ArcGIS Online (www.arcgis.com) or your Portal for ArcGIS, and sign in.**

3. **Click Map to open the map viewer, and start a new map.**

4. **Click the Add button, click Add Layer from File, locate the Top_50_US_Cities.csv file on your computer, and click Import Layer.**

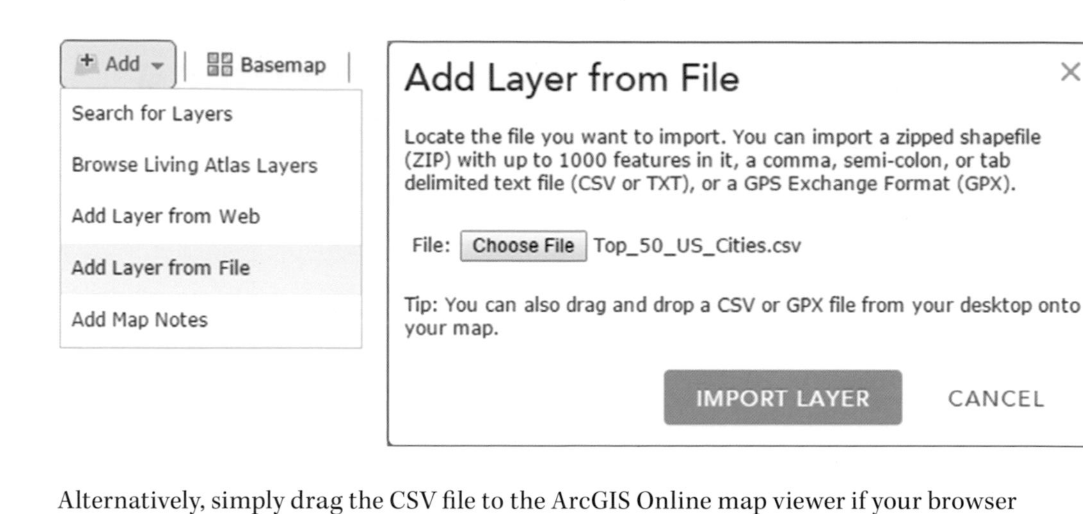

Alternatively, simply drag the CSV file to the ArcGIS Online map viewer if your browser supports drag-and-drop operations.

5. **In the Add CSV Layer window, review the geocoding settings for the Country and Location fields.**

Because the CSV file does not have latitude and longitude fields, the map viewer will geocode your data to map the points. Geocoding converts addresses and other identifiers (such as place-names and postal codes) into coordinates. ArcGIS Online geocoding services include more than one hundred countries. ArcGIS Online for Organizations and Portal for ArcGIS can also be configured to use your own geocoding services.

If your data contains addresses for a single country, select its name from the **Country** list. If the addresses refer to multiple countries or to a country not on the **Country** list, select **World**. Then review the location fields. To correct a field, click its cell.

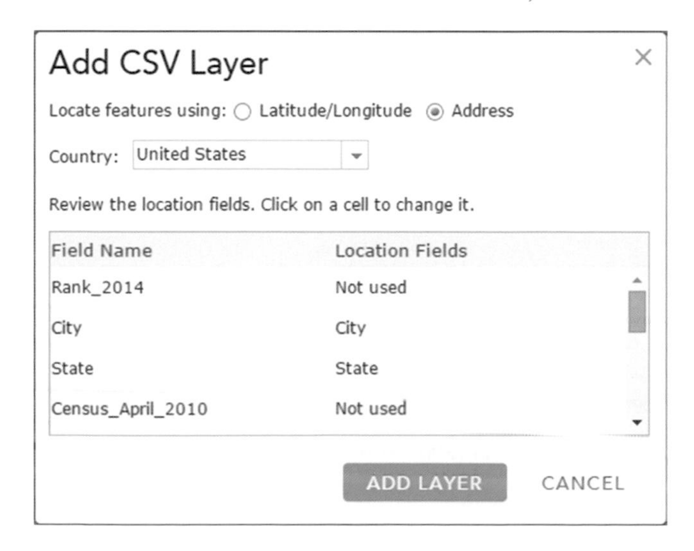

6. **In the Add CSV Layer window, click Add Layer. Notice that the cities are then geocoded and displayed on the map.**

7. **Save your web map by clicking the map viewer Save button. In the Save Map window, enter appropriate title and tags, and then click Save Map.**

The information specified in the window constitutes your web map's metadata. Other users can use this information to discover your web map.

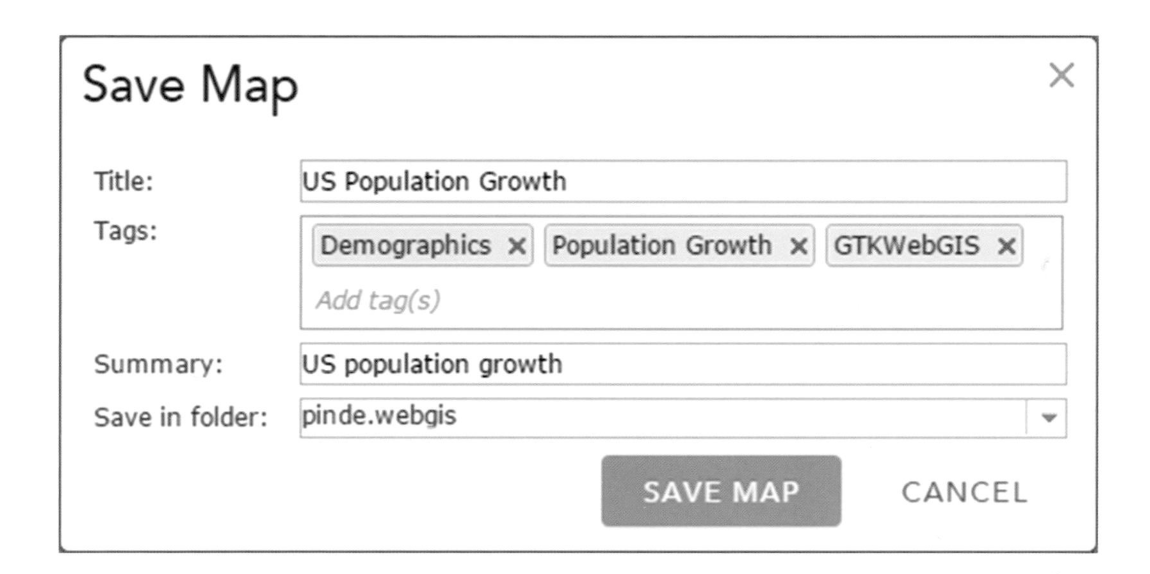

You have created a basic web map. You will edit the data and enhance the map in the following sections of this chapter.

2.2 Edit data in the map viewer

You may find problems or errors in your CSV file or shapefile after you have created a web map. You do not always have to go back to the source data and redo your work. Knowing how to edit your data directly in the portal map viewer can save you time.

In this section, you will add a missing city (New Orleans, Louisiana) and add a missing value (for Los Angeles).

1. **As you continue from the last section, in the Contents pane, click Cancel to close the Change Style pane.**

You should see all the cities displayed in the default single symbol. You will configure the actual drawing style in the next section.

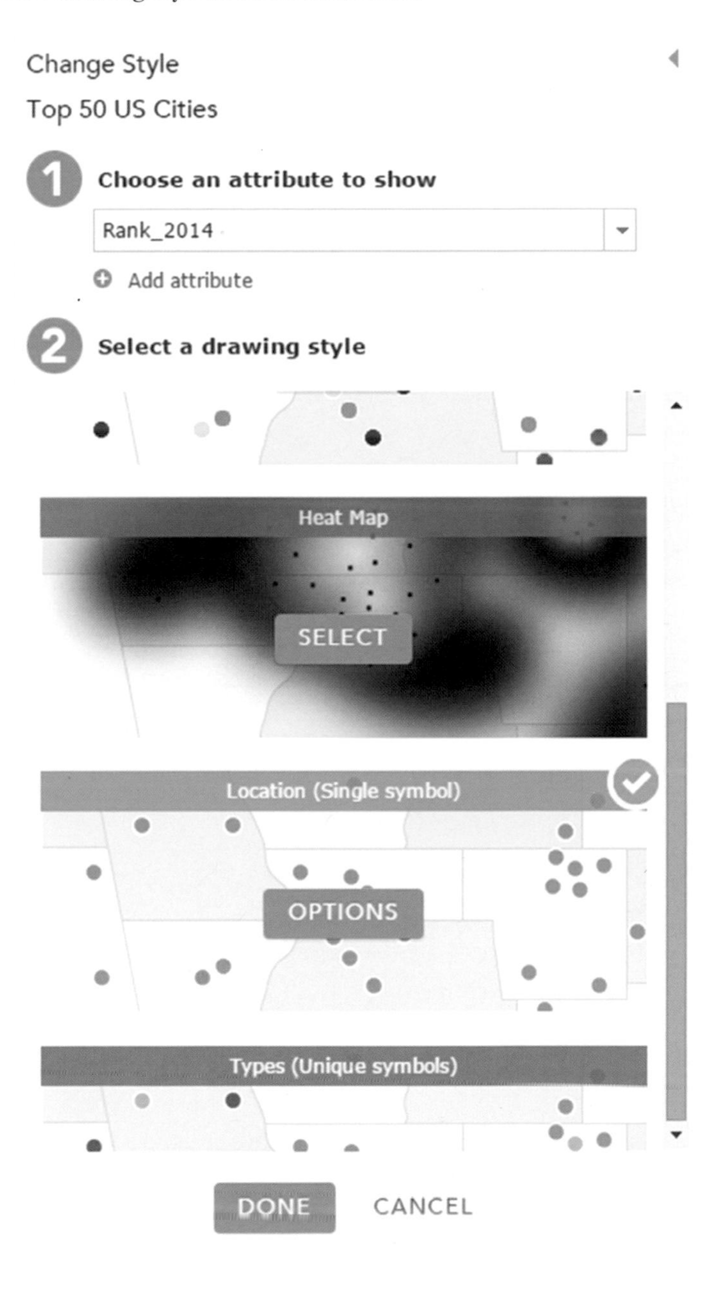

2. In the **Contents** pane, point to the layer **Top_50_US_Cities**, click the **More Options** button ···, and then click **Enable Editing**.

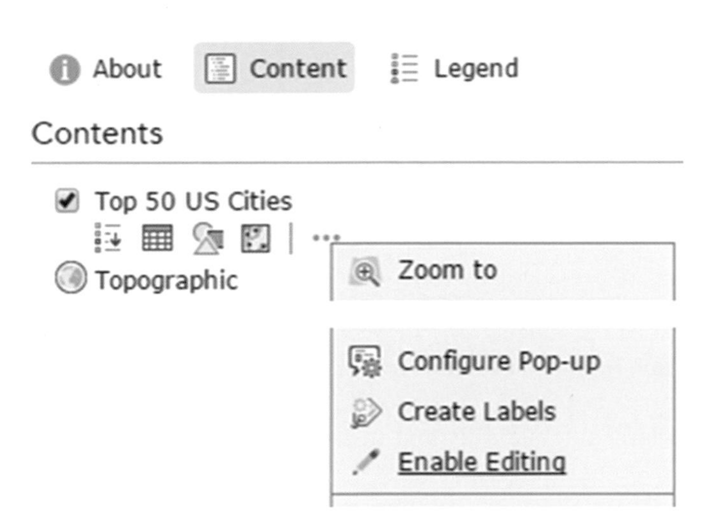

3. In the map viewer search box, type **New Orleans, Louisiana**, and then press **Enter** or click the **Search** button.

The **Search** function finds the city and zooms there. A **Search result** pop-up appears, indicating the location.

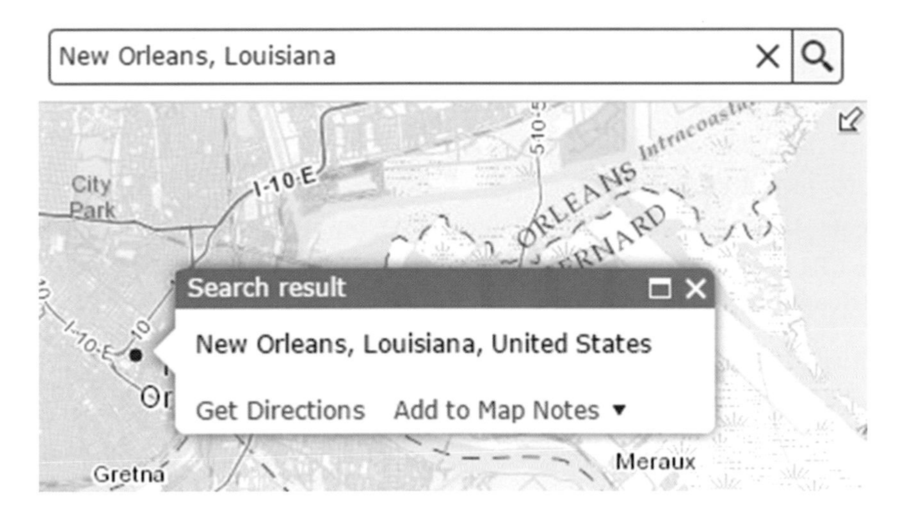

4. **On the menu bar, click the Edit button. Then click New Feature to select the symbol.**

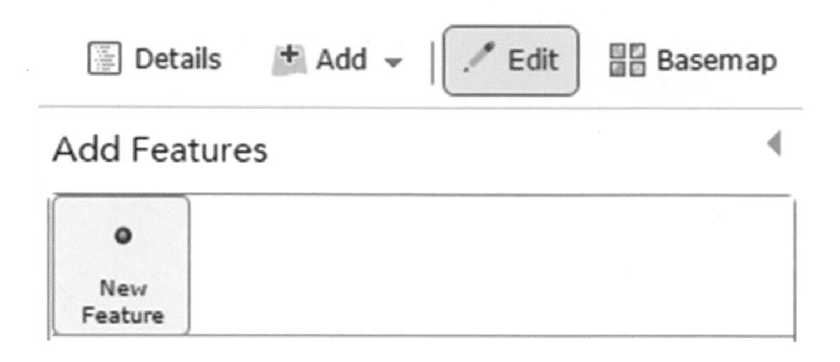

5. **On the map, click any location in the City of New Orleans.**

This city location is added to your layer automatically, and an attribute editor pop-up appears.

6. **In the attribute editor, specify the following attribute values:**

- **Rank_2014: 50**
- **City: New Orleans**
- **State: Louisiana**
- **Census_April_2010: 343,829**
- **Estimate_July_2010: 343,829**
- **Estimate_July_2010: 347,987**
- **Estimate_July_2011: 360,877**
- **Estimate_July_2012: 370,167**
- **Estimate_July_2013: 379,006**
- **Estimate_July_2014: 384,320**
- **Wikipedia_URL: https://en.wikipedia.org/wiki/New_Orleans**
- **Picture URL: https://upload.wikimedia.org/wikipedia/commons/7/75/Seal_of_New_Orleans%2C_Louisiana.png**

You can copy the values from **C:\EsriPress\GTKWebGIS\Chapter2\New_Orleans.txt.**

7. Click Close.

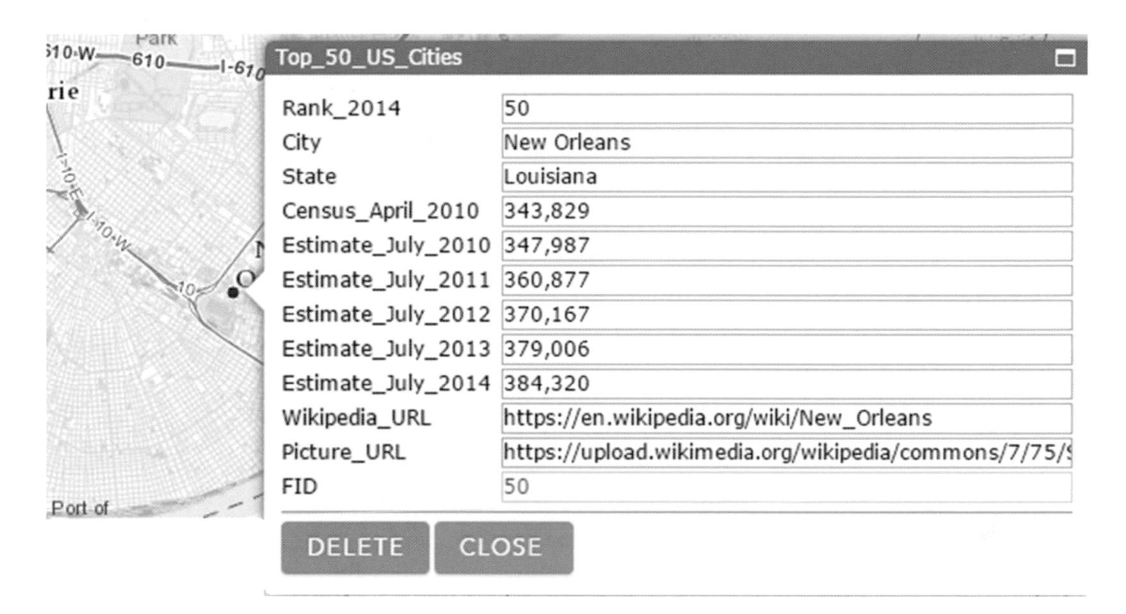

8. Navigate the map to Los Angeles.

If you do not know the location of Los Angeles, you can search for it as you did in step 3.

9. Click the existing point symbol of Los Angeles so the attribute editor appears. Fill in the missing value for Estimate_July_2014 as 3,928,864, and then click Close.

10. **On the menu bar, click the Details button to bring up the Contents pane. Point to the layer Top_50_US_Cities, click the More Options button ···, and then click Disable Editing.**

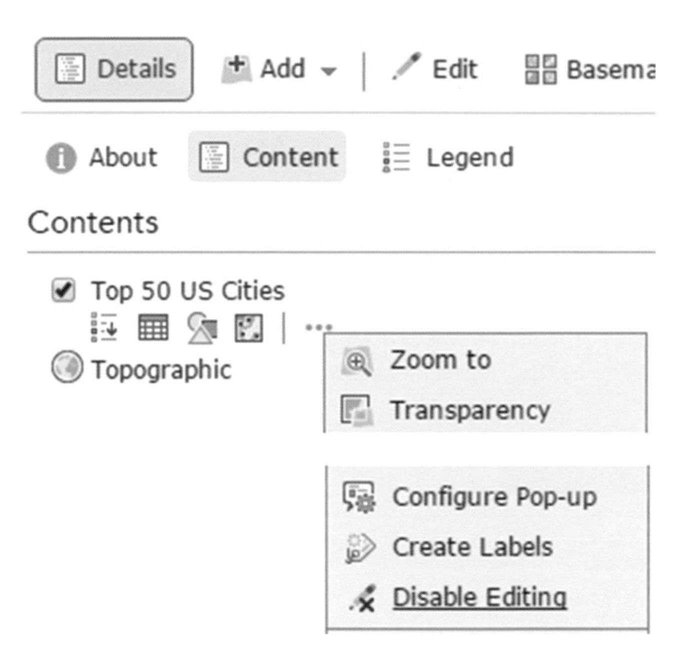

You have finished editing your data and have exited the editing mode, which will prevent unintended data changes.

2.3 Configure layer style

In this section, you will explore smart mapping and configure the drawing style of your layer.

1. **With the map viewer open, in the Contents pane, point to Top 50 US Cities, and then click the Change Style button.**

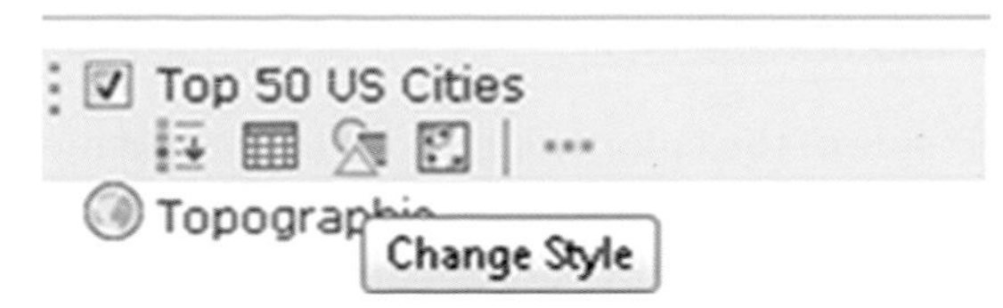

2. **In the Change Style pane, select Heat Map, and then zoom the map out to include all the cities.**

You will see the checkmark next to **Heat Map** indicating your current choice. You can make additional configurations using the **Options** button.

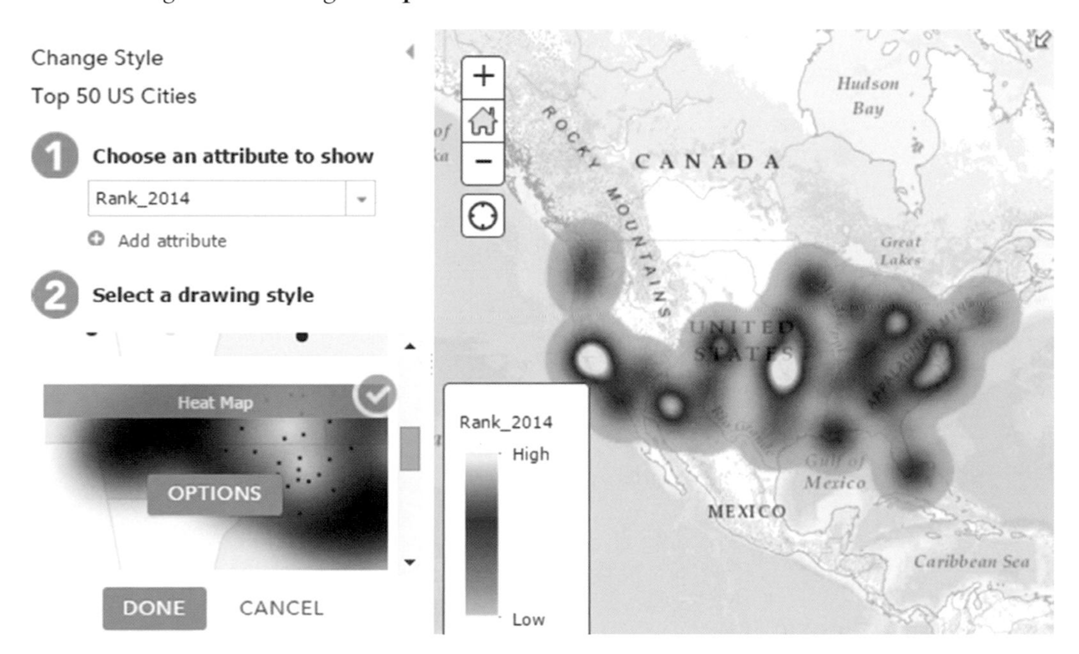

Heat maps are especially useful when you map many points that are close together and cannot easily distinguish the points. Heat maps calculate the relative density of these points and display the density as smoothly varying sets of colors ranging from cool (low density of points) to hot (many points).

In this example with only 50 points, a heat map is not the best way to present the data. However, the heat maps still clearly show that the West and East Coasts have higher densities of large cities, as indicated by the stronger colors appearing to glow hotter.

3. **In the Change Style pane, click Options on the Heat Map drawing style to experiment with and understand the effects of various options.**

   - **Adjust the two handles on the left side of the color ramp slider to change how the colors get applied to the density surface.**
   - **Adjust the Area of Influence slider to make the clusters become larger and smoother or smaller and more distinct.**
   - **Click Symbols to choose a different color ramp.**
   - **Click OK to exit the Options mode.**

Top 50 US Cities

Heat Map

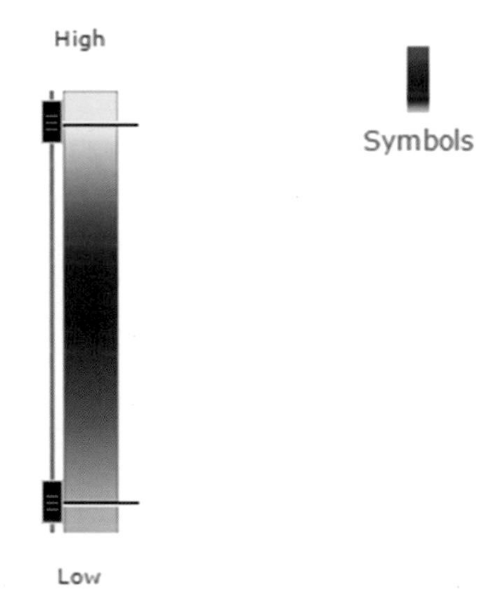

Area of Influence

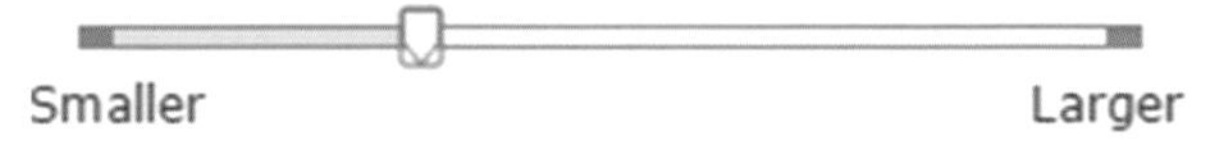

Transparency

Next, you will use smart mapping to display US populations of cities and population changes.

4. **In the Choose an attribute to show text box, choose Estimate_July_2014 as the field to show in the Change Style pane.**

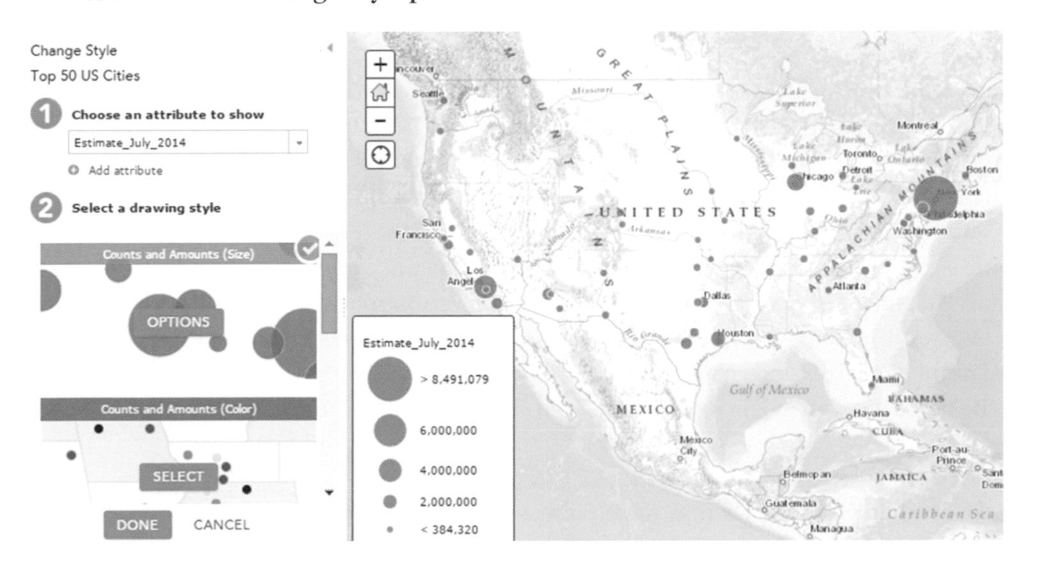

Smart mapping automatically selects **Counts and Amounts (Size)** as the default style for this numeric field. This style uses an orderable sequence of different sizes to represent your numeric data or ranked categories. You can draw points, lines, and areas using this approach. The proportional symbols make the map intuitive. On the map, can you easily tell the top three most populated cities in the United States?

5. **Click Options in the Counts and Amounts (Size) style to further configure the layer style. Set Divided By to Census_July_2010.**

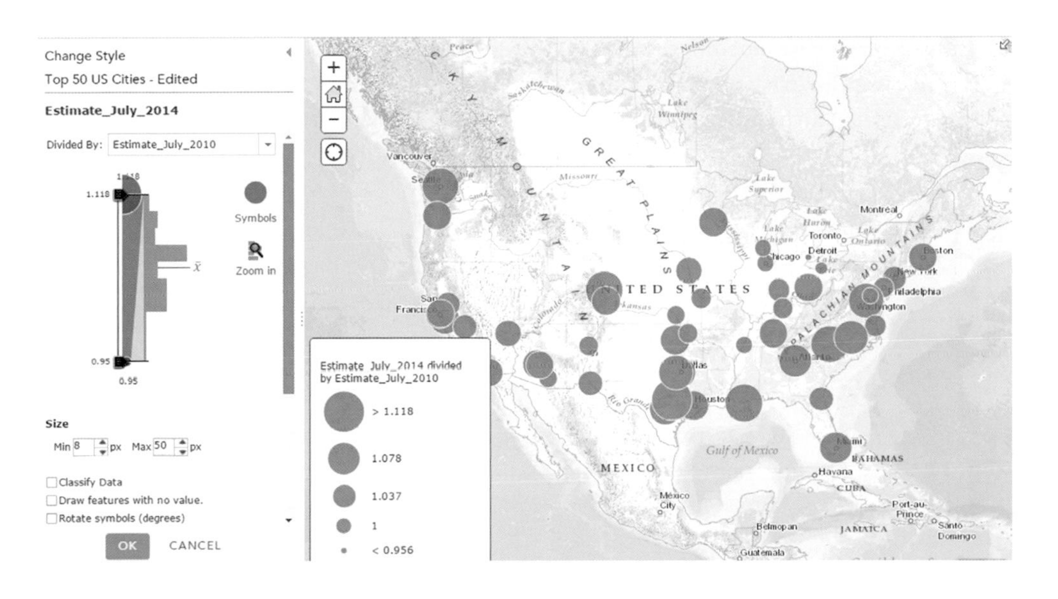

The map now displays the ratio between the 2014 population and the 2010 population, which is a good indicator of population growth over the period of four years. Bigger circles indicate bigger increases, and smaller circles indicate smaller increases and even decreases. A value of 1.0 means there was no change, numbers greater than 1.0 indicate growth, and numbers less than 1.0 indicate population loss. This ratio is a good approach to reveal how things change over time and which locations change the fastest, without having to create multiple maps. This approach works not only for population data but for any two numeric fields in a layer.

You may also change the **Min** and **Max** symbol sizes and adjust the two handles on the histogram slider to exaggerate the cities with the most and the least population growth. Smart mapping easily emphasizes certain ranges of the data and uncovers subtle details.

Smart mapping also uses continuous sizes and colors automatically, but it does not remove any of the traditional methods to break features into a set of classes and show them with a limited number of sizes and colors.

In the next step, you will manually break the cities into classes and assign symbols for these classes.

6. **Scroll down in the Change Style pane, and check the box next to Classify Data. Select manual breaks with 4 classes and Round classes to 0.01 (in other words, round the class break values to 2 decimal points).**

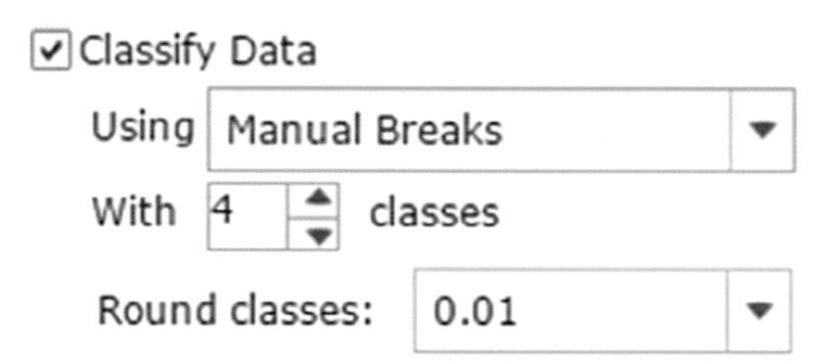

7. **Starting from the bottom, set the class break values to 1, 1.05, and 1.1. You can do so by clicking the existing break values and typing in the new values, as illustrated, or by dragging the handles on the histogram slider in Change Style.**

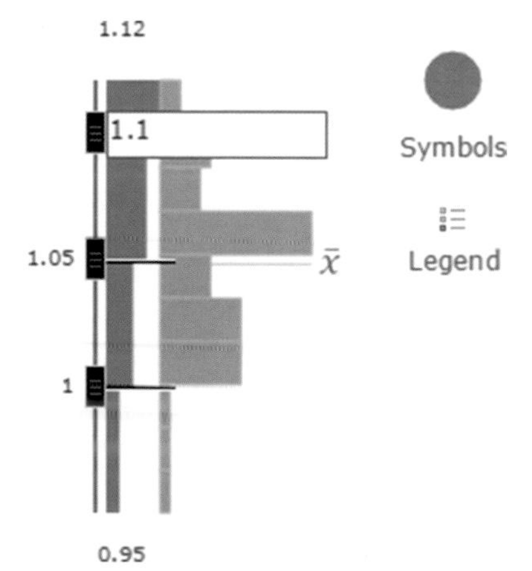

Note the **Symbols** and **Legend** icons in the figure. Clicking the **Symbols** icon allows you to set a symbol for all the classes. Clicking the **Legend** icon allows you to set an individual symbol and label for each class.

Currently, all of the classes are using the same symbol with gradual sizes. Next, you will set each class to use a different symbol.

8. **Click Legend to edit the symbol for each individual class.**

9. **Click the biggest circle—the one next to > 1.1 to 1.12. A window pops up showing the available point symbols. Click the list to choose Arrows, click the solid red up arrow, set its size to 36, and click OK.**

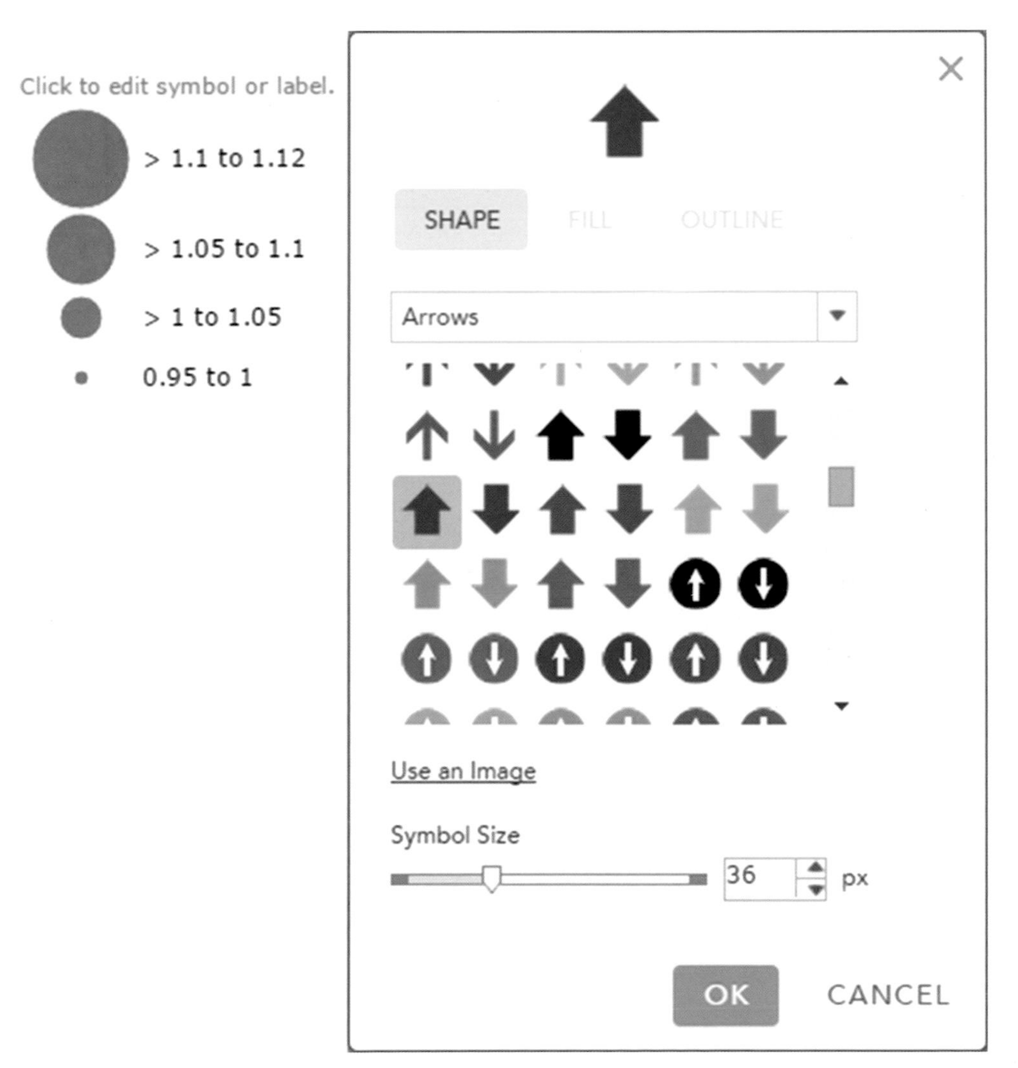

The map viewer provides a collection of symbols for business, damage, disasters, infrastructure, recreation, people, places, points of interest, health, transportation, and other categories, as shown in the list of options. You can browse through these categories and get a sense of the variety of available symbols. To browse, you can click the symbol list to access the options.

10. **Similarly, change the symbol for the > 1.05 to 1.1 class to a brown up arrow ⬆ in the Arrows group, set its size to 24, and click OK.**

In addition to the collections of symbols, the map viewer allows you to use your own images as symbols. You will do so in the next steps.

11. **Click the symbol for the > 1 to 1.05 class, click Use an Image, specify http://esrimapbook.esri.com/GTKwebgis/chapter2/yellow_up.gif** (or the short equivalent, **http://arcg.is/1ryHfTU**) as the URL, click the **Plus** icon, set its size to **16**, and then click **OK**.

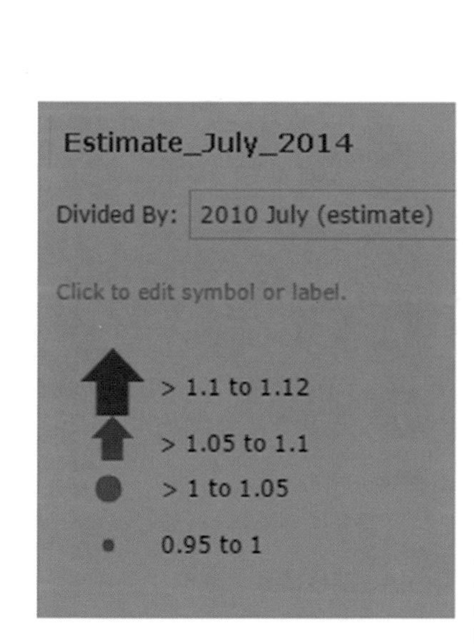

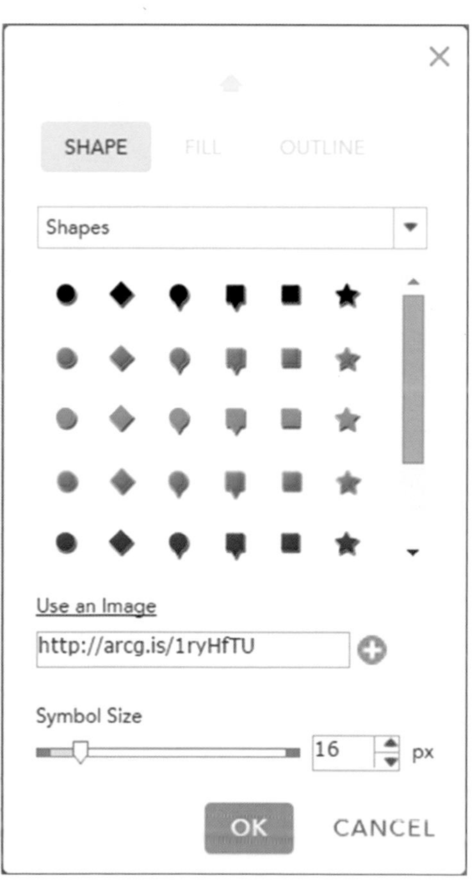

12. **Similarly, change the symbol for the 0.95 to 1 class to a blue arrow: specify http://esrimapbook.esri.com/GTKwebgis/chapter2/blue_down.gif (or the short equivalent http://arcg.is/1A2gdXp), and set its size as 16.**

The last class had a population loss from 2010 to 2014, so the class was symbolized using a solid down-arrow icon.

13. **Click OK, and then click Done to exit the Change Style pane.**

Next, you will change the basemap to a neutral background.

14. **To change the basemap, click the Basemap button on the map viewer toolbar, and click Light Gray Canvas.**

This basemap provides a neutral background with minimal colors, labels, and features. The neutral background allows your operational data layer to stand out clearly.

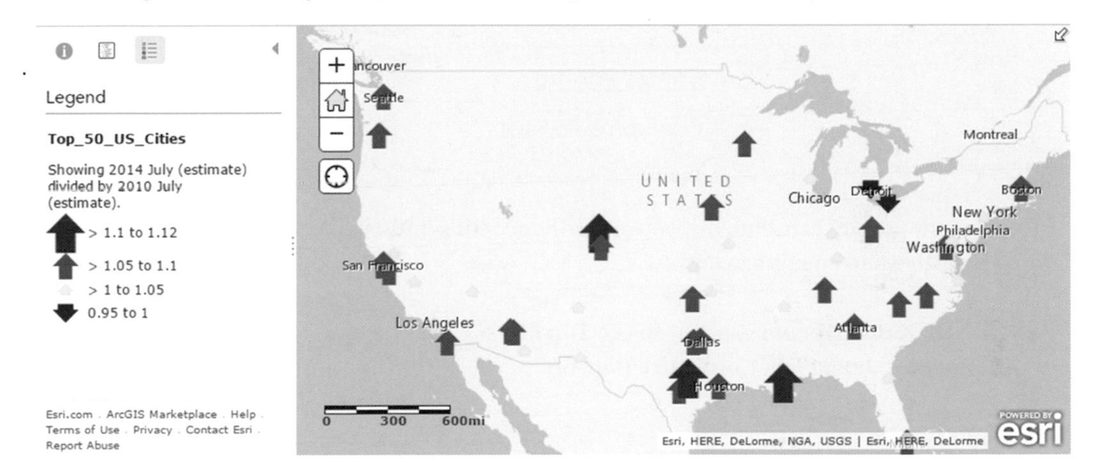

15. **Save your web map.**

You have configured the style of your operational layer. With this map, you can quickly and easily see which cities increased or decreased in population.

2.4 Configure layer pop-up

In this section, you will decide what information to show and how to display the information in the pop-ups of the cities.

1. **While you have your web map open in the map viewer, click one of the 50 cities to see the default pop-up.**

If a pop-up window does not appear, it may be disabled for this layer. You can enable the pop-up by pointing to the layer, clicking the **More Options** button and clicking **Enable Pop-up**.

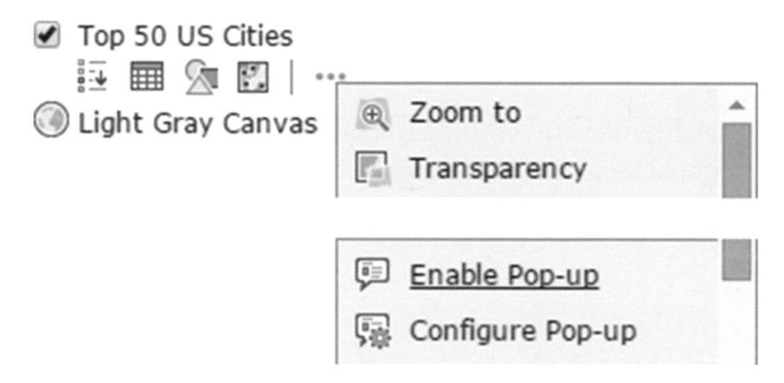

This default pop-up is useful, but you can enhance the window to communicate information in more intuitive and engaging ways.

2. **In the Contents pane, point to the Top 50 US Cities layer, click the More Options button ···, and click Configure Pop-up.**

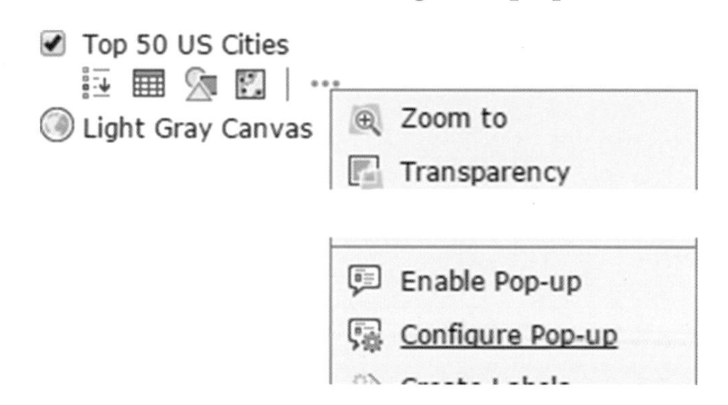

In the **Configure Pop-up** pane, you can configure the pop-up's title, contents, and media sections.

Next, you will configure the title, which can be static text, a set of attribute field values, or a combination of the two.

3. **Click the Plus button ⊞ under Pop-up Title to select the City field. Type a comma and a space, and click the Plus button again to select the State field.**

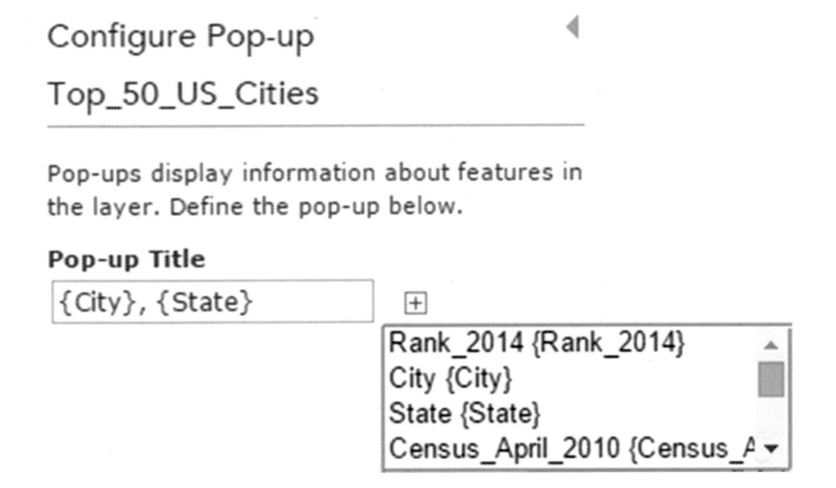

Next, you will configure the pop-up contents, which can include a list of attribute fields or a custom description based on attribute values.

4. In the **Display** menu under **Pop-up Contents**, click the arrow, and choose **A list of field attributes**, and then click the **Configure Attributes** link.

In the **Configure Attributes** window, you can choose which fields to show or hide and define their aliases, order, and formats.

5. In the **Configure Attributes** window, make the following choices:

   - **Select {Rank_2014}, and set its alias as Rank 2014.**
   - **Clear the check boxes for {City} and {State}. They already appear in the title, so you do not need to repeat them.**
   - **Keep {Census_April_2010}, {Estimate_July_2010}, {Estimate_July_2011}, {Estimate_July_2012}, {Estimate_July_2013}, and {Estimate_July_2014}, and set their aliases as 2010 April (Census), 2010 July (estimate), 2011 July (estimate), 2012 July (estimate), 2013 July (estimate), and 2014 July (estimate).**
   - **Clear the check boxes for the rest of the fields.**
   - **Click OK to close the Configure Attributes window.**

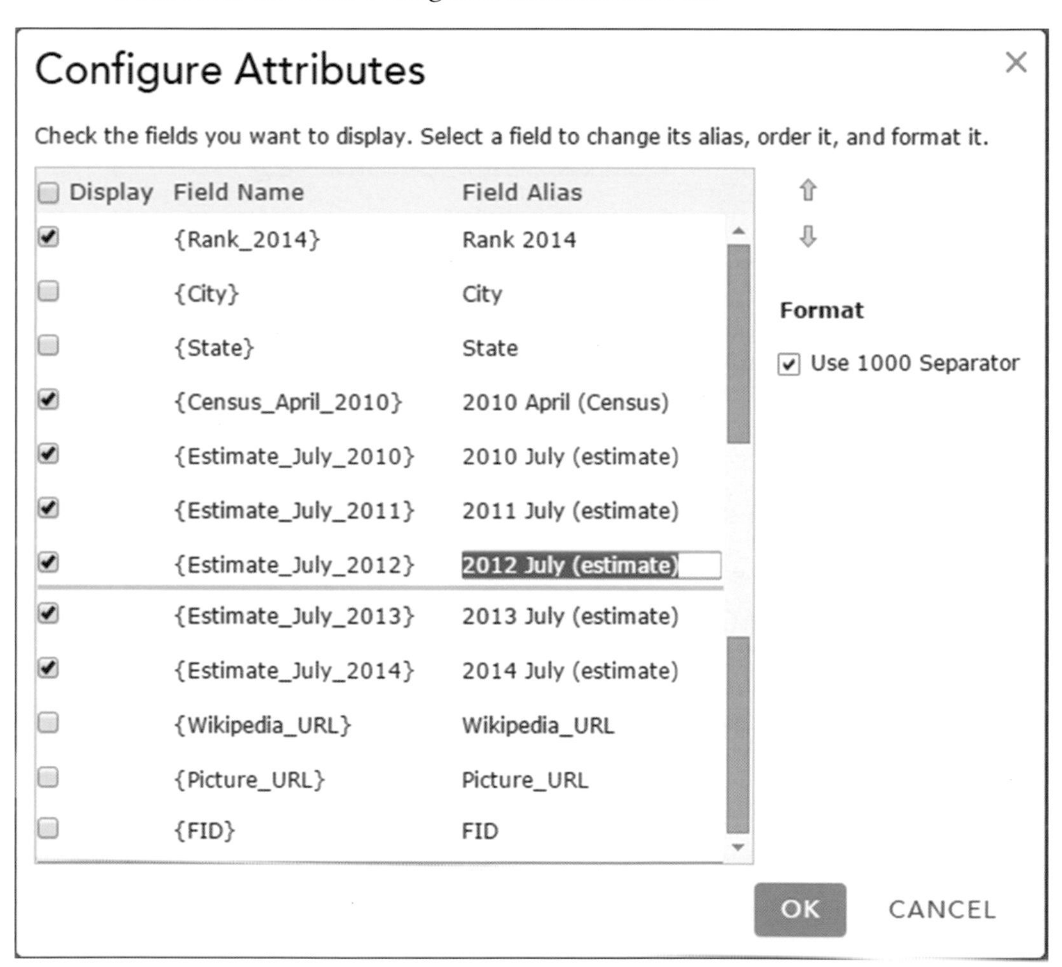

6. **In the Configure Pop-up panel, click Save Pop-up to apply your pop-up configuration.**

7. **Click any city on the map to review the new pop-up.**

You will see that the pop-up window is simpler and easier to read.

8. **On the map viewer menu bar, click the Save button to save your changes.**

2.5 Add images and charts to pop-up windows

Media, such as images and charts, can engage users more effectively than straight text and can improve their understanding of your data.

- The tutorial CSV file contains two URL fields, Wikipedia_URL and Picture_URL. You will use the first URL to add a picture to the city layer's pop-up and the second URL to link the picture with the city's Wikipedia page so that users can gather supplementary information about the city's population changes.
- Charts require numeric attribute fields. The US cities layer contains several of these fields. You will display them in appropriate charts to exemplify the population trends of the cities.

1. In the **Contents pane**, point to the **Top 50 US Cities** layer, click the **More Options** button, and click **Configure Pop-up.**

2. In **Pop-up Media**, click **Add**, and then click **Image.**

Pop-up Media

Display images and charts in the pop-up:

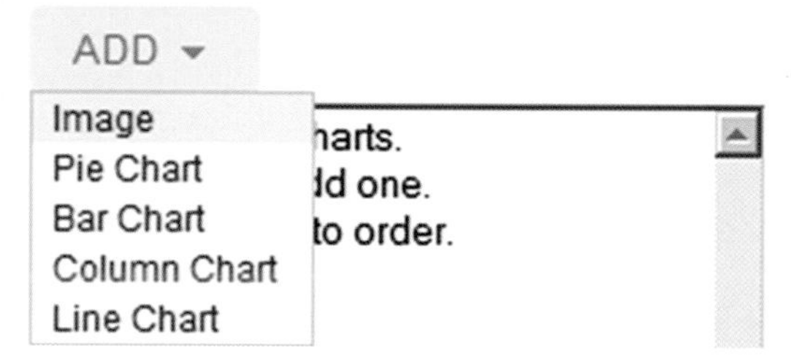

3. In the **Configure Image** window, make the following changes:

   - **Enter About the city as the title (leave the Caption window blank).**
   - **In the URL box, click the Plus button, and click the Picture_URL field.**
   - **In the Link (optional) box, click the Plus button, and click the Wikipedia_URL field.**
   - **Click OK to close the Configure Image window.**

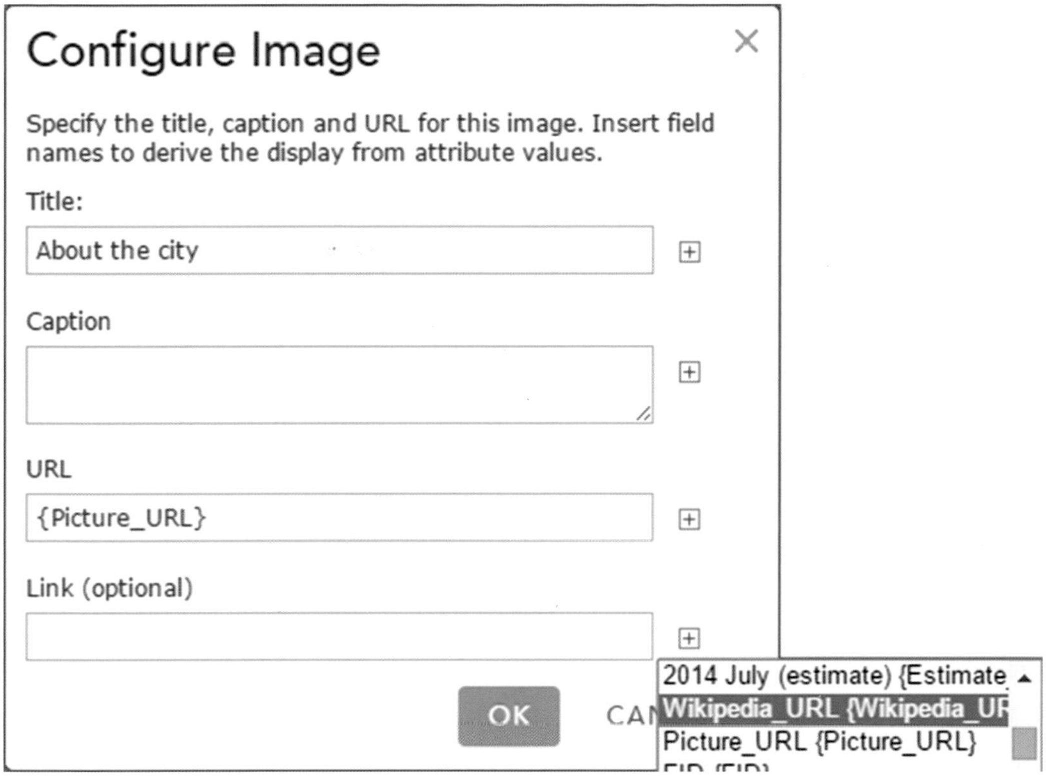

The image title, caption, image URL, and link URL can all take the form of static text, attribute field values, or a combination of the two. If you lack image and link URL fields when you do the assignment, specify a static URL instead. For example, you can use **http://www.census.gov/history/img/Census_Logo.jpg** as the image URL and **http://www.census.gov** as the Link URL. This way, the pop-up windows for all cities display the same image and link to the same web page.

You can add more than one image to your pop-up window simply by repeating steps 2 and 3.

4. **In the Configure Pop-up pane, click Save Pop-up to apply your pop-up configuration.**

5. **Click a city on the map to observe the new pop-up.**

The pop-up window displays the city's seal or flag. If you click on the image, the city's Wikipedia page will appear.

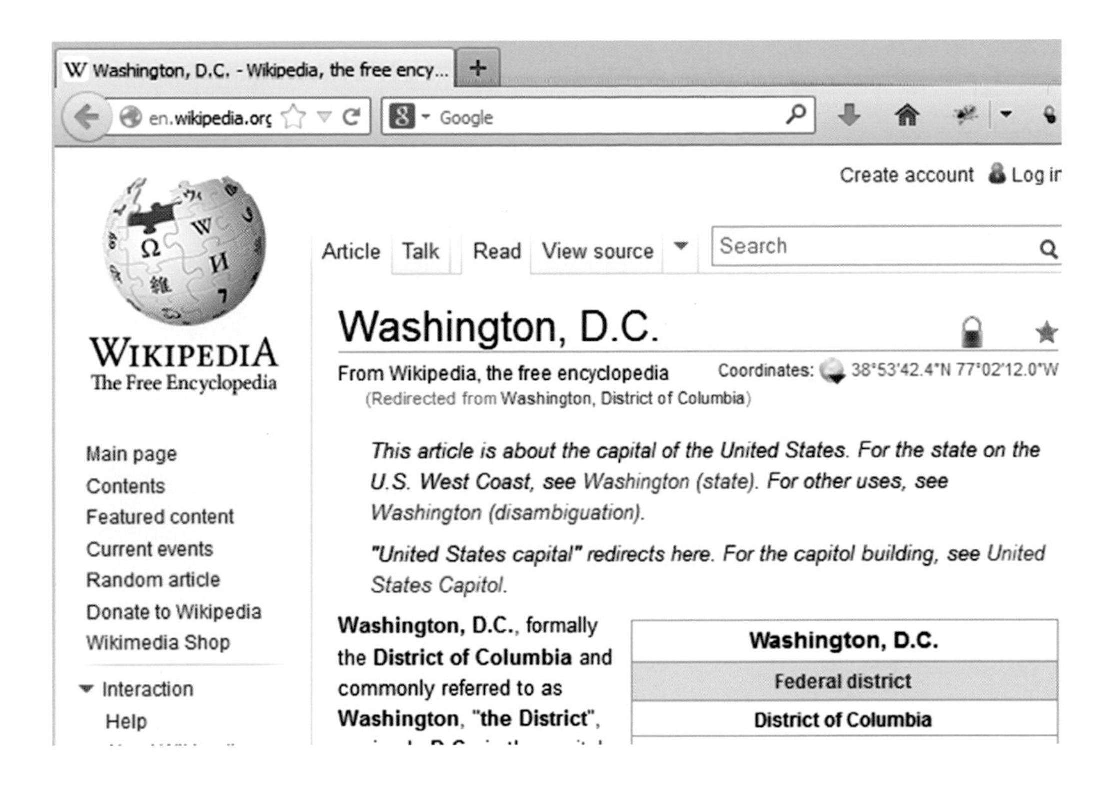

6. In the **Contents** pane, point to the **Top 50 US Cities** layer, click the **More Options** button, and click **Configure Pop-up** again. In the **Pop-up Media** section, click **Add**, and then click **Line Chart**.

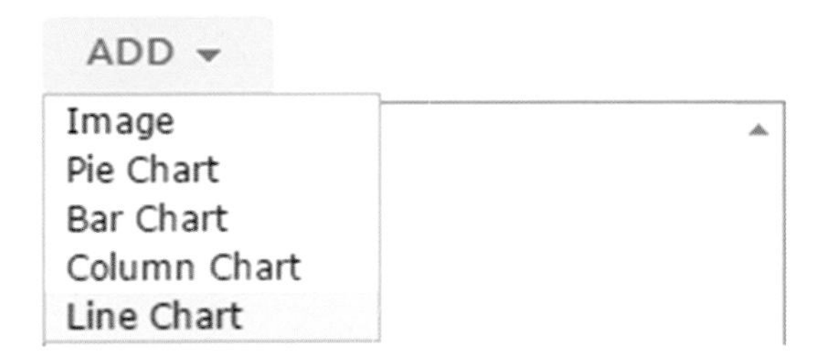

7. In the **Configure Line Chart** window, make the following changes:

   - **For Title, specify Population Change (2010–2014).**
   - **For Chart Fields, check the 2010, 2011, 2012, 2013, and 2014 estimated population fields.**
   - **Click OK to close the Configure Line Chart window.**

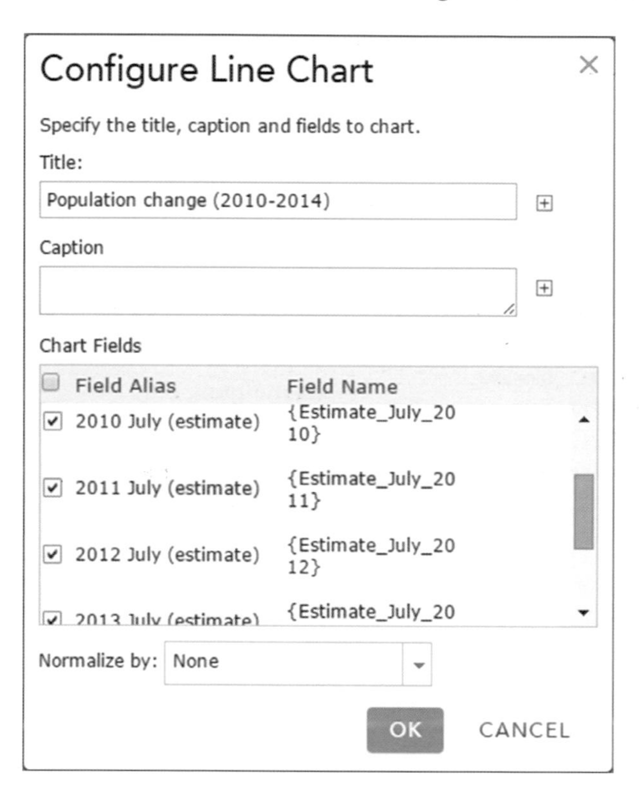

8. Click **Save Pop-up.**

9. **Click a city—for example, Detroit, Michigan—to see the new pop-up (use the Search box to find Detroit and other cities if necessary). To the right of the city seal or flag image, click the right arrow ▶ to see the chart you configured.**

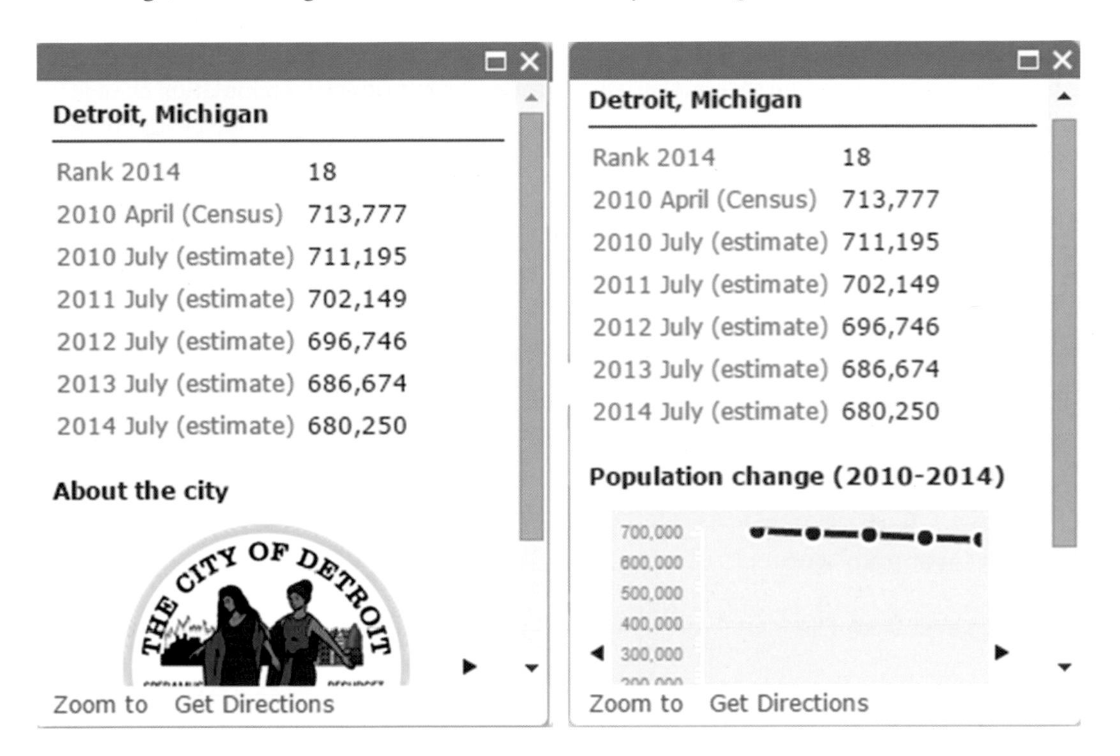

You will notice that the population of Detroit has been decreasing.

10. **Save your web map.**

The new pop-up you configured does more than display raw attribute values. The pop-up also leverages the URLs stored in the attributes to display and link a picture to a website that provides details. The pop-up also charts the numeric fields to provide a visual interpretation of population changes.

2.6 Use layers from the Living Atlas

In addition to showing the population changes of the major US cities, you will add additional operational layers to further display US population change patterns so users can explore the reasons behind the patterns. In this section, you will add a layer to show US population change from 2015 to 2020 and a layer to show 2015 unemployment rates. You are not provided with the data, but you will find these layers in the Living Atlas.

1. **On the map viewer menu bar, click the Add button, and from the list, click Browse Living Atlas Layers.**

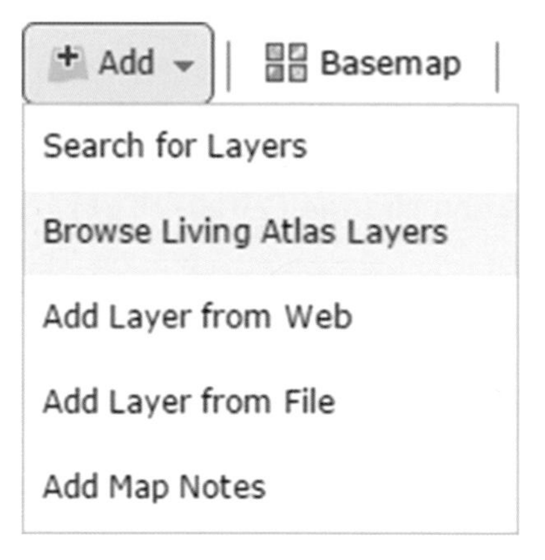

In the **Browse Living Atlas Layers** window, you will explore the extensive categories and subcategories of layers available.

2. **In the Browse Living Atlas Layers window, click the arrow to see the list of categories.**

3. **In the Demographics & Lifestyle category, click Population & Housing.**

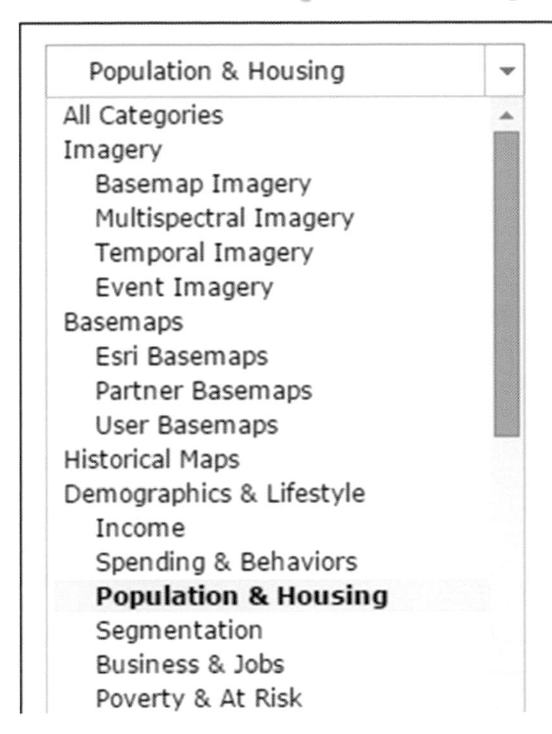

4. Find and point to the **2015–2020 USA Population Growth** layer to see the layer description. Click **Add layer to map**, click **As Layer**, and then close the **Browse Living Atlas Layers** window.

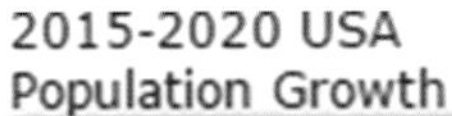

Add layer to map▾
As Layer
As Basemap

Notes:

- If you cannot find the **USA Population Growth** layer, your map extent may not cover the United States. You will need to unselect the **Within map area** check box at the upper-right corner of the **Browse Living Atlas Layers** window.
- If you added a wrong layer, you can remove that layer from the **Contents** pane. Just point to the layer, click the **More Options** button, and click **Remove**.

5. **In the Contents pane, point to the 2015–2020 USA Population Growth Rate layer you just added, click the More Options button ···, and click Show Item Details.**

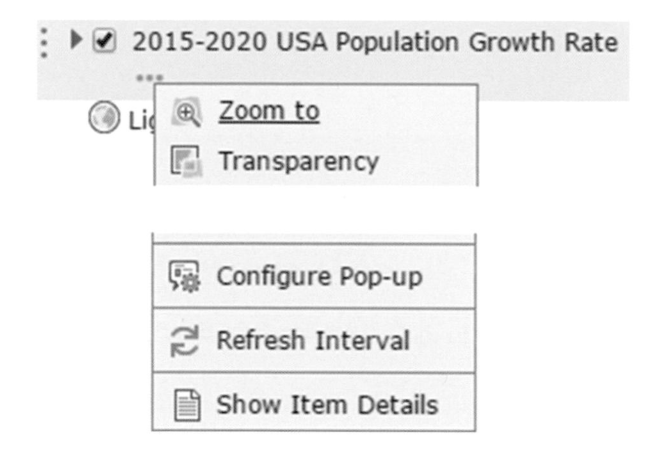

This action brings up the item details, or metadata (in other words, data about data) about the layer. This map layer shows the estimated annual growth rate of the population in the United States from 2015 to 2020 in a multiscale map by state, county, ZIP Code, tract, and block group. This layer is actually a map service. You will learn how to create such a service layer in later chapters.

6. **Go back to the map viewer by clicking the corresponding tab of your web browser. Zoom in and out of the map to examine the population change patterns. Click a state, county, ZIP Code, tract, or block group to see the information in the pre-configured pop-ups.**

You should see that the population change trends of the 50 major cities generally agree with the trends of their counties, ZIP Codes, tracts, and block groups.

Note: You may see multiple features at the location you clicked. In such cases, the pop-up will show one feature a time. You can click the arrow in the pop-up header to navigate through the features. You may also see pop-ups of multiple layers. You can disable the pop-up for a layer in the layer's **More Options** context menu.

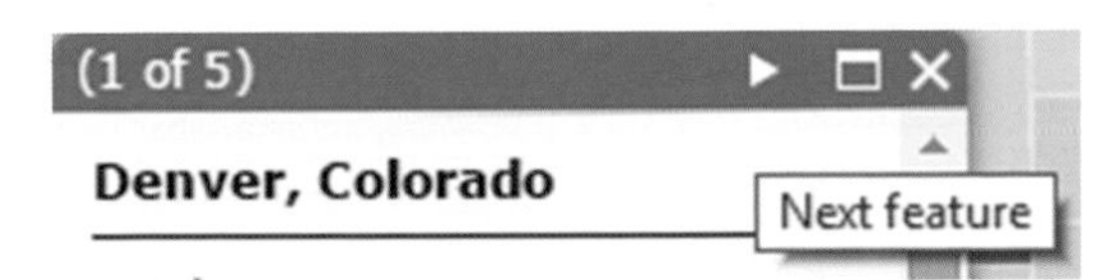

Next, you will add a layer about the unemployment rate. You can add the layer by repeating the first three steps or by searching, as illustrated in the following steps.

7. **On the map viewer menu bar, click the Add button, and click Search for Layers.**

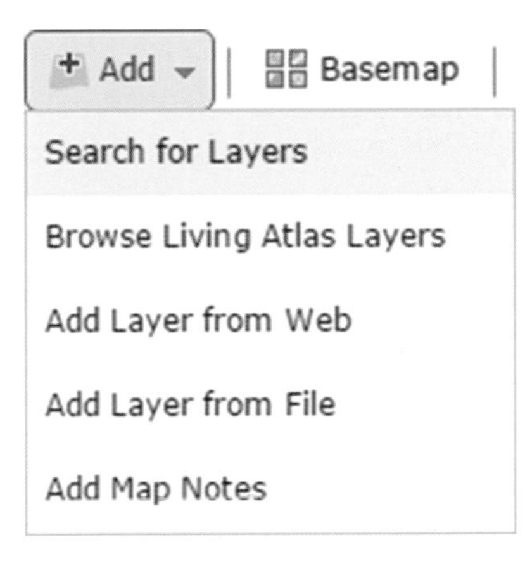

8. In the **Search for Layers** pane, make the following changes:
   - **In the Find text box, type US unemployment.**
   - **In the In box, click the arrow, and then click Living Atlas Layers.**
   - **Leave the Within map area check box selected if your current map extent covers the United States.**
   - **Click Go.**

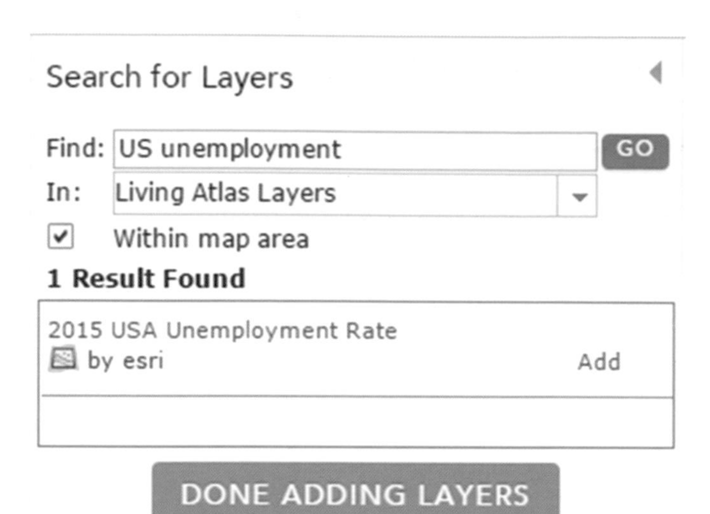

9. **In the results list, find the 2015 USA Unemployment Rate layer. On the right side of the layer, click Add to add the layer to your map.**

You typically will need to click the layer name and read its details to evaluate whether it provides what you need before adding the layer to your map.

10. **Click Done Adding Layers to close the Search for Layers pane.**

11. **Click the map to see the unemployment rates in Denver, Detroit, and other cities (you can use the Search box to find these cities if necessary).**

You will see unemployment is likely one of the causes for the population increase or decrease in these cities.

12. **Zoom the map to show all the 50 cities, and save your web map.**

The map extent saved is the default extent of your web map and web app.

2.7 Create and configure your web app

1. **On the map viewer menu bar, click the Share button.**

2. **In the Share window, share your web map with everyone (public).**

You are provided with a button to embed your web maps in your organization's existing website, your personal blog, and even Facebook. This embed approach is typically used for simple mapping apps. Next, you will create a more complete web app using the **Create A Web App** button.

3. **Click Create A Web App.**

4. **In the app gallery, find the Basic Viewer app. Click the app, and then click Create App.**

Alternatively, you can click **Preview** to see if the app meets your requirements before publishing it.

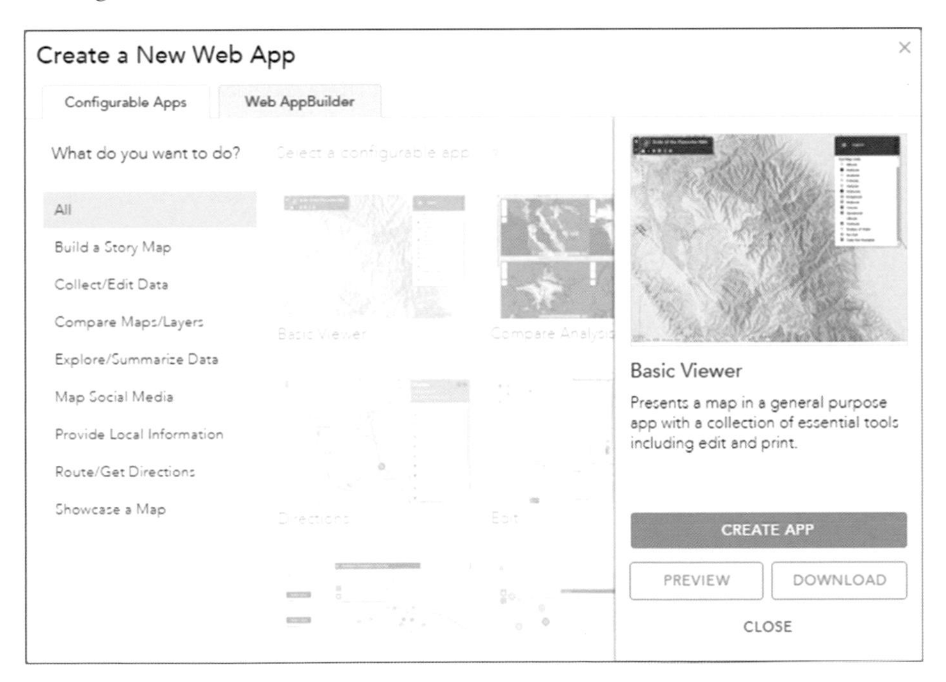

5. **Fill in the appropriate title, tags, and summary information of your app, and then click Done.**

You will be directed to the configuration mode of your app.

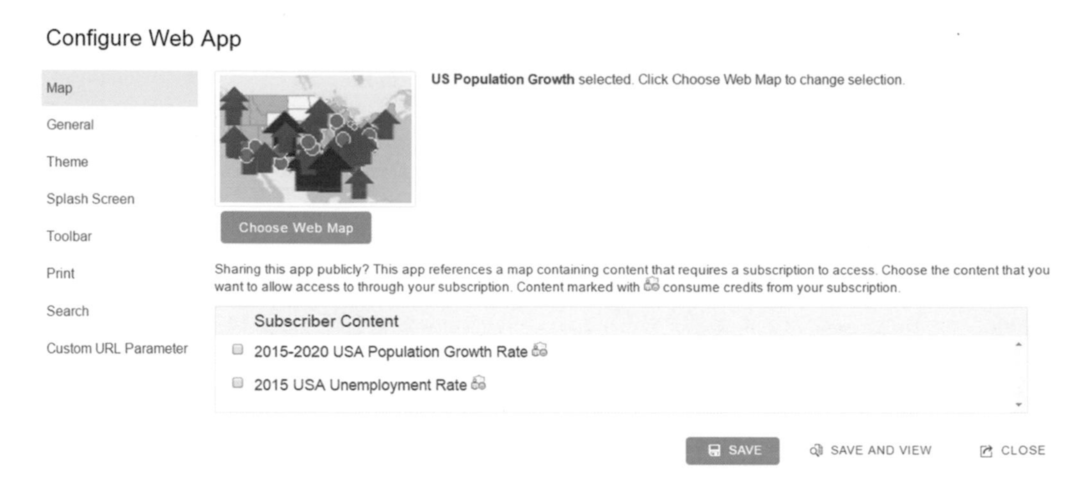

6. In the **Configure Web App** window, study the settings regarding Subscriber Content.

The **Subscriber Content** layers are from the Living Atlas. Some Living Atlas layers are premium content that require credits for access. You can select their check boxes if you want to allow your app users to access these layers through your subscription (using your credits). Otherwise, your users will be prompted to log in with their own ArcGIS Online subscription accounts, and the usage will charge their credits, not yours.

7. Study the options in the **General, Theme, Toolbar,** and **Search** sections.

You will notice that some settings require certain types of layers in your web map. For example, the **Editor** tool demands a feature service layer (discussed later in the book) and does not apply in this app.

The search settings allow users to find a location or data in the map using the default geocoding search or using searches configured in your web map.

8. **In the Print section, check the Print Tool, Display all Layout Options and Add Legend to Output boxes.**

With **Display All Layout Options** selected, all available print layouts—such as A3, A4, Letter, and Portrait/Landscape—supported by the printing service you use will display in the list of the print tool in your app.

9. **Click Save and View, and preview your app.**

After you click **Save**, your settings are applied to your app, and you can preview your app immediately. If needed, adjust the settings, save, and preview your app again.

10. **Click Close.**

This action exits the app configuration mode and directs you to the item details page of your app.

11. **Familiarize yourself with the item details page.**

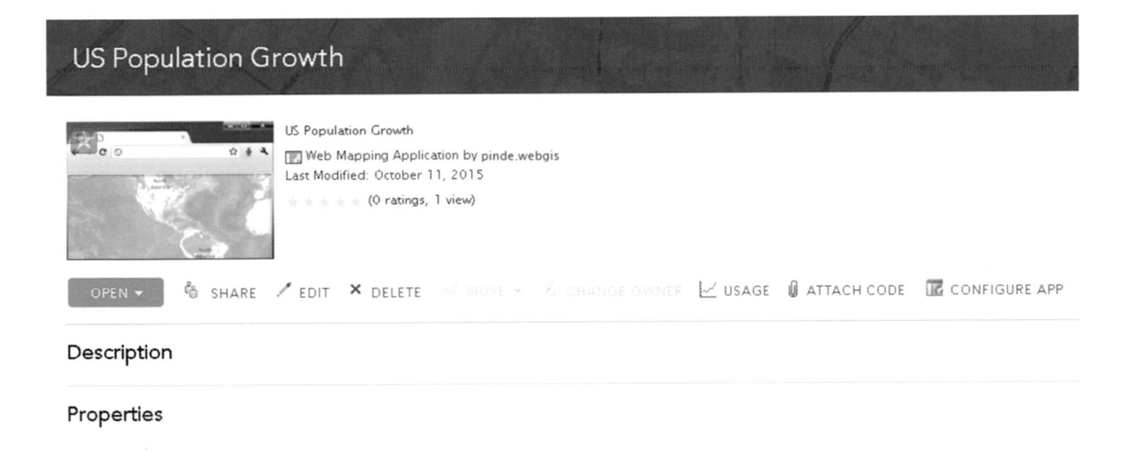

The item details page displays metadata about the contents of a portal item (here, your web app): title, thumbnail, description, tags, owner, ratings and comments, last modified date, share level, credits, and so on.

The menu bar on the item details page includes the following buttons:

- **Open** provides a list of ways to view the item (if it is a web map, scene, web app, hosted layer, and so forth) or download the item (if it's a CSV file, shapefile, map package, and so on).
- **Share** enables control over who has access to this item.
- **Edit** lets you edit the metadata or properties of this item.
- **Delete** lets you remove the item.
- **Move** allows you to move an item to a different folder under **My Content**.
- **Change Owner** allows you to change the owner of this item to another user. This option is available for administrator accounts.
- **Usage** charts the views, usage, and downloads of the item to help you gauge its popularity.
- **Attach Code** allows you to attach code (such as a ZIP file) to the item. This is helpful if you are sharing a sample or a configurable app and want others to have access to your code.
- **Configure App** directs you to the configuration mode of your web app.

Next, you will change the app's thumbnail into something more meaningful.

12. Click **Edit**.

As the owner of this item, you can update its metadata. Here you will change the app's thumbnail.

- **Click on the thumbnail image.**
- **In the Upload Thumbnail window, click Choose File and select C:\EsriPress\GTKWebGIS\Chapter2\images\thumbnail.png.**
- **Click OK.**
- **Click Save.**

The new thumbnail helps people who find your web app to quickly get a sense of what the web app is about and what the app looks like before they open the app.

13. **On the item details page, click the thumbnail image, or click Open, and click View Application.**

Your web app appears in a new window or tab. You can also view this app on tablets and smartphones.

14. **Explore your web app.**

Navigate through the map, check the pop-ups, turn layers on and off, create PDFs using the print function, examine the legend, study the patterns of US population change, and think about the causes behind the patterns you observe.

15. **To let your instructor and others use your web app, email them the app URL.**

You can point to each of the buttons on the toolbar in your web app to see the ToolTips for each button. If you click the **Share** button, you will see that you can share your web app using social media (such as Facebook, Twitter, and Google), copy and email the web app URL, or embed the app in other web pages.

In this tutorial, you created a web app with basemaps, operational layers, and a few tools (for example, the pop-up and print tools). You also experienced the flexibility of smart mapping and the rich contents of the Living Atlas.

QUESTIONS AND ANSWERS

1. **After I created my web app using my web map, I updated the layer styles and pop-ups of my web map. Will the changes be reflected automatically in my web app?**

   Answer: Yes.

   A web app maintains a link with its web map and depends on the web map as its source. Changes made to web map content, such as adding or removing layers, changing symbology, or configuring pop-up windows, are reflected automatically in the app.

2. **CSV files can hold point features. How can I add line or polygon features to my web map?**

   Answer: You have several options:

   - Use shapefiles. A shapefile is an Esri vector data storage format. It is not just one file but contains .shp, .shx, .dbf, and .prj files. Just zip these files into one ZIP File. Then, in the map viewer, click **Add > Add Layer from File**, find the ZIP File, and add the file to your map.

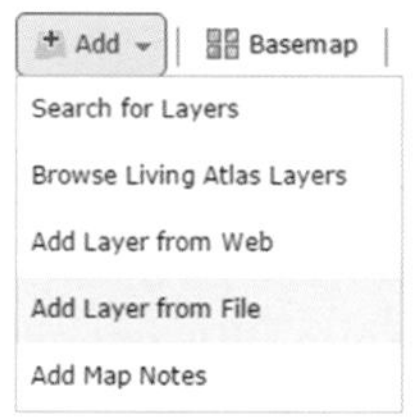

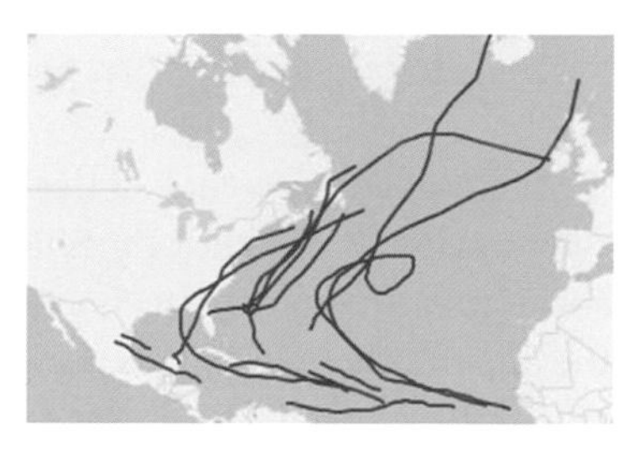

Note: Your line and polygon features may be generalized for better performance.

   - Use the Esri geodatabase and a feature service or map service, and then add the service to your web map.
   - Use a map notes layer.

3. **When I went to upload my own city's tree inventory (more than 30,000 features), I discovered that I could upload CSV files and shapefiles with only 1,000 or fewer features. How can I work around this limit?**

   Answer: This limit is a rational choice; you have better solutions for your data size.

   If you have more than 1,000 points in a CSV file or shapefile, you can split the file into smaller parts, with each one containing no more than 1,000 features, which you can then add to the web map.

   Be cautious when you split files into smaller parts. Layers from CSV files and shapefiles will be drawn by the browser side. Web browsers cannot handle large quantities of geometries and attribute data. In particular, some of your end users may still use older computers that cannot efficiently draw a few thousand points. Thus, your web app may perform slowly or may not work for some of your users.

   You will learn better solutions in the following chapters. The best practice in this case is to publish your data as a feature service, configure it with appropriate scale dependency, and then add the feature service to your web map. Other solutions include dynamic map service and a tiled map-service approach. All these solutions allow you to handle large datasets and still provide web apps with fast performance. You will learn these solutions in other chapters.

ASSIGNMENTS

Assignment 2: Create a web app to display one of the following examples:

1. The real estate for sale in your city, along with pictures and a chart of price histories in the pop-up window
2. The recent crime sites in your city
3. The recent earthquakes in the world
4. The upcoming garage sales in your city
5. Community-supported agriculture (CSA) sites and community gardens in your county
6. Other data you have or can collect

Data:

- For option 3, you can obtain the data from the US Geological Survey (USGS) website. USGS provides an earthquake feed in CSV and other formats. The CSV feed containing last week's earthquakes with magnitudes greater than 2.5 is located at **http://earthquake.usgs.gov/earthquakes/feed/v1.0/summary/2.5_week.csv**.
- For others, you must collect your own data using Microsoft Excel or some other spreadsheet tool. Whenever possible, add images, links, and charts to make your app interesting.

Requirements:

- Configure appropriate layer styles.
- Configure appropriate layer pop-ups. Include at least one chart or image with a link in the pop-up. (If there are no image URLs in your data, find an image URL.)
- Choose an appropriate configurable app template to transform your web map into a web app.
- Share your web map and app.

Tip:

- For option 3, you can make the URL link in your pop-up open the USGS web page with details about whichever earthquake users clicked to use. The URL pattern is **http://earthquake.usgs.gov/earthquakes/eventpage/{id}** (for example, **http://earthquake.usgs.gov/earthquakes/eventpage/ak10796120**).

1 2 3 4 5 6 7 8 9

What to submit: Email your instructor with the subject line **Web GIS Assignment 2: Your name**, and include the URL of your web app.

Resources

Help documents and tutorials

"Add Layers," https://doc.arcgis.com/en/arcgis-online/create-maps/add-layers.htm.

"Change Style," https://doc.arcgis.com/en/arcgis-online/create-maps/change-style.htm.

"Configure Pop-ups," https://doc.arcgis.com/en/arcgis-online/create-maps/configure-pop-ups.htm.

"Create Apps from Maps," https://doc.arcgis.com/en/arcgis-online/create-maps/create-map-apps.htm.

"Get Started with ArcGIS Online," http://learn.arcgis.com/en/projects/get-started-with-arcgis-online.

Living Atlas of the World, http://doc.arcgis.com/en/living-atlas/about.

Esri blogs

"Adding Geotagged Photos to Your Web Map," by Bern Szukalski, http://blogs.esri.com/esri/arcgis/2013/08/14/adding-geotagged-photos-to-your-web-map.

"Crafting Custom Attribute Displays in Pop-ups," by Bern Szukalski, http://blogs.esri.com/esri/arcgis/2013/07/25/custom-attribute-display-pop-ups.

"Introducing Smart Mapping," by Mark Harrower, http://blogs.esri.com/esri/arcgis/2015/03/02/introducing-smart-mapping.

"Maps and More: Discover the Living Atlas of the World," by Shane Matthews, http://blogs.esri.com/esri/arcgis/2015/09/14/maps-and-more-discover-the-living-atlas-of-the-world.

"Smart Mapping Part 2: Making Better Size and Color Maps," by Mark Harrower, http://blogs.esri.com/esri/arcgis/2015/03/17/smart-mapping-part-2-making-better-size-and-color-maps.

"Smart Mapping Part 3: Rounding classes for Color and Size Drawing Styles," by Adelheid Freitag, http://blogs.esri.com/esri/arcgis/2015/03/20/smart-mapping-part-3-rounding-classes-for-color-and-size-drawing-styles.

"Smart Mapping Part 4: Pairing Data with Maps," by Mark Harrower, http://blogs.esri.com/esri/arcgis/2015/05/12/smart-mapping-part-4-pairing-data-with-maps.

"Smart Mapping Part 5: Tips and Tricks," by Mark Harrower, http://blogs.esri.com/esri/arcgis/2015/06/16/smart-mapping-part-5-tips-and-tricks.

Videos

"ArcGIS Online: Smart Mapping—Make Brilliant Maps Quickly and with Confidence," by Jim Herries and Mark Harrower, http://video.esri.com/watch/4720/arcgis-online-smart-mapping-_dash_-make-brilliant-maps-quickly-and-with-confidence.

"ArcGIS Online Steps for Success—A Best Practices Approach," by Bern Szukalski and Jeff Archer, http://video.esri.com/watch/4718/arcgis-online-steps-for-success-_dash_-a-best-practices-approach.

"Living Atlas of the World," http://video.esri.com/series/224/living-atlas-of-the-world.

"Smart Mapping—A Closer Look," http://video.esri.com/watch/4618/smart-mapping-a-closer-look.

Chapter 3
Hosted feature layers and volunteered geographic information

In this chapter, you will learn about feature layers and feature services, which provide a more scalable approach than the approach you learned in chapters 1 and 2, when you added comma-separated values (CSVs) and other local files directly to your web map. This new approach allows you to go beyond the limit of 1,000 features per layer. More importantly, you can configure a feature layer to be editable. Web client users can add, update, and delete the geometries and attribute fields underlying the layer. This functionality is especially useful for collecting data via crowdsourcing and enterprise users.

Learning objectives

- *Understand the concepts of web layers, hosted web layers, and feature services.*
- *Understand the various ways to create hosted feature layers.*
- *Create hosted feature layers using ArcGIS Online or Portal for ArcGIS.*
- *Add and delete fields after a feature layer is created.*

- *Define feature templates.*
- *Create web apps that can collect volunteered geographic information (VGI) and authoritative data.*

Chapter overview

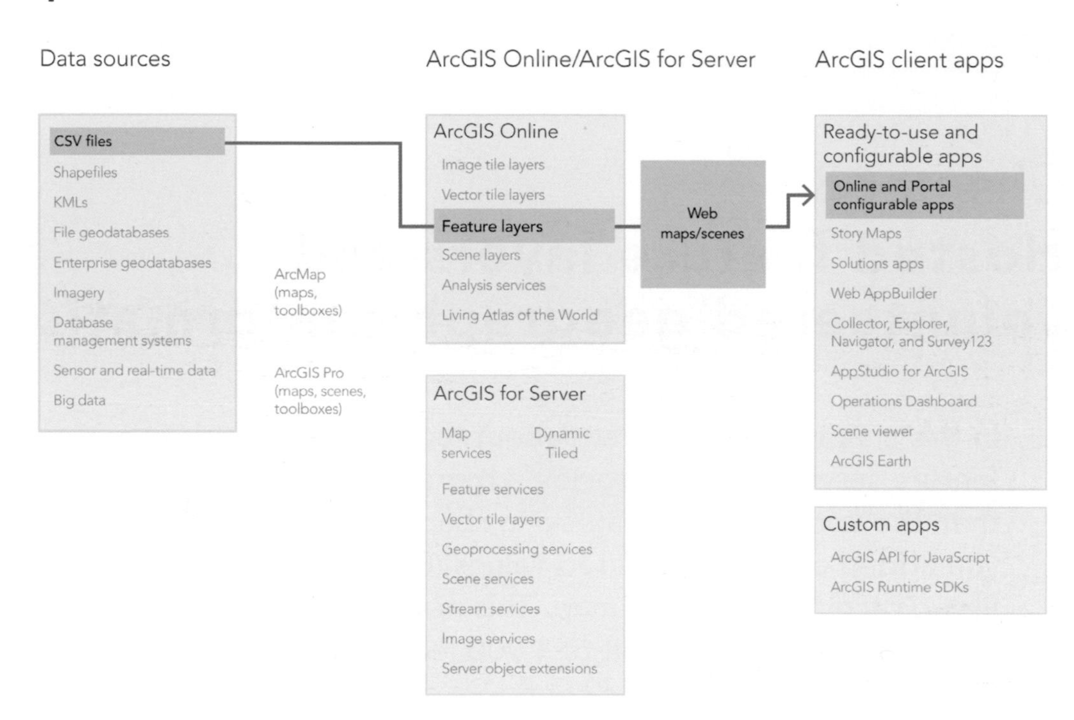

ArcGIS offers many ways to build web apps. The green line in the figure highlights the technology presented in this chapter.

Features vs. hosted feature layers

Earlier in the book, you added CSV files (also applies to shapefiles, and so on) to ArcGIS Online or Portal for ArcGIS map viewer. When you add data this way, the features in the source CSV file or shapefiles are imported and stored in your web map. ArcGIS Online limits the number of features in such layers so that you can efficiently manage and display them. The limit is 1,000 features from a shapefile, .csv, or .txt file using latitude/longitude coordinates per layer. For addresses using .csv or .txt files, the limit is 250 features per layer.

Hosted feature layers are a more scalable way to publish features. Hosted feature layers allow you to create and use layers efficiently in many web maps instead of in a single web map. You can also exceed the limit of 1,000 features per layer when you use hosted layers. More important, you can configure feature layers so they are editable and thus usable to support apps that collect GIS data.

VGI and web-based geospatial data editing

VGI (volunteered geographic information) is digital spatial data produced voluntarily by citizens rather than by formal institutional data producers. The term "VGI" was coined in 2007 to refer to user-generated content in the geospatial field. Various navigation and traffic apps, for example, collect VGI about highway traffic, accidents, and police car locations that mobile users contribute voluntarily. Other examples of VGI include georeferenced tweets collected by Twitter, georeferenced photos uploaded to Flickr and Instagram, and the annual Christmas bird count by the National Audubon Society (**https://www.audubon.org/conservation/science/christmas-bird-count**).

VGI marks a research frontier of significant practical value. With the help of crowdsourcing, in which large numbers of citizens act as sensors, VGI can enhance early warning systems for natural disasters or real-time social event monitoring. VGI supports tremendous new business opportunities. VGI is also an important source of big data.

From a technical perspective, VGI is collected via web-based geospatial data editing. Such editing capability allows users to add, remove, and update geometries and attributes. ArcGIS feature services and feature layers provide this capability. Feature services and feature layers support the collection of VGI as well as authoritative data in many ways, including the following examples:

- A utility company can have its personnel collect and edit data in the field.
- A city can invite citizens to report emergency and nonemergency incidents.
- A planning department can invite bus riders to propose locations for new bus stops and lines.
- Law enforcement can encourage citizens to report when and where they spotted crime suspects and even upload photos and videos of suspects.

Note: ArcGIS Online provides web editing capabilities to support simple feature editing. ArcGIS Pro and ArcMap are the best options for more complex editing operations such as topologies and geometric networks.

Feature services and hosted feature layers

Feature services are a type of web services that allows clients to request geographic features, which can include vector coordinates and attributes. Feature services is a broader term than hosted feature layers. In the ArcGIS platform, you can publish feature services to ArcGIS for Server, ArcGIS Online, and Portal for ArcGIS. Feature services published to the latter two are called hosted feature layers. Feature services and hosted feature layers can support both read and write data access and can track who added, updated, or deleted the features in a layer.

- **Read access:** Web clients can query and retrieve feature geometries, attributes, and symbology so that features can be drawn and changed on the client side.
- **Write access:** Web clients can edit feature geometries and attributes and save edits to the server database. You can disable and enable this type of access for hosted feature layers and hosted web layers.

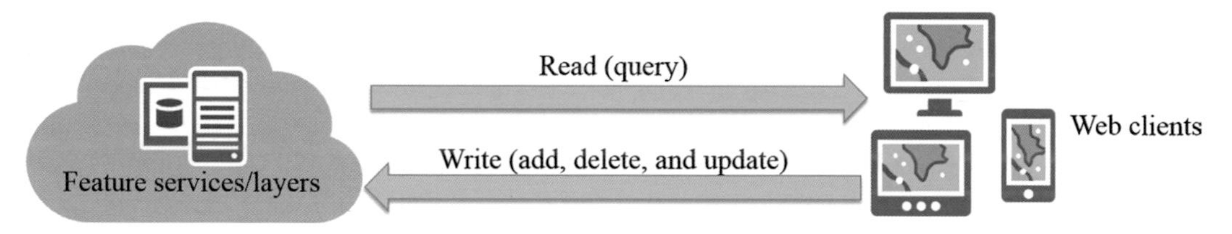

Feature services or layers can provide web clients with both read and write access to the server database.

You can publish your maps and data as hosted web layers on ArcGIS. When you do this, ArcGIS Online takes care of hosting your layers (as services) and scaling to meet demand. As a result, these layers are often called hosted layers or hosted web layers. Hosted web layers include the following types of layers:

- **Feature layers**: Hosted feature layers support queries, visualization, and editing. Hosted feature layers are most appropriate for visualizing data on top of your basemaps. In web apps, the browser draws feature layers that support interactive highlighting, queries, and pop-ups.
- **Tile layers**: Hosted tile layers support fast map visualization using a collection of predrawn map images, or tiles. These tiles are created and stored on the server after you upload your data. Hosted tile layers are appropriate for basemaps that give your maps geographic context.

Create hosted feature layers

You can create hosted feature services using ArcMap, ArcGIS Pro, or a web browser with ArcGIS Online or Portal for ArcGIS. If you use a web browser, you can choose among the following options:

- Create a feature layer from your own data (for example, CSV, shapefiles, GeoJSON, and file geodatabase). You can go to the **My Content** page and click **Add Item** if you want to create a feature layer this way.
- Create an empty feature layer by duplicating an existing template or layer, without needing your own data: Duplicating an existing layer will create a new empty layer of the same schema. In other words, the empty layer will contain the same attribute fields as the parent layer. This functionality is useful if you can find a parent layer that matches your needs or if you want to use the parent feature layer again and again. To duplicate an existing template or feature layer, you can go to **My Content** page, click **Create -> Feature Layer**, and choose **From Template** or **From Existing Layer** or **From URL**. You can find a parent layer from the templates, your own collection, your organization's collections, or anywhere else as long as you have the service URL of the feature layer you would like to copy.

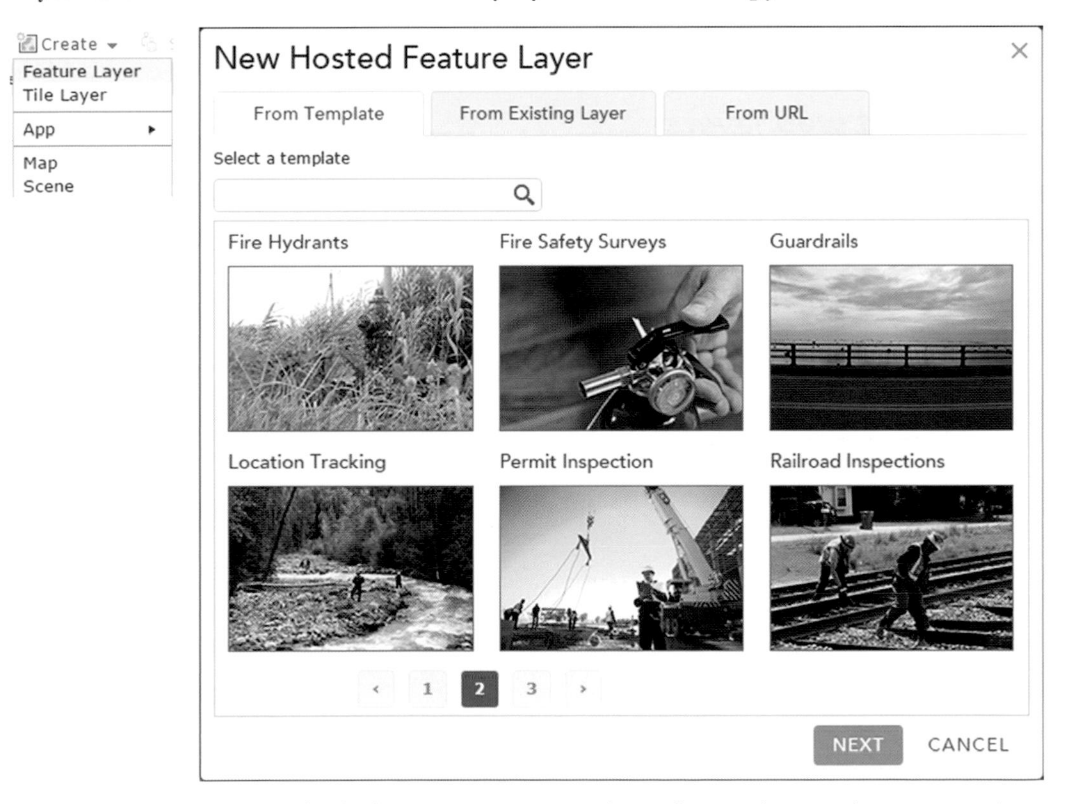

Create an empty feature layer by duplicating an existing template or layer, without needing your own data.

1 2 3 4 5 6 7 8 9 10

- Create an empty feature layer, and define your own fields interactively using the ArcGIS for Developers website. You can go to **http://developers.arcgis.com**, log in with your ArcGIS Online account, then click the **Hosted Data** button at the top, and click the **New Feature Service** button. The website will walk you through the steps to create a new feature layer. You will have the chance to specify the layer title, feature type (points, lines, or polygons), extent, attribute fields, and the layer's default symbols.

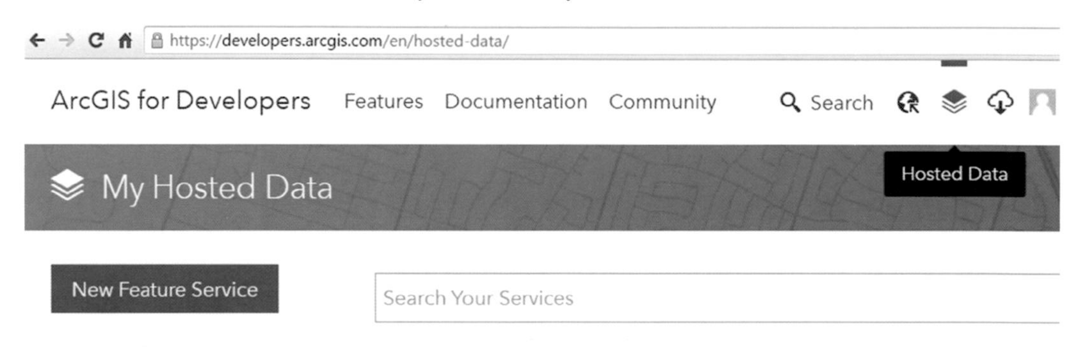

Create an empty feature layer and define your own fields interactively using ArcGIS for Developers site.

Use hosted web layers

You can add your hosted layers, including hosted feature layers, to the map viewer, save them to a web map, and then create web apps.

You can also reference the service URL of a web layer directly in your web map or web app, just like other types of web services that ArcGIS provides. The service URL is also known as the REST URL or REST (Representational State Transfer) end point. REST is the most commonly used type of web service interface today. With REST, every web resource has a URL. You can find the service URL of a hosted web layer on its item details page.

Feature template

A feature service typically contains one or more feature templates, which you can define in ArcMap (if used to author the map document) or portal map viewer (for hosted feature services).

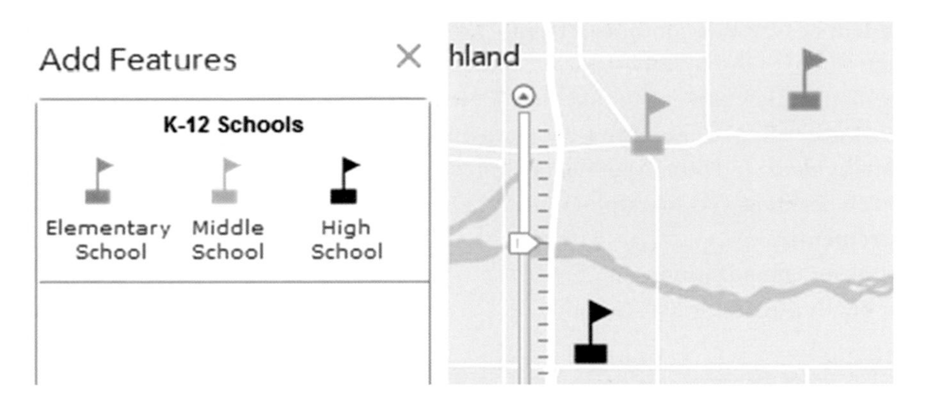

The feature template for K–12 schools defines the types of schools that users can add.

A feature template defines the types of data items that users can add to a layer. A template ensures data integrity and makes the editing easier for your end users:

- **Data integrity:** A feature template for a layer that represents schools, for example, might allow editors to classify a new feature as an elementary, middle, or high school. In a school feature template, you might preset these three options so that users can choose only one of these three types of schools. Presetting the options prevents users from entering invalid values for such an important attribute.
- **Ease of editing:** A feature template can have preset symbol and default values for one or multiple fields. Preset symbols make it easier for users to know what type of feature they are using or adding. With the default attribute values, users do not have to type these values manually, which is particularly convenient for mobile users.

Editor tracking and ownership-based editing

Editor tracking is the ability to track who has changed the data of a feature service, and when. Editor tracking can help create more accountability and quality control over the edited data. Tracking also can support ownership-based access control, which allows you to limit access so that only the user who created a certain feature can access that feature.

This tutorial

Your city wants to create a web app that helps citizens report nonemergency issues, such as potholes and graffiti, thus ensuring that the appropriate city departments can respond. The web app should allow citizens to describe the issue and report the location, as well as attach photos, videos, and other documents to help city staff better understand the details.

Data: 311Incidents.csv, which contains attribute fields that the city wants to collect.
Requirements:
Your web app should allow citizens to take the following actions:

- Report events according to predefined categories of incidents/issues.
- Attach photos, videos, and other types of documents.
- Use the app on desktops and smartphones.

System requirements:

- ArcGIS Online for Organizations
- A publisher or administrator user account

3.1 Prepare your data

You can publish hosted services using a variety of data formats. This tutorial uses CSV, the easiest one to work with.

Note: CSVs can store only point features. To publish a feature service of line or polygon feature types, use other formats, such as shapefiles or file geodatabases. (You will need to compress your shapefile or file geodatabase into a ZIP File; see the two ZIP Files under C:\EsriPress\GTKWebGIS\Chapter3\Assignments_data for examples.)

1. **In Microsoft Excel, open C:\EsriPress\GTKWebGIS\Chapter3\311Incidents.csv, and study its data fields.**

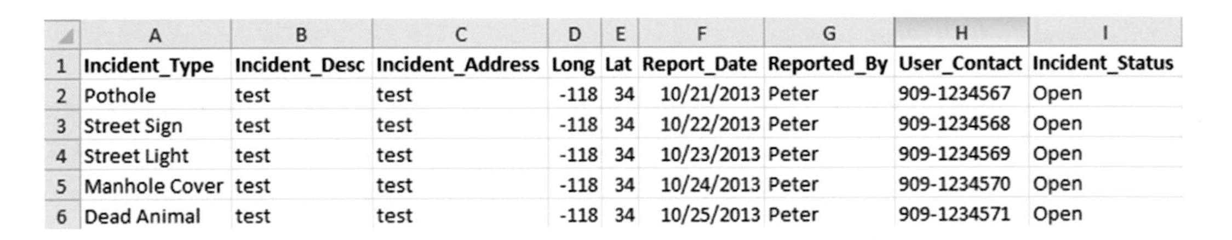

	A	B	C	D	E	F	G	H	I
1	Incident_Type	Incident_Desc	Incident_Address	Long	Lat	Report_Date	Reported_By	User_Contact	Incident_Status
2	Pothole	test	test	-118	34	10/21/2013	Peter	909-1234567	Open
3	Street Sign	test	test	-118	34	10/22/2013	Peter	909-1234568	Open
4	Street Light	test	test	-118	34	10/23/2013	Peter	909-1234569	Open
5	Manhole Cover	test	test	-118	34	10/24/2013	Peter	909-1234570	Open
6	Dead Animal	test	test	-118	34	10/25/2013	Peter	909-1234571	Open

If you have existing point features to start with, you can put them in the CSV file in the format as illustrated. The incidents already in the sample CSV file are merely placeholders, which you will delete once you have created your feature service. Keep them for now, because they serve two purposes:

- The incidents help define attribute field types. Unlike databases and shapefiles, in which you can explicitly define attribute field types, ArcGIS determines the field types in CSV files on the basis of their values. For example, if a field contains the value "10/21/2013," ArcGIS will recognize the entry as a date value.

- The incidents help define attribute domain values. For the **Incident_Type** field, the values allowed are limited to the following: **Pothole**, **Street Sign**, **Street Light**, **Manhole Cover**, and **Dead Animal**. (This list of values serves as an illustration, so it is incomplete). You will use these domain values to create feature templates. You can define feature templates and add new domain values using ArcGIS Online or Portal for ArcGIS. However, it is easier to first input all the possible values in the CSV file, and then create the feature templates all at once.

2. **Close Excel. Do not save the CSV if you made any changes.**

3.2 Publish a hosted feature layer

1. **Open a web browser, and go to ArcGIS Online (http://www.arcgis.com) or your Portal for ArcGIS. Sign in with a publisher or administrator account.**

The trial account you created earlier is an administrator account and allows you to create hosted feature layers.

2. **Click My Content on the main menu bar, click Add Item +, and then click From my computer.**

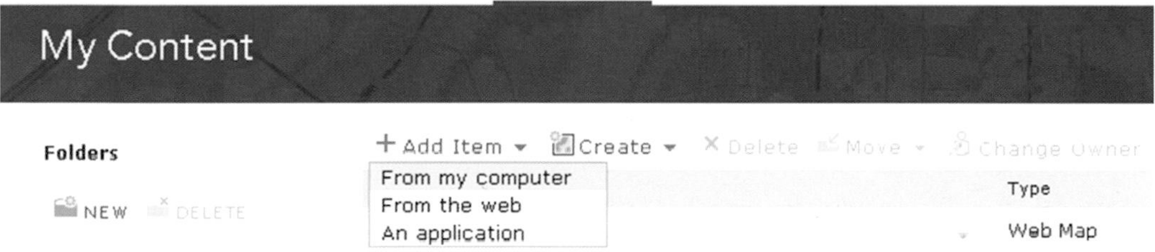

3. In the **Item from my computer** window, perform the following tasks:

   - For **File**, browse to **C:\EsriPress\GTKWebGIS\Chapter3\311Incidents.csv**, and click it. If you have published a file of the same name to your content before, rename your **311Incidents.csv** file to a unique name, and then click the file.
   - For **Title**, use the default, or specify a new one.
   - For **Tags**, specify keywords, such as **311**, **Incidents**, **Service Requests**, **VGI**, and **GTKWebGIS**, as illustrated. Separate the keywords with commas.
   - Make sure the check box next to **Publish this file as a hosted layer** remains selected.
   - Leave the **Use Latitude/Longitude** option selected.
   - Review the field types and location fields.
   - Click **Add Item**.

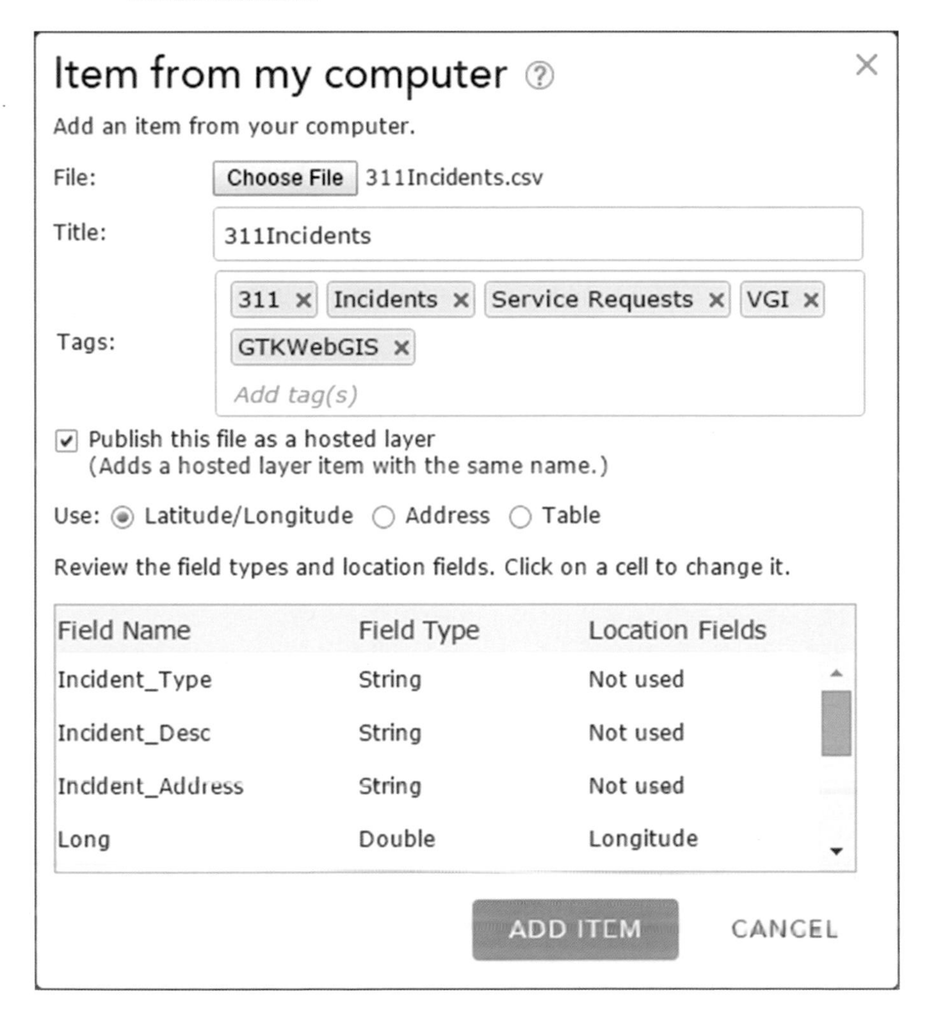

The item details page appears as your CSV file is being published as a hosted feature layer. Publishing a CSV file may take a few seconds.

4. **On the item details page, look for the Layers section. Click the arrow next to 311Incidents, and click Enable Attachments.**

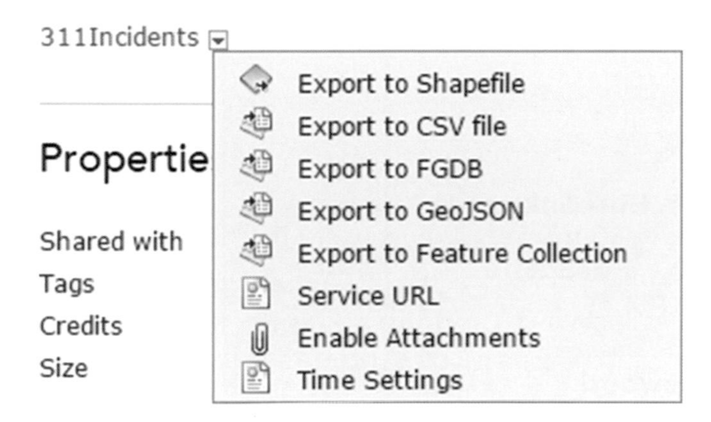

This step allows your users to attach photos, videos, and other files with the incidents your users report. Users can attach multiple files for one incident, and each file can be up to 10 MB.

5. **While still on the item details page, click Edit.**

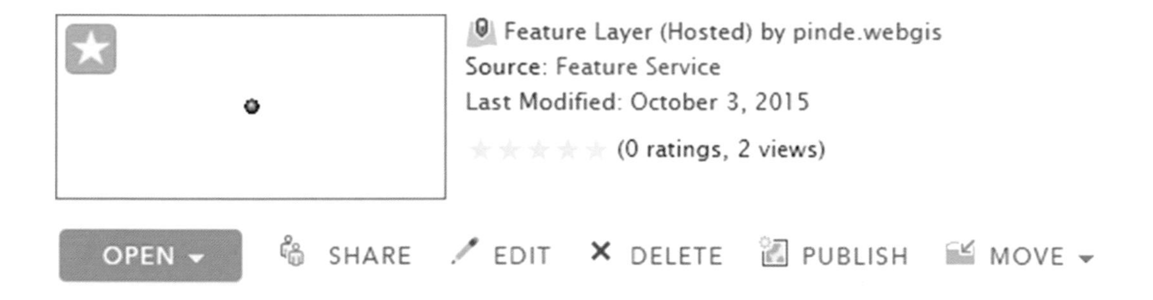

On the editor settings page, scroll down, and click **Set Extent**.

6. **In the Set Extent window, define an area to cover your city or the area you expect to collect data. You can define an area by searching for an address or place, for example Los Angeles or your city, or draw an extent on the map, or simply type in the left/right longitudes and top/bottom latitudes.**

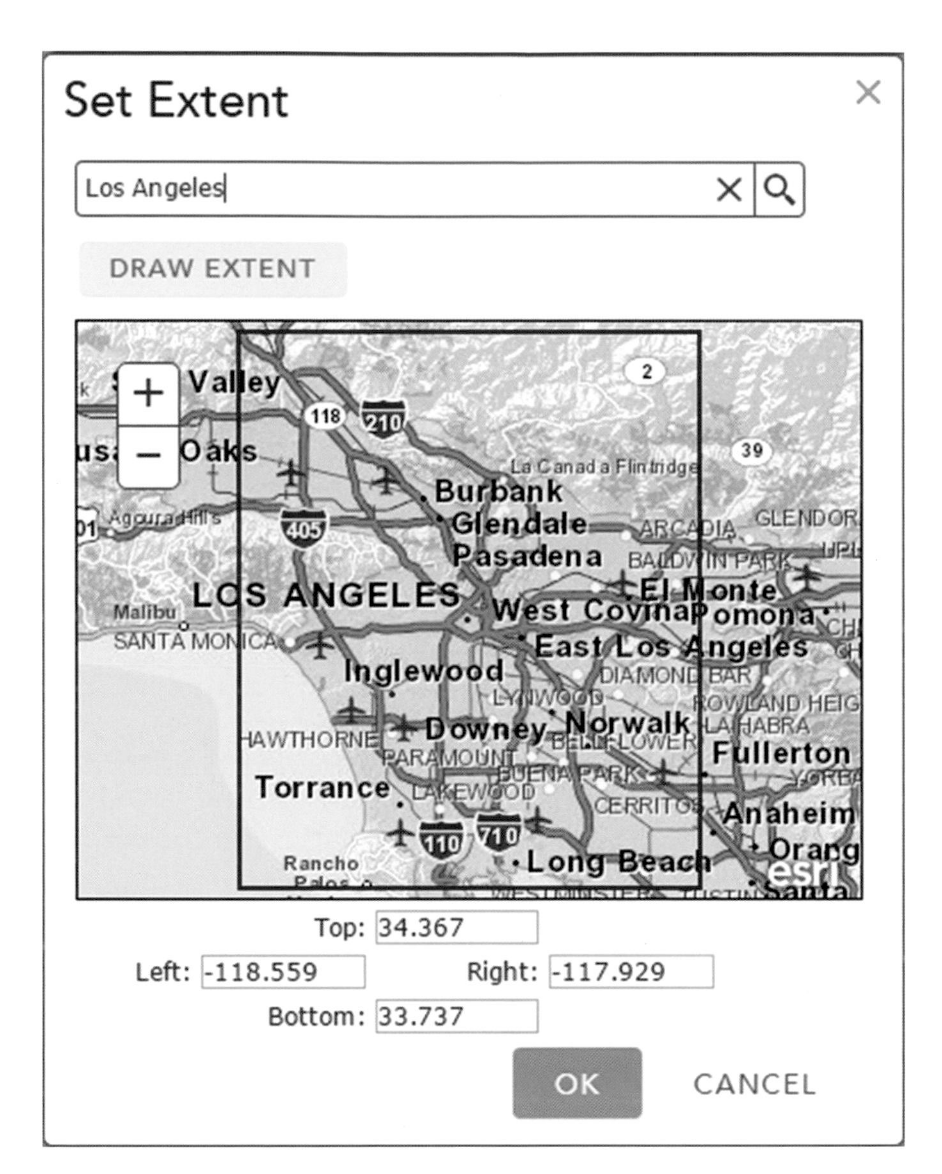

7. Click OK to close the **Set Extent** window.

8. While you are still on the editor settings page, select the check box for **Enable editing.**

9. Leave **Add, update, and delete features** as selected.

Editing ☑ Enable editing and allow editors to:
◉ Add, update, and delete features
○ Update feature attributes only
○ Add features only

Of the three options, **Add, update, and delete features** gives users the most privileges but also allows users to update and delete incidents that others have reported. Do you want to trust that your users will respect each other's posts? If you choose this option, a user can delete other posts, either by accident or on purpose.

The second option, **Update feature attributes only**, allows users to enter attribute information but does not allow them to change feature geometries. For example, your service may contain a set of observation towers. In this case, you may not want users to edit the locations of these towers, but this option still allows users to update their attributes.

The final option, **Add features only**, allows users to report something new but not to delete or update existing features.

10. **Select the check boxes next to Export Data and Sync.**

Export Data ☑ Allow others to export to different formats.

Sync ☑ Enable Sync (disconnected editing with synchronization).

Enabling **Export Data** allows others to export your feature layer into a CSV, shapefile, file geodatabase, or GeoJSON format. Enabling **Sync** supports offline editing. End users who need to work offline can export the latest data, use the data offline, edit the data, and synchronize with the hosted feature layer when they connect again.

11. **Review the Track Edits options, and leave them unselected.**

Track Edits ☐ Keep track of who created and last updated features.
☐ Editors can update and delete only the features they add.
☐ Editors can view, update, and delete only the features they add.

If you wish to track and restrict user edits, you can select these options. If you select these options, your app will require all users to log in so that the feature service knows who is using the app. For tutorial purposes, all users are allowed to report incidents without logging in.

12. **Click Save.**

This step will save your settings and direct you back to the item details page.

13. **On the item details page, click Share, and share your service with Everyone (public).**

This way, your web users can use your feature layer without having to log in.

3.3 Define feature templates

In this section, you will define feature templates to improve the editing experience of your end users.

1. **On the item details page, under the item thumbnail, click Open, and click Add layer to new map with full editing control.**

A map opens in the map viewer.

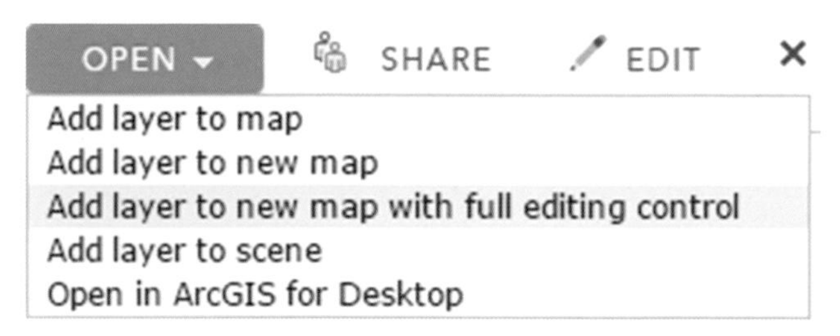

2. **In the Contents pane, point to the 311Incidents layer, and click the Change Style button.**

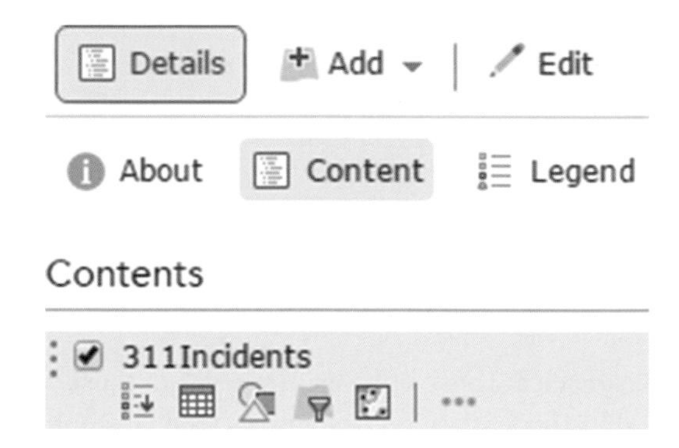

3. **In the Change Style pane, for the attribute to show, click Incident_Type.**

Once you choose the attribute, a number of different drawing styles appear. Typically, the most suitable style is applied automatically and indicated by a checkmark.

4. **For Types (Unique symbols), click Options.**

Change Style

311Incidents

1 Choose an attribute to show

Incident_Type

2 Select a drawing style

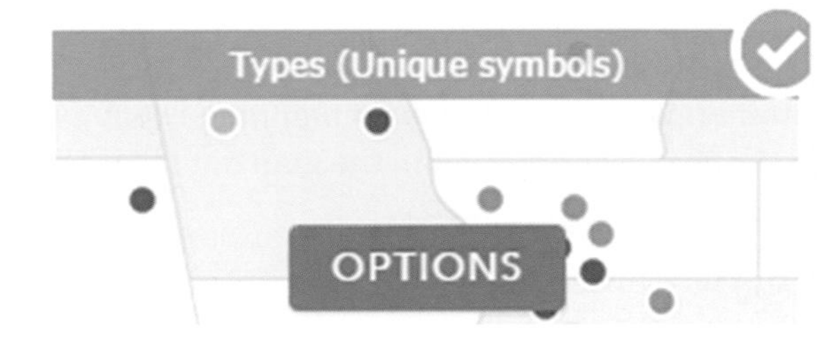

5. Click the symbol of each incident type, change it to create a symbol that users can intuitively understand, and set its size to **24**.

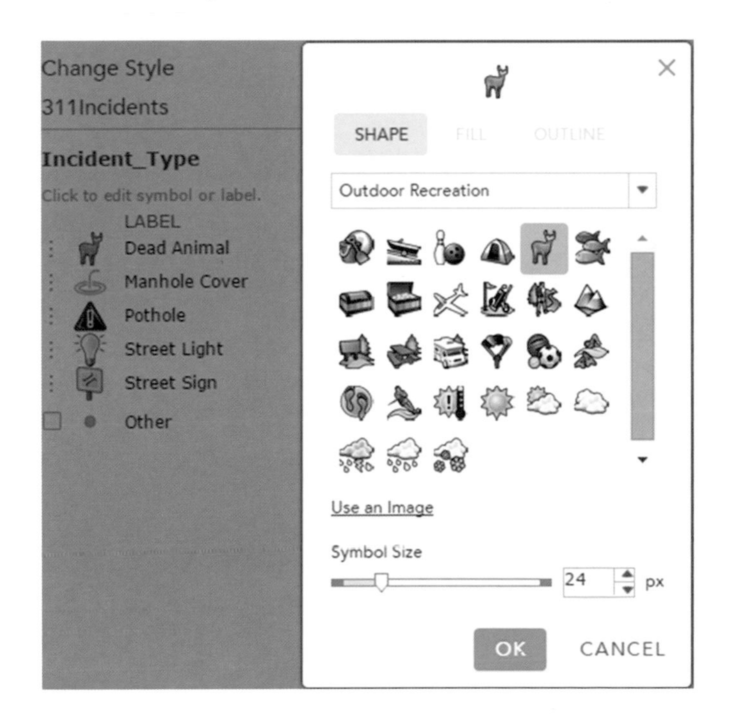

6. **For this tutorial, use the symbols provided in ArcGIS Online or Portal for ArcGIS, according to the following list:**
    - **Dead Animal from the Outdoor Recreation set**
    - **Manhole Cover from the Cartographic set**
    - **Pothole from the Transportation set**
    - **Street Light from the People Places set**
    - **Street Sign from the Transportation set**

For an actual project, you may need to create custom symbols to better match the incident types. To create a custom symbol, you would select the **Use an Image** option shown in the figure.

7. **Note that the Visible Range is automatically set by the smart mapping capability. Do not be surprised if you do not see the layer when you zoom the map out fairly far.**

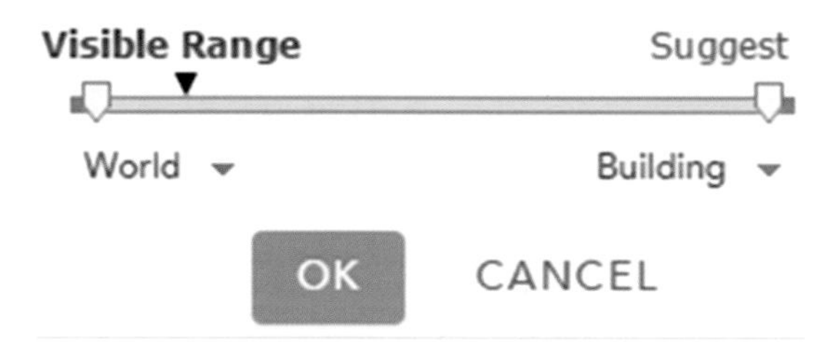

8. **Click OK and then Done to exit the Change Style pane.**

The new symbols appear on the map.

9. **In the Contents pane, point to the 311Incidents layer, click the More Options button ···, and click Save Layer.**

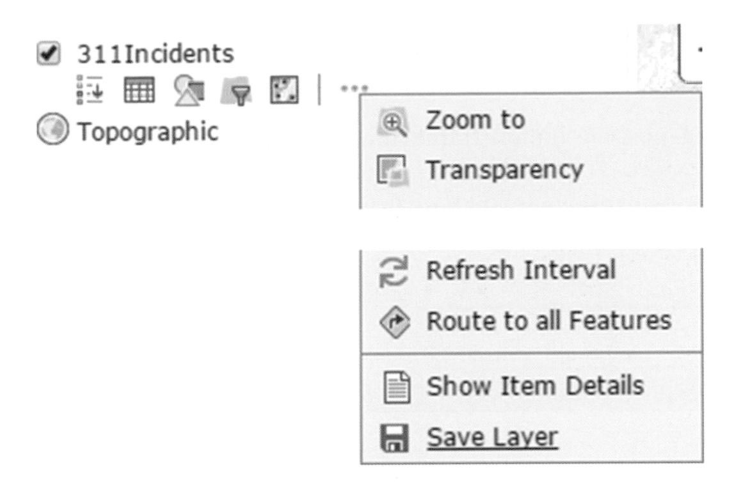

This step saves the symbols you just defined to the feature layer. In other words, these symbols will go with this layer. If you add this layer to another web map, you will not need to repeat the above steps to define its style again.

10. **On the map viewer, click the Edit button to see the feature templates in the Add Features pane.**

The feature templates are actually the incident types and symbols you defined earlier. If you included all the possible values for the incident type field in your CSV file, your feature templates would be complete. If you missed some values, you can add them here. To illustrate this task, you will add a new type of incident, **Graffiti**.

11. **Under the feature templates, click Manage so that you can manage the feature template (for example, to add a new incident type).**

Only the service owner or an organization administrator can see the **Manage** button.

12. **In the Manage New Features pane, click Add New Type of Feature.**

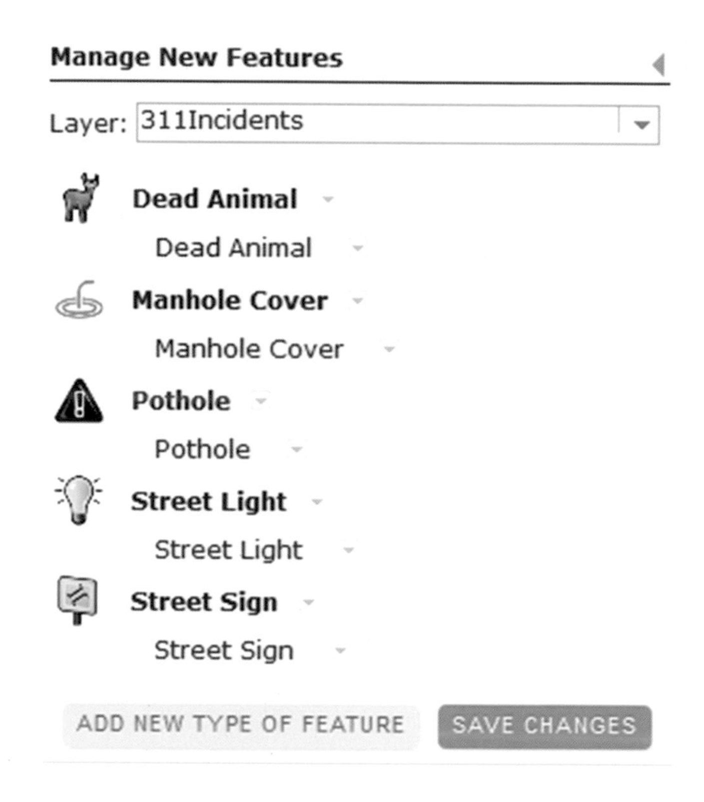

13. **In the Type Properties window, perform the following tasks:**

- **For Label, specify Graffiti.**
- **For Attribute: Incident_Type, also specify Graffiti.**

You have now defined a new value for the **Incident_Type** field.

- **For Symbol, choose an appropriate symbol (from the Safety Health set)**
- **Set the Symbol Size to 24.**
- **Click Done.**

Type Properties
Edit the properties of the feature type
Label:
Graffiti
Attribute:
Incident_Type
Graffiti
Symbol:
SHAPE
FILL
OUTLINE
Safety Health
Use an Image
Symbol Size
24
px
DONE
CANCEL

14. **Click Save Changes.**

This step saves the new template to the feature layer.

Next, you will delete the latitude and longitude fields. You needed the fields to create the feature layer, but you no longer need the fields afterward. The feature layer internally manages the geometries of your points. Your end users will not need to fill in the latitude and longitude fields.

15. **Click the Details button.**

16. **In the Contents pane, point to 311Incidents, and click the Show Table button.**

17. **In the table header, point to the Long field, click the Settings button, and delete the field. When prompted, confirm the deletion.**

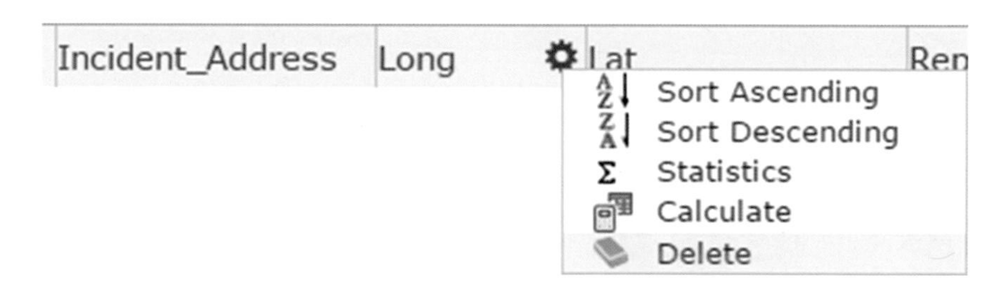

18. **Repeat the previous step to delete the Lat field.**

19. **Click the X in the upper-right corner of the table to close the table.**

Now that you have created the feature layer and feature template, you no longer need the filler points imported from the CSV. Next, you will remove these points.

20. Click the **Edit** button on the map viewer menu bar. Click each of the sample incidents on the map. As each pop-up window appears, scroll down and click **Delete** to remove the points.

21. In the upper-left corner of the page, click **Home**, and click **My Content**.

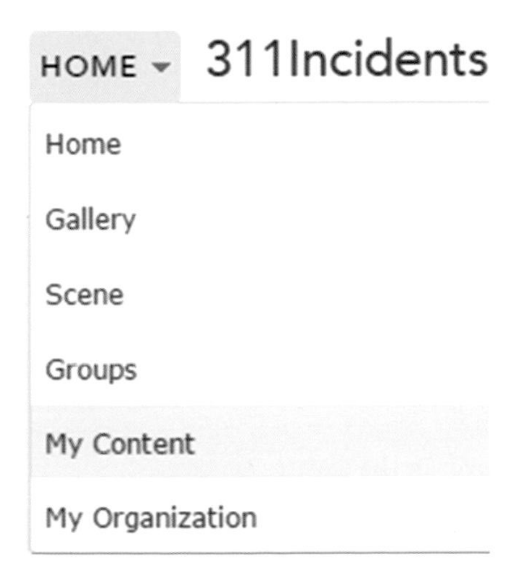

22. **On the My Content page, click the feature layer you just created to see the details.**

If you see a lot of content items, click **Modified** in the content table header to sort your contents in descending order. The feature layer you just created should be the first in the list.

23. **On the item details page, locate the Layers section. Click the arrow next to 311Incidents layer, and choose Service URL.**

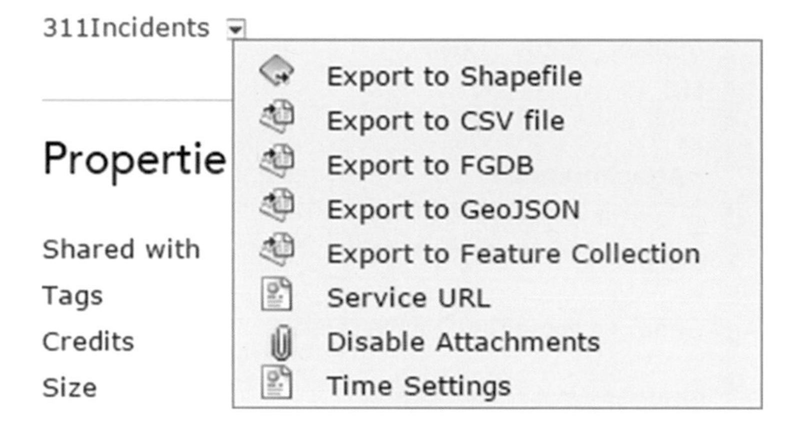

This choice will direct you to the service URL or the REST URL of the layer, which will open in the **ArcGIS REST Services Directory**. This directory lists the web services hosted on a GIS server and displays the metadata of these web services.

In your browser's URL box, you will see that the service URL of your hosted layer has the following pattern:

http://server/organization_id/arcgis/rest/services/service_name/FeatureServer/layer_number.

- The protocol can be **http** or **https**. The **https** protocol is a secured **http** that encrypts the information communication between a server and its web clients.
- The **server** is a server name of ArcGIS Online, for example, services2.arcgis.com, or your ArcGIS for Server name.
- The **organization_id** is a long string that uniquely identifies your organization.
- **/arcgis/rest/services** means this is the REST URL of the feature layer.
- The **service_name** (in this case, **311Incidents**) is the name specified when you created your feature layer.
- **FeatureServer** indicates that this is a feature service.
- The **layer_number**, for example, **0**, is often the first layer in the feature service.

24. Review the **ArcGIS REST Services Directory** page of the **311Incidents** layer and find the following sections:

- **For Drawing Info, verify that the renderer is uniqueValue on the Incident_Type field.**
- **For Has Attachments, notice that the value is true.**
- **For Fields, notice this section contains all the fields from your CSV except the Lat and Long fields you deleted earlier.**
- **For Types, notice that this section contains all the feature templates you defined.**

This metadata confirms that the style of your feature layer is correct and that the feature layer templates are defined. The feature layer accepts attachments and all the attribute fields you would like to collect.

ArcGIS REST Services Directory

Home > services > 311Incidents(FeatureServer) > 311Incidents

JSON

Layer: 311Incidents(ID:0)

View In: ArcGIS.com Map

Name: 311Incidents

Types:

ID: Dead Animal
Name: Dead Animal
Domains:
Templates:
Name: Dead Animal
Description:
Drawing Tool: esriFeatureEditToolNone
Prototype:
Attributes:
- *Incident_Type*: Dead Animal

ID: Graffiti
Name: Graffiti
Domains:
Templates:
Name: Graffiti
Description:
Drawing Tool: esriFeatureEditToolNone
Prototype:
Attributes:

Now you are ready to use your published feature layer to create web maps and web apps.

3.4 Use your layer in a web map and define editable fields

In this section, you will define which fields you will allow your users to edit in your web app.

1. **Go to the item details page of the 311Incidents feature layer.**

If you continue from the previous section, the item details page should be open in a tab of your web browser. Otherwise, you can log in to ArcGIS Online or your Portal for ArcGIS and find and click the feature layer in your content list.

2. **Under the item thumbnail, click Open, and click Add layer to new map with full editing control.**

3. **Click Yes, Open The Map.**

4. **In the Contents pane, point to the 311Incidents layer, click the More Options button ···, and choose Configure Pop-up.**

5. **In the Configure Pop-up pane, click Configure Attributes to open a window.**

Next, you will configure the aliases of the attribute fields and the fields you want to allow users to edit in the **Configure Attributes** window.

6. **In the Incident_Status field, clear the Edit check box.**

This field is restricted to internal use only—for example, a public works dispatcher who needs to track an incident's status as open, assigned, or closed. End users can see the field but cannot change the status of the incident.

7. **Also in the FID field, clear the Display check box.**

When you clear this internal feature identification field, users cannot see or edit the field.

8. **Specify aliases for the following attribute fields:**

- **Incident_Type: Type**
- **Incident_Desc: Brief Description**

- **Incident_Address: Incident Address**
- **Report_Date: Report Date**
- **Reported_By: Reported By**
- **User_Contact: Email or Phone**
- **Incident_Status: Status**

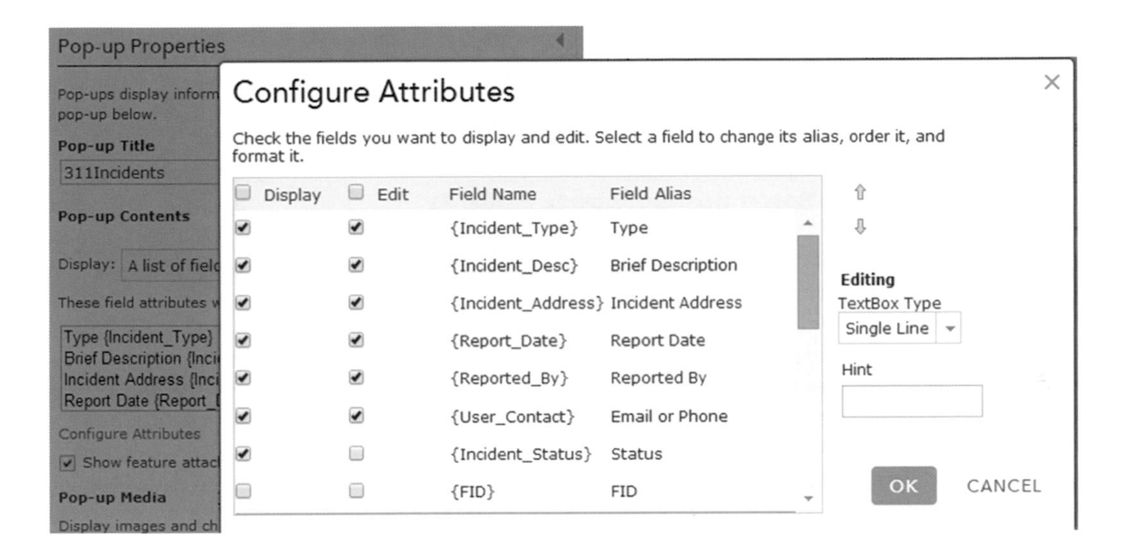

9. Click **OK** to close the **Configure Attributes** window.

10. In the **Configure Pop-up** pane, click **OK** to save your pop-up configuration.

11. Zoom and pan the map to your city or intended study area—that is, the area for which you are creating the app.

12. Save your web map. On the menu bar, click **Save** > **Save**.

13. Configure the **Save Map** window with the following information:

    - Title: **Report 311 Incidents.**
    - Tags: **311, Incidents, VGI,** and **GTKWebGIS.**
    - Summary: **Report non-emergency incidents.**
    - Save in folder: Leave as the default value, or choose one of your folders.

1
2
3
4
5
6
7
8
9
10

14. Click **Save Map.**

Save Map

Title: Report 311 Incidents

Tags: 311 × Incidents × VGI × GTKWebGIS ×

Add tag(s)

Summary: Report non-emergency incidents

Save in folder: pinde.webgis

SAVE MAP CANCEL

Note: The web map you have just created is for the public. If you wish, you can create a separate web map available only to internal or management users. This internal web map should allow internal users to edit the **Incident_Status** field.

You will leave your web map open if you are going directly to the next section.

3.5 Create a web app for data collection

You will create a web app using an ArcGIS Online configurable app. However, not all apps support editing, and not all of the apps respect the editable fields you defined earlier.

This section will use the GeoForm configurable app, and you will learn to define the editable fields because this app does not reuse the editable field settings in the web map.

1. Continuing from the previous section, on the map viewer menu bar, click the **Share** button.

2. In the **Share** window, share your web map with **Everyone (public).**

3. **Click Create A Web App.**

4. **In the list of categories on the left, click Collect/Edit Data, and look for a configurable app that supports editing.**

The following configurable apps support editing. You can click each of the apps, read their descriptions, and click **Preview** to experiment with the apps:

- **Basic Viewer:** Provides common tools including editing.
- **Edit:** Provides simple editing capabilities for editable layers in a web map.
- **Find, Edit, and Filter:** Support viewing and editing layers based on a filtered field value.
- **GeoForm:** Provides a form-based editing experience (for point feature types only). This particular app does not respect the editable fields you defined in the previous section.

5. **Click GeoForm, and click Create App.**

6. **Fill in the appropriate Title, Tags, and Summary information (this may already be done for you), and then click Done.**

You will be directed to the builder mode of your app.

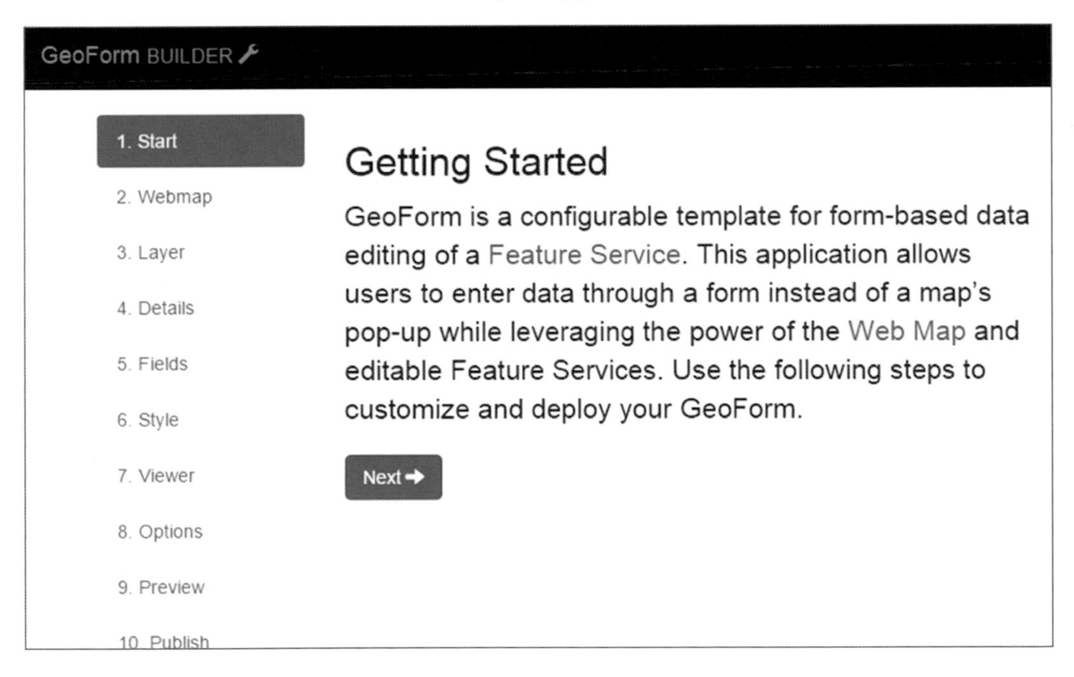

7. **For the Start step, read the information after Getting Started, and click Next.**

8. **For the Webmap step, click Next so that you can continue to use your current web map.**

9. **For the Layer step, click Next. This action tells the app to build its form from your web map layer, which is the feature layer you created earlier.**

10. **For the Details step, choose Use Small Header, and set the Title as Report a non-emergency incident.**

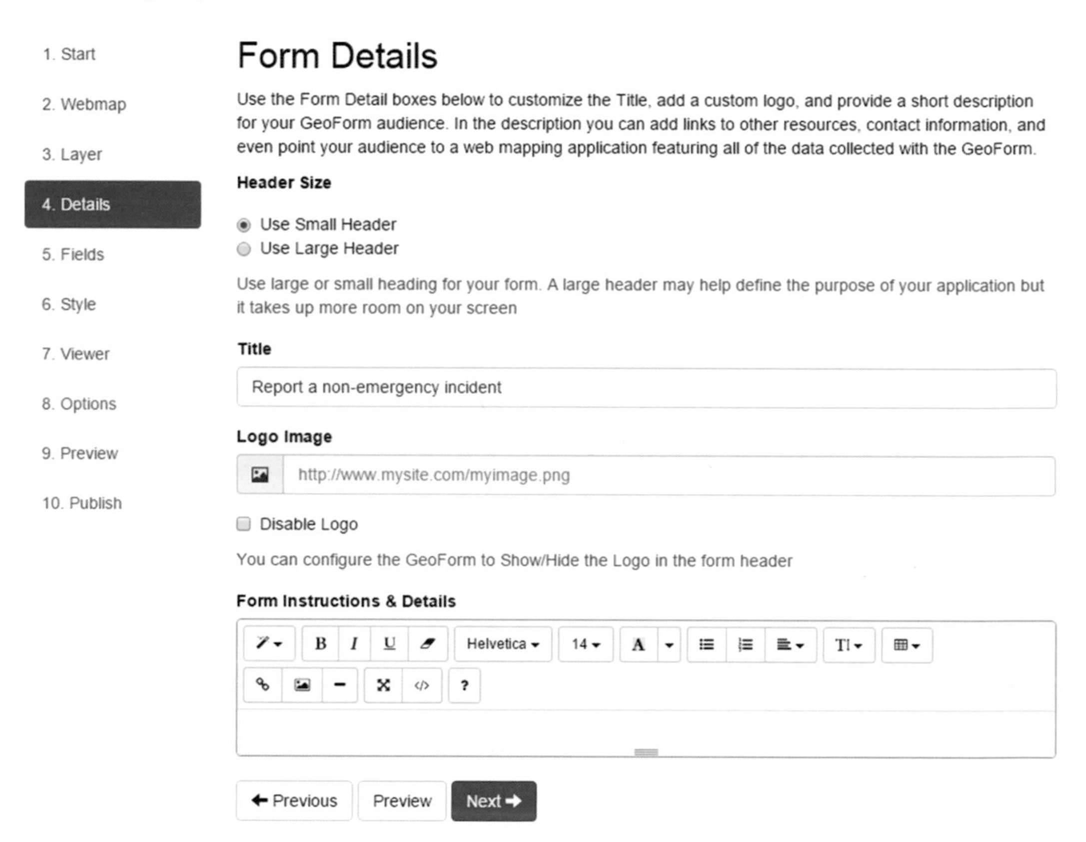

Optionally, you can add a custom logo and provide some instructions about how to use your app and details about your app for your end users.

If you want to use a custom logo, you will need to specify a URL to your logo image. If you don't yet have your logo image online, you will need to upload your logo image to a web server and then obtain the image URL for use here.

11. **Click Next.**

12. For the **Fields** step, set the **Enabled** and **Label** fields according to the following directions:

    - **In the Enabled field, select the boxes for Incident_Desc, Report_Date, Report_By, and User_Contact.**
    - **Clear the remaining check boxes in the Enabled field.**
    - **Starting at the top of the Label fields, rename the labels to Type, Description, Incident Address, Date, Your Name, Your Email or Phone, and Incident Status.**

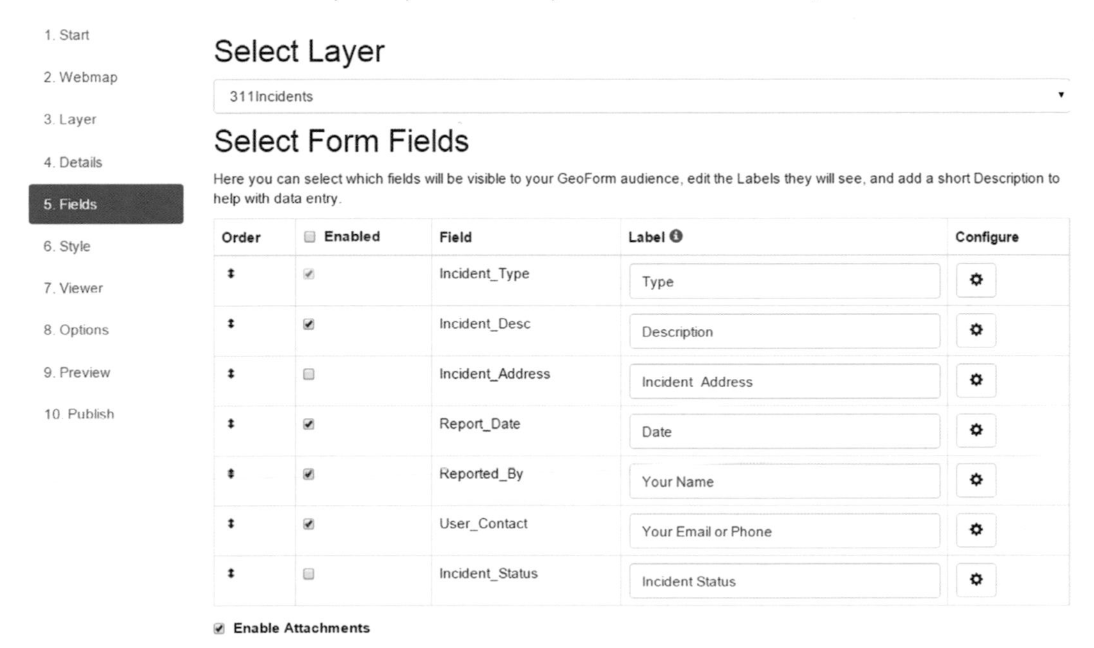

13. Click **Next**.

Unlike most other configurable apps, the GeoForm app does not take the pop-up configuration you defined earlier. For this reason, you must define the fields again here.

14. For the **Style** step, select a style (for example, **Superhero**) from the list, and click **Next**.

Styles define the color scheme, font family and sizes, and so on.

15. For the **Viewer** step, leave **Incident_Type** as the display field, which will appear in the viewer as a title, and click **Next**.

16. For the **Options** step, make sure the check box next to **My Location** is selected, and click **Next**.

Users can click **My Location** and zoom the map to the current location of the current user. This function is especially convenient for mobile users who are using small hand-held devices.

17. **For the Preview step, preview your app (if necessary, click Previous to change some settings), and click Next.**

18. **For the Publish step, click Save.**

A pop-up window appears notifying you that the item has been successfully saved.

19. **In the pop-up window, copy the URL under Form Link.**

This is your app URL. You can also find the app URL in the **Properties** section of the app's item details page.

20. **Paste your app URL in a web browser, and press Enter to open your app.**

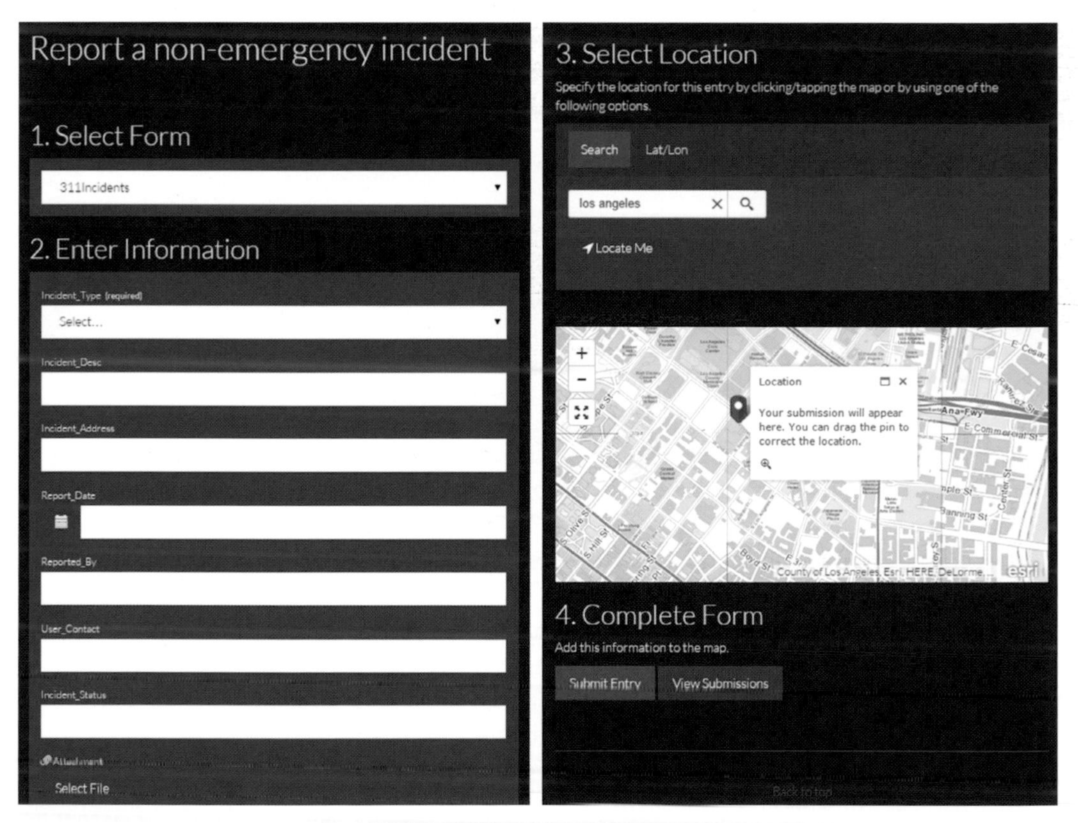

You can also open your app by clicking on the thumbnail on the item details page of this app.

21. **Test your web app by reporting an incident.**

   - **For Attachment, choose a photo or other document.**
   - **For Select Location, click Locate Me to find your current location, which is especially useful for mobile users.**
   - **Click View Submissions to see all the incidents submitted.**

You now have a web app that can collect data about non-emergency incidents for your city or study area. For an actual project, you can share the app's URL with the public via social networks or your organization's home page. The URL is the app URL in step 19.

Note: Users typically submit VGI at the site of the incident using mobile devices. The GeoForm app is optimized for smart devices. You can try this app on a smart mobile device by opening the app in your mobile device in a couple of different ways:

- You can find the app URL on your desktop computer, email the URL to yourself, check your email on your mobile device, click the URL to open the app, and test the app there.
- You also can start the web browser on your mobile device, go to ArcGIS Online or Portal for ArcGIS, sign in, find the app in your content list, and open the app.

QUESTIONS AND ANSWERS

1. **How does a map notes layer differ from a hosted feature layer?**

   Answer: A map notes layer is stored in the web map in which the layer was created. You cannot add a map notes layer to other web maps. Anyone with whom you share your web map can view this layer, but only the owner can edit the layer. Edits are saved to the containing web map. If you edited the data but are not the owner, you will need to save a new copy of the web map.

   In contrast, an owner can configure a hosted feature layer as read-only or editable. Anyone who has been granted permission from the owner can edit the layer. Edits are saved to the feature layer itself rather than to the web map. A feature layer is not limited to one web map. It can be added into in many web maps.

1
2
3
4
5
6
7
8
9
10

2. After I created my feature layer, I wanted to delete an existing field and add a new field. Do I have to change my CSV before I re-create my feature layer?

Answer: No, you do not need to start over from your CSV. You can directly modify your feature layer in the map viewer.

To delete a field, refer to section 3.3, step 17.

To add a new field, show the table of the feature layer, click **Table Options** in the map viewer, and choose **Add Field**. Follow the instructions to add a new field.

3. What kinds of editing can you perform with lines and polygons?

Answer: ArcGIS configurable apps, ArcGIS API for JavaScript, and Web AppBuilder for ArcGIS provide flexible capabilities for editing lines and polygons, including defining shapes with mouse clicks or freehand; moving, rotating, scaling shapes; adding, deleting, moving vertices; and even feature snapping.

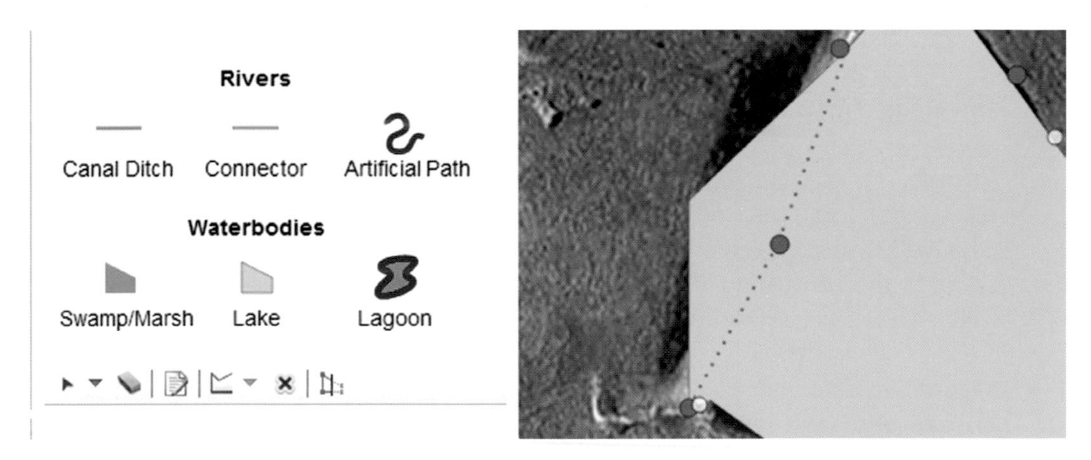

4. You can attach photos and other documents to host feature layers. Where are the attachments stored?

Answer: Attachments are stored in a database table underlying your hosted feature layer. A feature can have zero to many attachments; thus, a feature can have zero to many corresponding records in the table.

5. **For data that is collected by a hosted feature layer, where is that data stored? Can I export the data to my own computer?**

   Answer: Data collected by a hosted feature layer of ArcGIS Online is stored in the cloud. Data collected by a hosted feature layer of Portal for ArcGIS is stored in the ArcGIS Data Store for your Portal for ArcGIS. ArcGIS Data Store is a complementary component of Portal for ArcGIS. This component is required if you want your Portal for ArcGIS to be able to host feature layers.

   You can export a hosted feature layer into CSV files, shapefiles, file geodatabases, GeoJSON (an open standard format designed for representing simple geographical features, along with their nonspatial attributes, based on JavaScript Object Notation) files, or feature collections, depending on the following conditions:

   - You own the feature layer.
   - You are an administrator for your ArcGIS Online organization or Portal for ArcGIS.
   - You are not the hosted feature layer owner or the administrator, but the owner or administrator has configured the hosted feature layer to allow others to export the data (you can change this setting by clicking **Edit** on the hosted feature layer's item details page in the website and clicking **Allow others to export to different formats under the Properties** list).

   You can export from a hosted feature layer on the item details page of the layer by clicking **Export** (at left in the illustration) or by clicking the arrow button next to the layer name (at right in the illustration).

ASSIGNMENTS

Choose one of the following two assignments:

Assignment 3A: Create a web app that allows citizens to report wanted suspects.

Law enforcement typically asks citizens to report wanted suspects. The police department in your city wants to enhance the workflow. The police would like to build a web GIS app that citizens can use to report where they have seen four highly dangerous suspects and upload photos and videos that include images of these suspects.

Data: No data is provided. You can create a CSV file with the following fields:

- Suspect_Name or Suspect_Number
- Description: Describe what the suspect was doing and where you saw the suspect.
- Long
- Lat
- Date_Saw_Suspect
- Report_Date
- Reported_By
- User_Phone

Requirements: Your app should work on smartphones, allow citizens to upload photos and videos, and include feature templates (so that each template represents a suspect).

What to submit: Email your instructor with the subject line **Web GIS Assignment 3A**: and **Your name**, and include the following information:

- The REST URL of your feature layer
- The URL of your web app

Tips:

- Use the four suspect names or numbers (for example, **Suspect #1**) as the domain values.

- In an actual project, use the photos or pictures highlighting known appearance characteristics of the suspects as their symbols. Here, simply use a different icon (head and shoulders icon preferred) for each suspect.

Assignment 3B: Create a web app to collect non-emergency incidents.

This assignment is similar to the tutorial in this chapter, but you will use either shapefiles or a file geodatabase.

Data: Use either Incidents_gdb.zip or Incidents.zip under C:\EsriPress\GTKWebGIS\Chapter3\Assignments_data. The first option contains a file geodatabase that contains three feature classes: **Incidents_Points, Incidents_Lines**, and **Incidents_Polygons**. The second option contains three shapefiles that are equivalent to the three feature classes (the extents of the data may be different from the extent of your study area. Refer to step 6 of section 3.2 to set your feature layer's extent to match your study area).

Requirements:

- Your app should allow citizens to report non-emergency incidents that are points, lines, and polygon types.
- Your feature layer should have feature templates, each template representing, for example, an incident.
- Your app should allow citizens to upload attachments.

What to submit: Send an email to your instructor with the subject line **Web GIS Assignment 3B:** and **Your name**, and include the following information:

- The REST URLs of your feature layers
- The URL of your web app

Resources

ArcGIS Online Help document site

"Manage Feature Templates," http://doc.arcgis.com/en/arcgis-online/share-maps/manage-feature-templates.htm.

"Manage Hosted Web Layers," http://doc.arcgis.com/en/arcgis-online/share-maps/manage-hosted-layers.htm.

"Publish Features," http://doc.arcgis.com/en/arcgis-online/share-maps/publish-features.htm.

Chapter 4

Story Maps and more configurable apps

Apps are important because they are the face of web GIS. They bring web GIS to life. You have used several ArcGIS configurable apps in previous chapters. ArcGIS as a platform offers far more apps. This chapter first presents an overview of ArcGIS configurable apps, which include apps from ArcGIS Online, Portal for ArcGIS, Story Maps, and ArcGIS Solutions. The tutorial then teaches how to create apps using additional configurable apps. In addition, this tutorial teaches how to find data in ArcGIS Open Data and how to use smart mapping to symbolize two attribute fields to better present the temporal patterns in your data.

Learning objectives

- *Understand the suite of ArcGIS configurable apps.*
- *Discover data in ArcGIS Open Data.*
- *Symbolize two fields using smart mapping.*
- *Use the Compare Analysis app to create your own apps.*
- *Use the Story Map Swipe and Spyglass app to create your own apps.*
- *Use the Story Map Journal to create your own apps.*

This chapter in the big picture

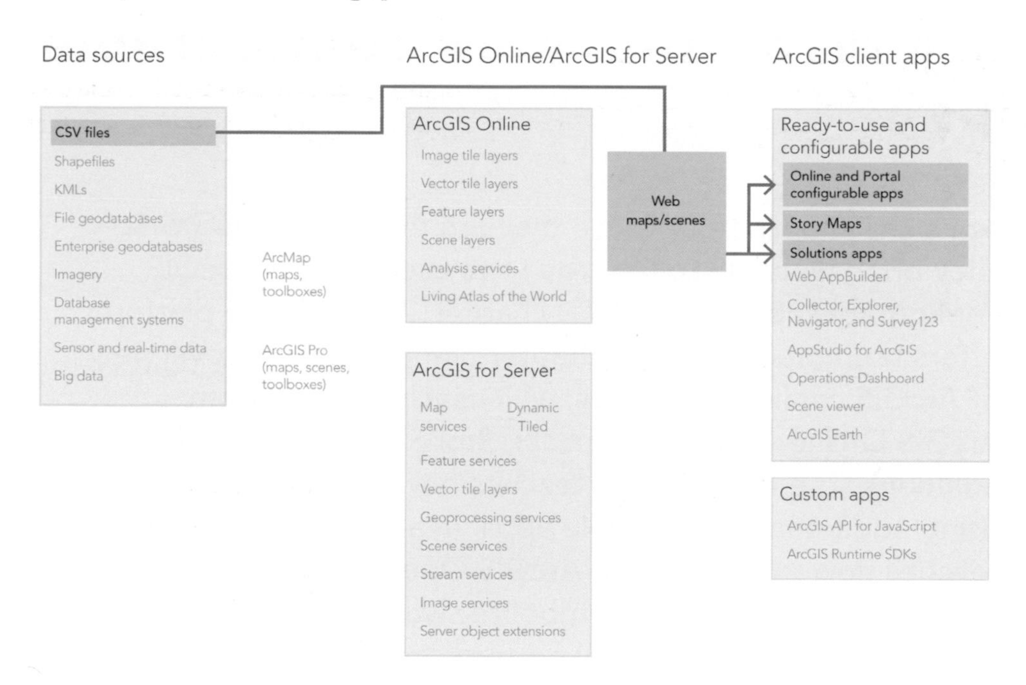

ArcGIS platform offers many web GIS apps and many ways to build apps. The green line in the figure highlights the technologies that this chapter teaches.

Apps, the face of web GIS

GIS apps are lightweight, map-centric computer programs that end users use on mobile devices, in web browsers, and on desktops. Web GIS end users directly interact with apps. The backend data, services, and server computation power of web GIS become live and useful through GIS apps.

Configurable apps enable the broad user community to build engaging apps with no GIS or web development skills. ArcGIS platform provides the following configurable apps:

- **ArcGIS Online and Portal for ArcGIS configurable apps:** You have learned several such apps in the previous chapters and will learn more of them in this and later chapters.
- **Story Maps:** Apps created from Story Maps combine maps and data with rich narratives and multimedia content so people feel connected and stay engaged. Some of the story apps have become ArcGIS Online and Portal for ArcGIS configurable apps.
- **ArcGIS Solutions apps:** These apps provide data and map and app templates for you to quickly jumpstart your applications projects.

App builders, including Web AppBuilder for ArcGIS and AppStudio for ArcGIS, enable users to build more sophisticated apps using a builder-user experience. You will learn more about them in other chapters.

Steps to use ArcGIS configurable apps

Using ArcGIS configurable apps generally involves three steps: choose, configure, and deploy.

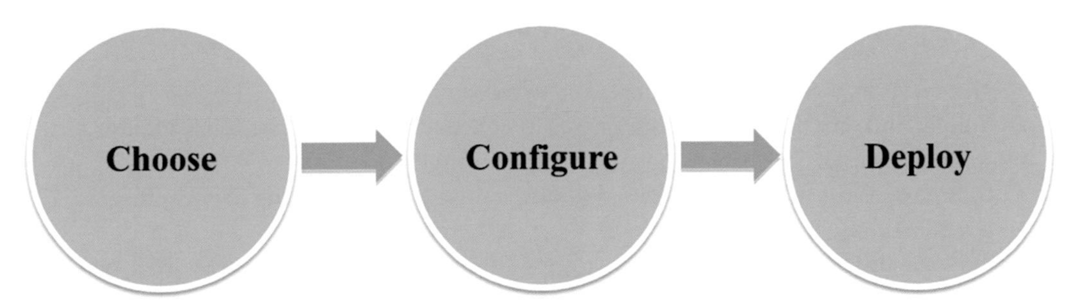

Using ArcGIS configurable apps generally takes three steps: Choose the app, configure the app, and deploy the app.

- **Choose**: In this step, you discover the data, maps, and configurable apps that match your app requirements. Consider the following factors:
  - **Purpose**: Who is your intended audience? Where and how will your audience use your app? What key points do you want to communicate?
  - **Functional requirements**: What critical functionalities support your purpose?
  - **Aesthetic**: How will the app's layout and color scheme support your brand or message?
- **Configure**: Configure the apps to use your data, and brand the apps for your organization.
  - ArcGIS Online and Portal for ArcGIS configurable apps typically have a configuration user interface.
  - Story Maps provide a builder-user experience.
  - ArcGIS Solutions apps provide a configuration file that often requires manual editing.
- **Deploy**: Deploy your new apps for your end users.
  - Esri automatically hosts the apps created with ArcGIS Online and Portal for ArcGIS configurable apps and Story Maps in the cloud. If needed, you can download the source code for these open-source apps and host the apps on your own web servers.
  - You must deploy and host apps created with ArcGIS Solutions templates on your own web servers.

ArcGIS Online and Portal for ArcGIS configurable apps

In the previous chapters, you have used several configurable apps. You can choose from several more apps. To help you to choose the right one, ArcGIS organizes these apps, according to their purposes, in the following categories:

- **Build a story map**: These apps are taken from Story Maps.
- **Collect and edit data**: These apps primarily collect data. These apps fall into the subcategories of crowdsourcing and general editing.
- **Compare maps and layers**: These apps are focused on comparing geographic phenomena.
- **Explore and summarize data**: These apps allow your users to interact with attributes and in some cases other services to facilitate a deeper exploration of the content of your map.

- **Make a gallery:** These apps create a gallery of maps, apps, or other content that you can use as a convenient access point for all of your geographic content. These apps require a group.
- **Map social media:** These apps include social media content to supplement your message with relevant contents.
- **Provide local information:** These apps highlight the resources available at a location. Options include highlighting all features within a certain distance of a location and informing users that their addresses are located within a certain geographic area.
- **Route and get directions:** Use these apps to provide driving directions from a user-defined starting point to the geographic features within your map.
- **Showcase a map:** Apps in this group include Basic Viewer, Map Tools, Minimalist, Simple Map Viewer, and Story Map Basic.

Story Maps

Every story has a location. Web GIS can enhance storytelling by visually and intuitively illustrating the "where" component of every story. Story Maps (**https://storymaps.arcgis.com**) are simple web apps that combine interactive maps, multimedia content—text, photos, video, and audio—and intuitive user experiences to tell stories about the world. Story Maps use geography as a way to organize and present information.

Story Maps combine interactive maps, multimedia content, and user experiences to tell stories.

Table 4.1 **Main types of configurable app templates provided by Story Maps**

Presenting a sequential, place-based narrative		
Story Map Tour		Uses the form of a series of geotagged photos and captions linked to an interactive map.
Story Map Journal		Uses a set of journal entries with text, maps, images, and video.
Presenting a series of maps		
Story Map Series Tabbed Layout		Uses a set of tabs.
Story Map Series Side Accordion Layout		Presents a series of maps, and accompanying text and other content for each map, in an expandable panel.
Story Map Series Bulleted Layout		Presents a series of maps via numbered bullets, one per map.
Comparing two maps or two layers of a single web map		
Story Map Swipe	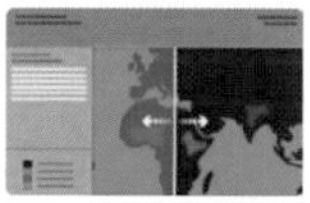	Users can slide the swipe tool back and forth to compare one map theme to a second map theme.
Story Map Spyglass		Similar to Swipe but enables users to peer through one map to another using a spyglass function.
A curated list of points of interest		
Story Map Shortlist		Presents a set of places organized into a set of tabs based on themes.
Presenting one map		
Story Map Basic		Presents a map via a simple minimalist user interface.

You can create a story map in either ArcGIS Online or Story Maps:

- **ArcGIS Online**: For example, you can share a web map, click **Create a Web App**, and choose a story map template.
- **Story Maps website**: You can select a type of story map and follow the builder wizard.

ArcGIS Solutions apps

ArcGIS Solutions provides a gallery of free templates to jumpstart your projects. The templates typically include apps source code. Some templates also include data models, layers, maps, and sample web services. The data model and map styles are created on the basis of industry best practices and emerging trends. ArcGIS Solutions apps cover almost all industries. The apps focus mostly on the web and mobile platforms and occasionally on the desktop platform.

You can search for templates in ArcGIS Solutions by products, industries, and keywords. For each template, you can read its introduction, requirements, and contents in the package. You can try the app live, download the template, configure the app, and deploy it.

ArcGIS Solutions provides a rich collection of configurable apps for almost all industries.

1
2
3
4
5
6
7
8
9
10

Configuring ArcGIS Solutions web apps typically involves editing the config.js file. In this file, you replace the default URLs of the maps and layers with the URLs of your maps and layers and replace the default attribute field names with the default attribute field names of your data. Deploying these web apps typically requires a web server such as Apache, Microsoft IIS (Internet Information Service), or other products. You can set up your own web server or purchase a web hosting service. Once you have a web server, deploying a JavaScript web app requires you to copy the files to a folder under your web server's web root, such as **c:\inetpub\wwwroot** for IIS.

ArcGIS Marketplace apps

ArcGIS Marketplace is another option for finding apps and data. These apps and data are provided by Esri Business Partners, Esri Distributors, and Esri. ArcGIS Marketplace aims to be an app store that explicitly serves the GIS market and user community.

To get apps from ArcGIS Marketplace, go to **http://marketplace.arcgis.com**, search for the app you need, request access to the app (this requires an administrator-level account), pay for the app if necessary, and obtain access to the item. All the apps in the Marketplace are built to work with ArcGIS Online, and you can easily share these apps among ArcGIS Online groups and users within your organization.

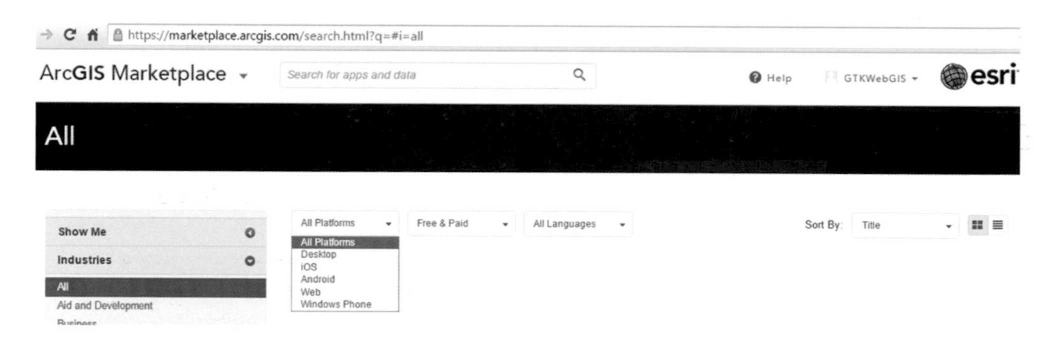

You can search for apps in ArcGIS Marketplace by industry, platform, and other criteria.

Find data for your apps in ArcGIS Open Data portal

Great apps need great data. In the previous chapters, you used your own data, layers from the Living Atlas of the World, and other content you discovered in ArcGIS Online. The ArcGIS Open Data portal (**http://opendata.arcgis.com**) is a good place to find rich, authoritative data for your apps.

ArcGIS Open Data allows organizations to use the ArcGIS platform to provide the public with open access to their authoritative data. You can enable ArcGIS Open Data within your ArcGIS Online My Organization settings page. See the **Questions and Answers** section for more information on how to enable ArcGIS Open Data. Organizations can configure the website with their own look and feel and designate groups that can share items to ArcGIS Open Data.

The general public can use ArcGIS Open Data sites to search for data by topic or location. ArcGIS Open Data allows you to explore the data through charts, tables, and maps to understand the data, download the data in several popular formats, and use the data directly via APIs.

This tutorial

This tutorial teaches ArcGIS Open Data, Story Maps, and more ArcGIS Online configurable apps.

- Section 4.1 searches, explores, and downloads NAEP (National Assessment of Educational Progress) data from ArcGIS Open Data portal.
- Section 4.2 creates a web map symbolizing two fields using smart mapping.
- Section 4.3 creates a web app using the ArcGIS Online Compare Analysis app. This app allows users to visually analyze the spatial and temporal patterns of NAEP scores.
- Section 4.4 creates a swipe story map that compares US 2015 average household income and median home value. This information can be helpful, for example, in comparing the salary of a new job offer to the average income and housing costs of the area in question.
- Section 4.5 creates an app from Story Map Journal to display the apps you have learned and the apps you will learn in this book.

System requirements:

- An ArcGIS Online publisher or administrator account.
- A desktop-based web browser: The 3D scene steps of section 4.5 require a video graphics card that supports WebGL (Web Graphics Library). WebGL is not yet supported on most mobile browsers. For more details, go to **http://arcg.is/1PQ9bKb** (the short URL for **https://doc.arcgis.com/en/arcgis-online/reference/scene-viewer-requirements.htm**). If your computer does not have this kind of video graphics card, you can still do most of section 4.5; simply skip the 3D scene steps.

4.1 Explore ArcGIS Open Data

Have you wondered where to find good data for your homework or for building your apps? This section will illustrate where you can find, evaluate, and download the data.

Note: The user interface of ArcGIS Open Data is undergoing significant changes. As you work through these steps, the actual interface may appear differently than the interface you see in this section. If you find major discrepancies between what you see in this section and the actual user interface, you can skip steps 4 through 13 so that you can find, download, and use the data in the subsequent sections.

1. **In a web browser, go to ArcGIS Open Data (http://opendata.arcgis.com), and sign in with your ArcGIS Online account.**

You can search for data within the Open Data site using the search box or the map. Here, you will search using keywords.

2. **In the Find text box, type in NAEP GTKWebGIS, and press Enter or click the Search button.**

The search result appears. The search result displays information about who shared the dataset, the number of features or rows in the dataset, and the first few lines of the dataset description. This NAEP dataset has 51 records. When you move your cursor to the search result, you will see that the spatial extent of the dataset displays on the map.

National Assessment of Educational Progress 2005-2015
Shared by GTKWebGIS
Since NAEP assessments are administered uniformly using the same sets of test booklets across the nation, NAEP results serve as a common metric for all states and selected urban districts. The assessment stays essentially the same from year to year, with only carefully documented changes. This permits NAEP to provide a

Custom License 4/5/2016 Spatial Dataset 51 Rows

3. **Click the dataset to see its details.**

You can see the full description, links to the licensing information, and links to access the data.

National Assessment of Educational Progress 2005-2015 Download Data ▾ APIs ▾
Custom License 1/21/2016 Spatial Dataset 51 Rows

4. **Read the description to understand what the data is about. You may need to expand the text to see the full description.**

5. **In the expanded description, click the field descriptions link to download and view the field aliases.**

The download is in the comma-separated value (CSV) file format. You can see in the CSV that this data layer has many attribute fields. For example, **M4_2005_AvSS** stands for "NAEP 2005, 4th Grade Math, Average Scaled Score," and **R8_2015_AvSS** stands for "NAEP 2015, 8th Grade Reading, Average Scaled Score."

6. **Click APIs, and then examine but do not select any of the options.**

The API (application programming interface) options are typically for developers to use to build applications. Here, you see the **GeoJSON** and **GeoService** formats in the list. GeoJSON is an open standard format designed for representing simple geographical features, along with their nonspatial attributes, based on JavaScript Object Notation. You will learn how to use the GeoService format in the JavaScript chapter.

You can use some datasets, like this one, directly via the API links. You can download all datasets found in Open Data, which is useful when you need to edit or enhance the data. We will use the download approach.

7. **Click Download Data, and choose Spreadsheet.**

You can download the data in either spreadsheet, Keyhole Markup Language (KML), or shapefile format.

8. **Locate and open the downloaded file in Excel or a text editor. Quickly review the data fields, and then close Excel or your text editor.**

You have downloaded the data. You will create a web map and a web app from this data in the following sections.

4.2 Map two variables with smart mapping

Smart mapping makes it easier for you to create attractive maps and reveal the meaning behind your data. You learned the basics of smart mapping in previous chapters. This section will teach it in further depth.

1. **In a web browser, navigate to ArcGIS Online (http://www.arcgis.com) or your Portal for ArcGIS, and sign in.**

2. **Click Map to open the map viewer.**

3. **Locate the CSV file you downloaded previously, drag the CSV file to the map area, and drop the file there.**

Notice that the states are displayed as points on the map. On the left side, the smart mapping interface appears.

4. **In the Change Style pane, for Choose an attribute to show, choose M4_2005_AvSS, which displays fourth-grade math scores from 2005.**

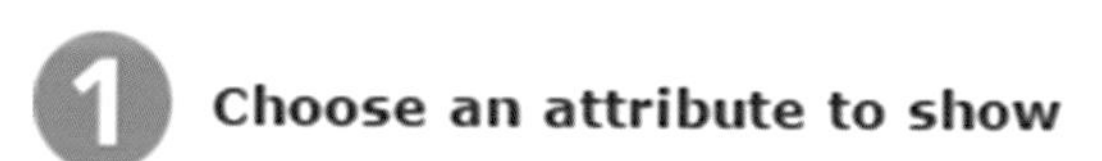

The map updates to display this attribute field by size. Next, you will further symbolize the layer with the math scores from 2015.

5. **Click Add attribute.**

6. **For the new attribute, choose M4_2015_AvSS, and leave Color & Size as the default style choice.**

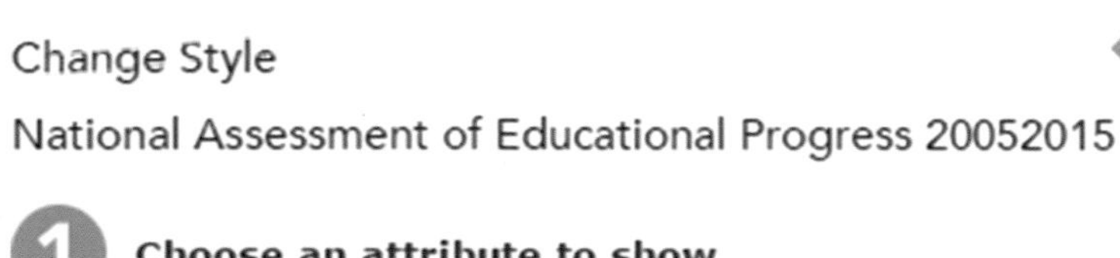

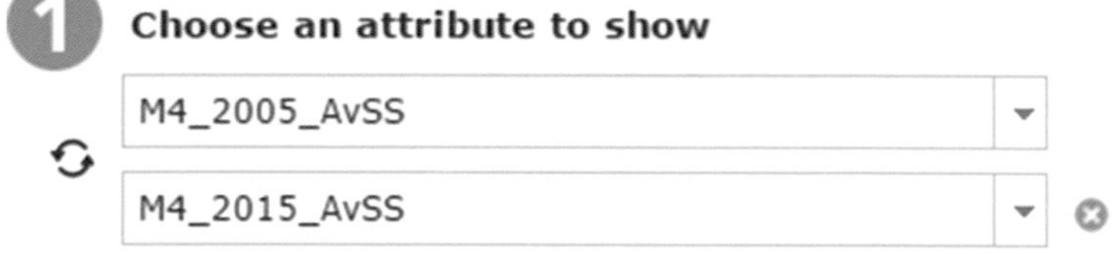

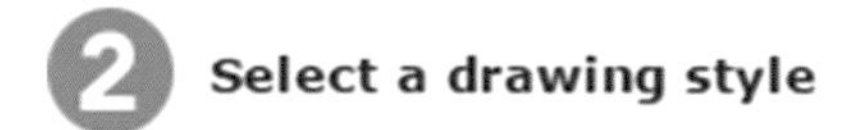

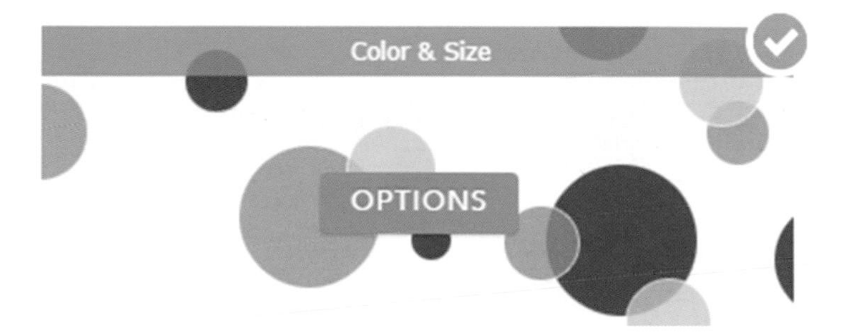

On the map, you will see that the point symbol of each state shows math scores from 2005 and 2015. A color ramp symbolizes the first field, M4_2005_AvSS, and circles of different sizes symbolize the second field, M4_2015_AvSS.

7. Click **Options** of the **Color & Size** style.

This action brings up a new interface, allowing you to further configure the style.

Change Style

National Assessment of Educational Progress 20052015

This layer has multiple styles.

M4_2005_AvSS

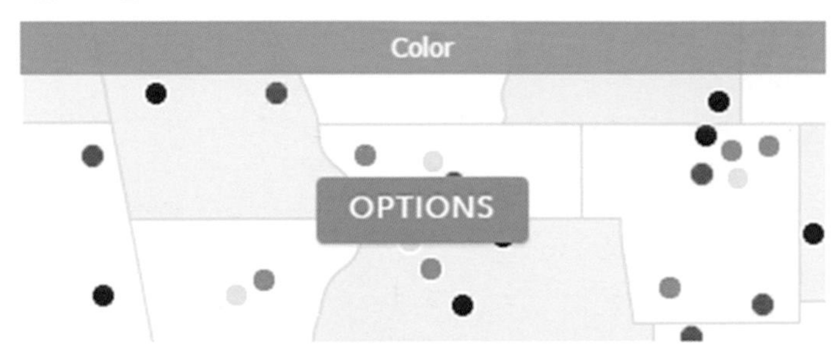

M4_2015_AvSS

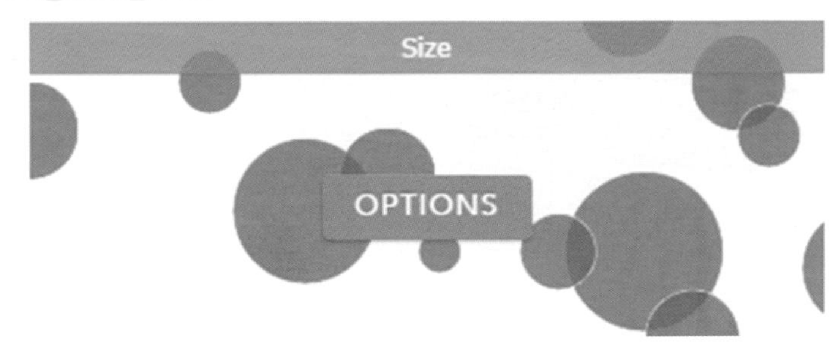

8. Click the **Options** of **Color**, and then click **Symbols**. In the gallery of symbols, click **Fill**, choose a different color ramp, such as the one that ramps from white on the bottom to dark red, and then click **OK** and **OK** again.

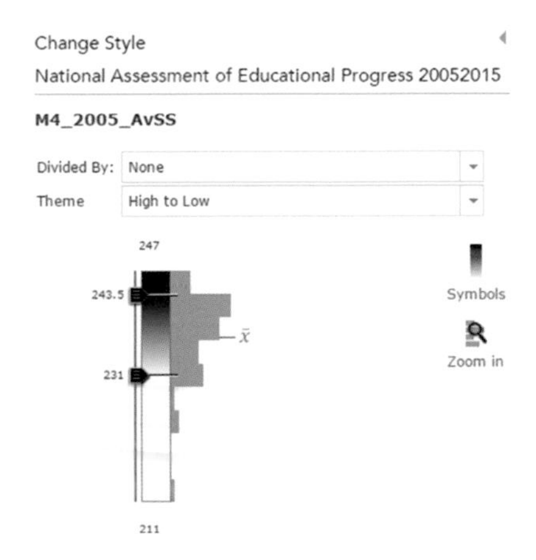

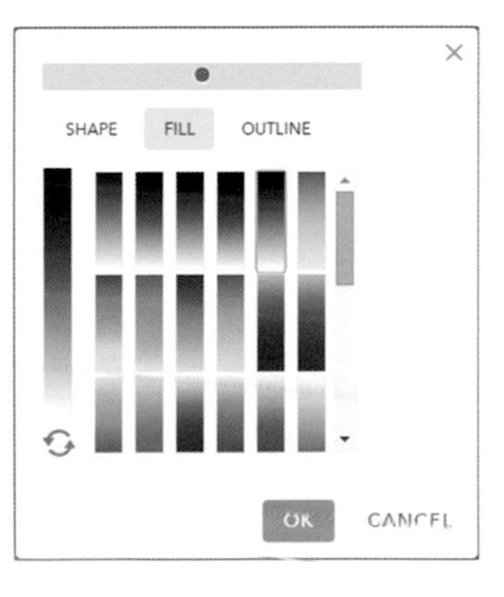

9. **Click Done and Done again to exit the Change Style pane.**

10. **Click the Basemap button on the map viewer toolbar, and change the basemap to Dark Gray Canvas.**

The NAEP layer should stand out better on this basemap, as illustrated. You should notice the following information:

- In a state such as California, a lighter color and a smaller circle indicate low scores in both 2005 and 2015.
- In a state such as Massachusetts, a darker color and larger circle indicate high scores in both years.
- A lighter color and a larger circle in a state indicate scores have increased from low in 2005 to high in 2015.
- A darker color and smaller circle in a state indicate scores have decreased from high in 2005 to low in 2015.

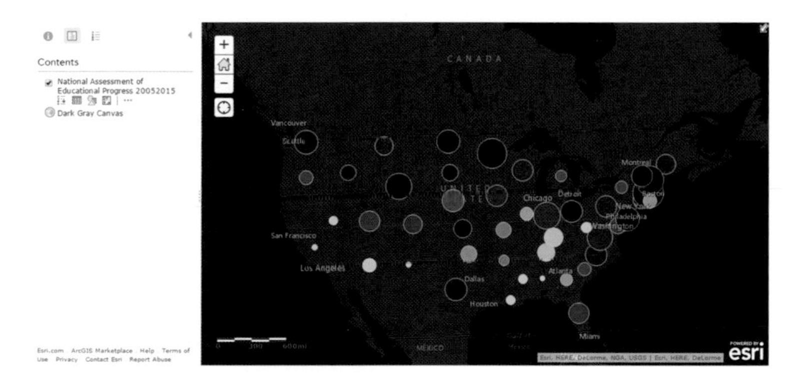

11. **In the Contents pane, point to the National Assessment of Educational Progress 20052015 layer, click the More button ···, and choose Rename.**

12. **Set the Layer Name as Math 2005 vs. 2015, and click OK.**

13. **Click Save on the map viewer toolbar, and choose Save from the list.**

14. **In the Save Map window, perform the following actions:**

- **Set the Title as NAEP 4th Grade Math 2005 vs. 2015.**
- **Set the Tags as NAEP, GTKWebGIS.**
- **Set the Summary as 2015 scores are displayed by color and 2005 scores are displayed by size.**
- **Click Save Map.**

In this section, you symbolized a layer with two attributes and created a web map for use in your web app in the next section.

4.3 Use Compare Analysis to create your own apps

In this section, you will use the Compare Analysis configurable app to display four maps side by side. This display allows you to analyze the spatial and temporal patterns of NAEP scores in multiple ways.

1. **Continuing from the previous section, on the map viewer menu bar, click the Share button.**

2. **In the Share window, share your web map with Everyone (public).**

3. **If a pop-up message prompts you about a shortened URL, click OK.**

4. **Click Create A Web App.**

5. **In the Create a New Web App window, click Compare Maps/Layers.**

6. **In the list of categories, choose the Compare Analysis app, and then click Create App.**

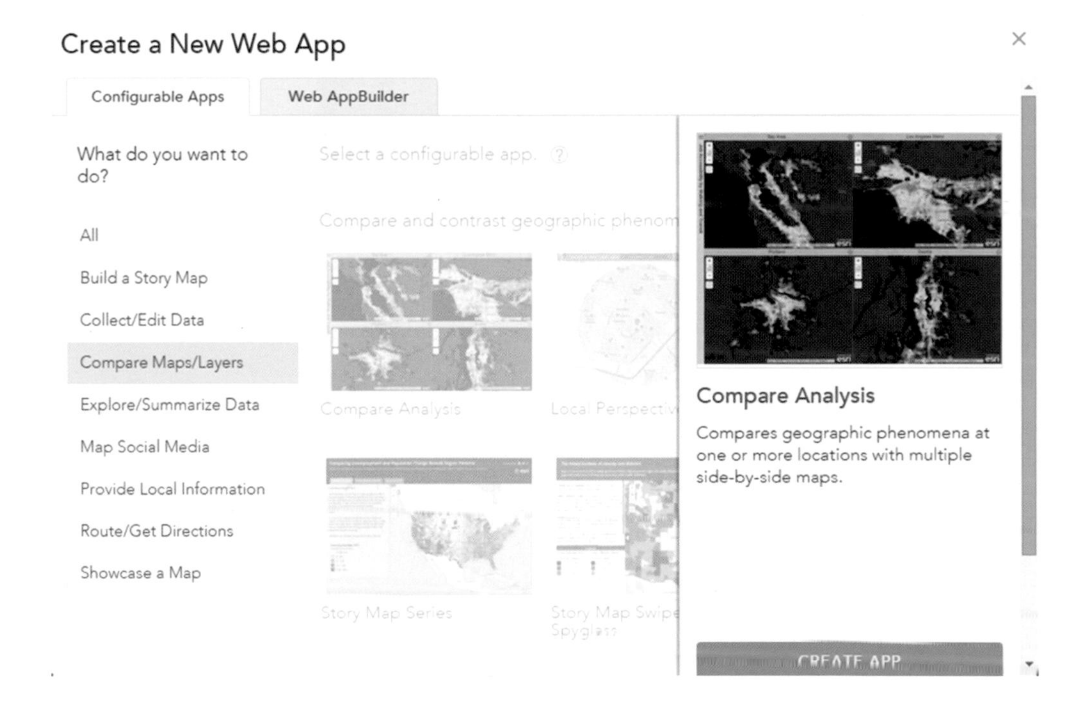

7. **Use the default values or fill in the appropriate Title, Tags, and Summary information for your app, and then click Done.**

The web app will be shared with everyone at the same level as the web map.

The Compare Analysis app needs two or more web maps. You started with a web map. You will select three more web maps in the next steps.

8. **In the Configure Web App page, click Choose Web Map(s).**

9. **In the Select Web Map window, search in ArcGIS Online for NAEP sample owner:GTKWebGIS, and press Enter.**

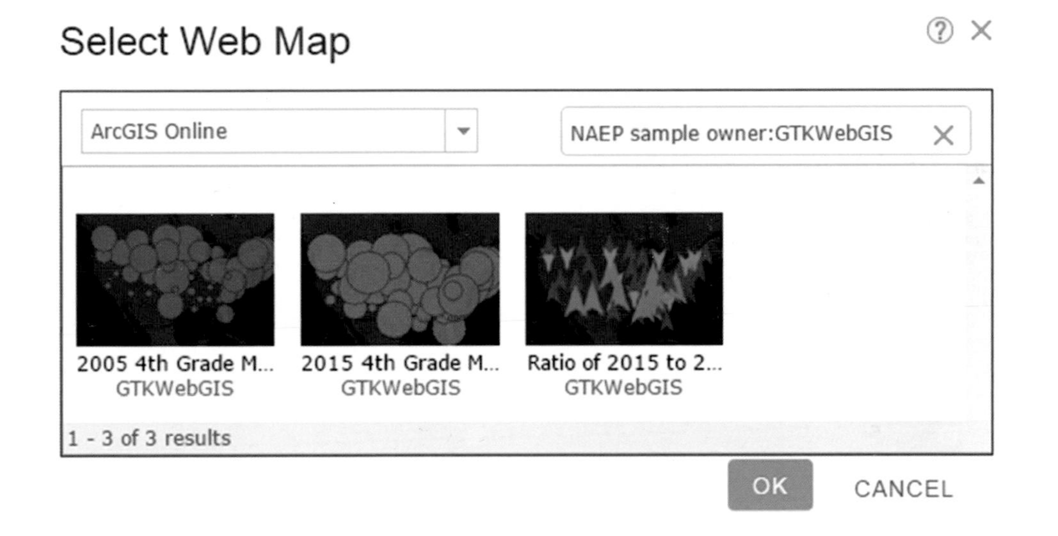

Next, you will select these web maps.

Note: The order of your web map selections will be the same order that your web maps are placed on the app.

10. **Click Ratio of 2015 to 2005, click 2005 4th Grade Math, and then click 2015 4th Grade Math. Click OK.**

If you need to unselect a web map, hold the **Control** key (for Windows) or **Command** key (for Mac) on your keyboard, and click the web map you want to unselect.

11. **Click the Side Panel, and clear the Show side panel. For Title, type NAEP 4th Grade Math 2005 vs. 2015.**

Configure Web App

Map

Theme

Side Panel

Options

Uncheck the 'Include side panel' box to remove the side panel from the app.

☑ Include side panel

☐ Show side panel

Title (Displays on side panel)

NAEP 4th Grade Math 2005 vs. 2015

12. **Click Save and View to preview your app.**

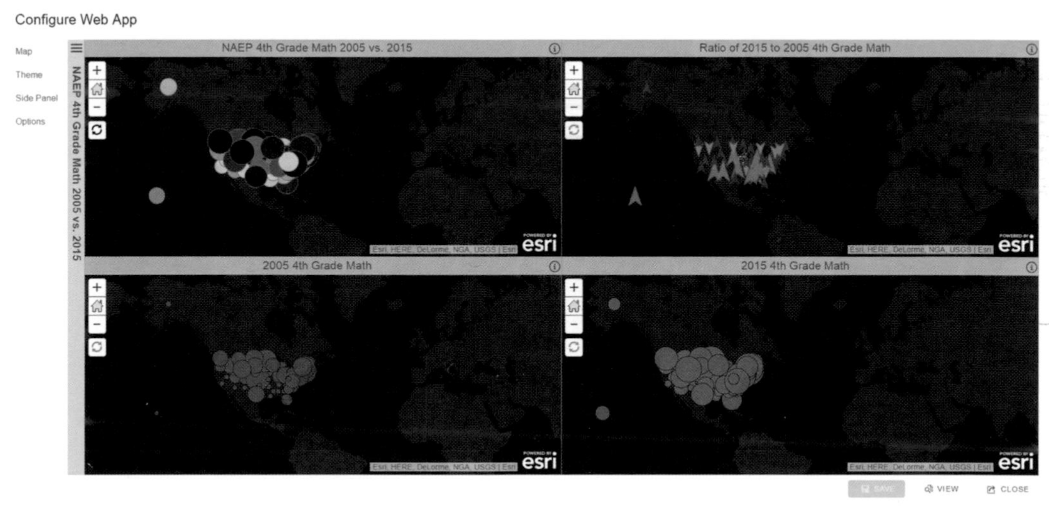

You will see the four maps displayed in one page.

13. **Pan or zoom the first map; note that the other maps follow.**

14. **Click Close.**

You will be directed to the **Details** page of the app.

15. **On the Details page, click the thumbnail to open and explore your app.**

In this section, you created a web app for comparison analysis. This app makes it easier for you to appreciate the spatial and temporal patterns of the NAEP scores.

- The 2005 and 2015 web maps allow you to see the spatial patterns of each year individually. In general, the states in the north and east scored higher than other states.
- Putting the 2005 and 2015 web maps side by side allows you to see the temporal pattern: Most states scored higher in 2015. But you still might have difficulty understanding which states increased or decreased and by how much.
- The 2015 versus 2005 web map symbolizes two fields in one map, with 2005 scores shown in color and 2015 scores shown in size. You can read the 2005 and 2015 scores and changes that appeared previously.
- The ratio map does not show the actual scores of each year but clearly presents the temporal pattern. As you can see on the map, the scores of most states increased. Washington, DC, had the largest increase, while scores in some mid-northern states decreased from 2005 to 2015.

4.4 Use Story Map Swipe and Spyglass to create your own apps

In this section, you will use the Story Map Swipe and Spyglass app, which is especially suited for the following purposes:

- Displaying changes of one theme over time, such as the difference between the current sea level and a projected rise in sea level, or visualizing an area at different times, such as before and after a tornado
- Displaying the similarity or contrast between two themes, such as the similarity of spatial patterns for diabetes and obesity or the contrast between unemployment rates and home value changes

Your web app will compare average household income with the median home values in a given area. This app can be helpful in many situations. For example, if you are offered a job in another city or state, you can use the app to find income and home value information for that area. On that basis, you can decide whether to accept the terms of the job offer or ask for a higher salary.

The swipe web app needs two layers in one web map or two web maps. A web map with two layers has been created for this tutorial. You will find and use this web map to create the web app.

You can create a swipe web app using ArcGIS Online/Portal for ArcGIS or Story Maps. You will use Story Maps in this section.

1. **Start a web browser, go to Story Maps (http://storymaps.arcgis.com), sign in with your ArcGIS Online account, and click Apps in the menu at the top.**

This page lists the available **Story Maps**, with intuitive graphics depicting the functions and general layout of each map.

2. **Find Comparing Two Maps, and under Story Map Swipe, click Build.**

3. **In the welcome window, click the Search button.**

4. **In the Select Web Map window, search in ArcGIS Online for Income Home value owner:GTKWebGIS.**

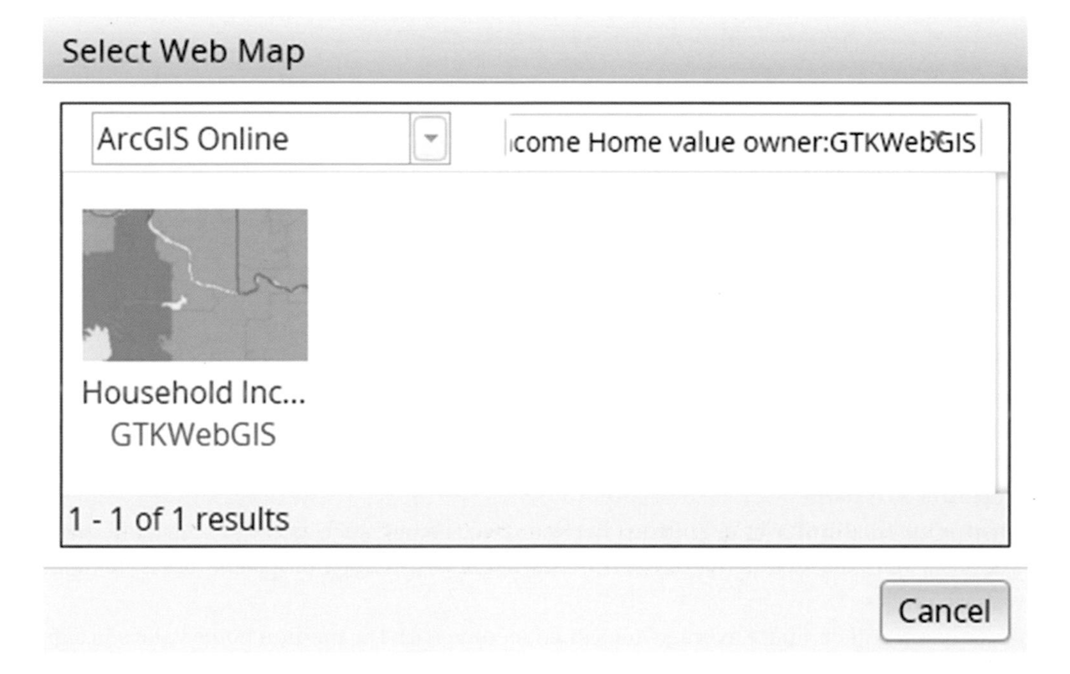

5. **Click the web map you find.**

6. **In the welcome window, click Next.**

The **Welcome to the Swipe/Spyglass Builder** window opens. The builder provides a step based wizard to guide you through the process of creating your app. The first step is **Swipe Style**, which allows you to choose between the vertical bar swipe style and the spyglass layouts. You already chose the swipe mode in step 2, so in the second step, you are presented with **Swipe Type**.

7. **In Swipe Type, select 2015 USA Median Home Value as the layer to be swiped. Click Next.**

Note: You must select the top layer in your web map. If the layer you want is hidden by upper layers, swipe will not have any visual effect.

8. **In the App Layout step, select Enable Legend, Enable Swipe series, Enable popup, Enable an address search tool, and Enable a 'Locate' button. Click Next.**

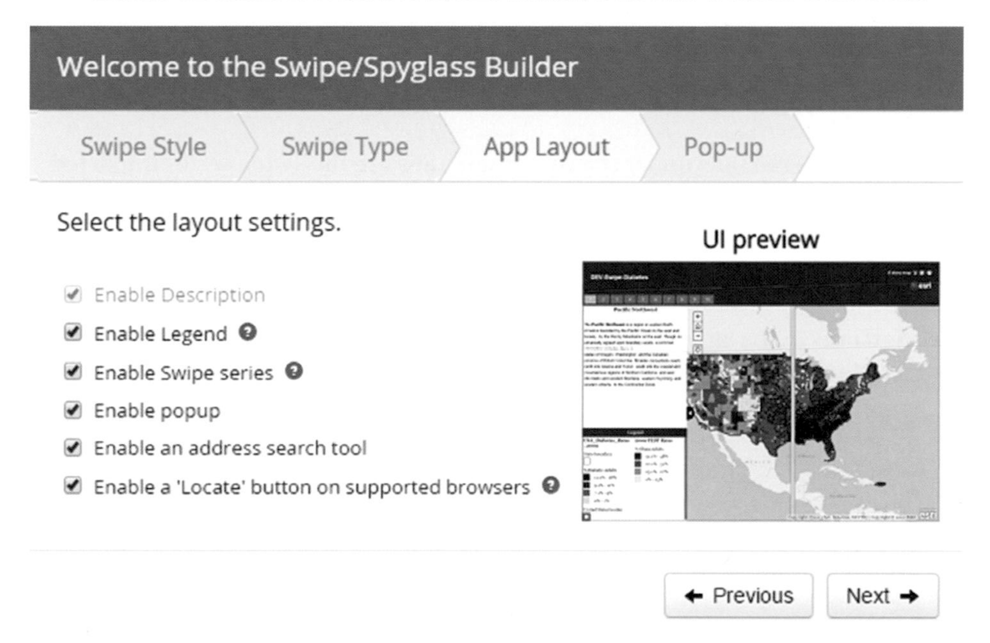

The swipe series allows you to create and edit a selection of locations with accompanying titles and text. This web map has three bookmarks already defined. These bookmarks will appear in the swipe series.

9. **In the Pop-up step, enter Average Household Income as the header title for the left map and Median Home Value as the title for the right map. Click Open the app.**

Your app opens in the configuration mode. You will see pencil buttons next to the title and subtitle. You can click these buttons to change the title and subtitle.

10. **Note the numbered tabs under the title. These tabs are the swipe series imported from the web map bookmarks.**

You can click the **Plus** button to add more tabs to the series.

11. **Click Settings in the banner. You can further configure the app theme and header. Click Apply if you make any changes; otherwise, click Cancel.**

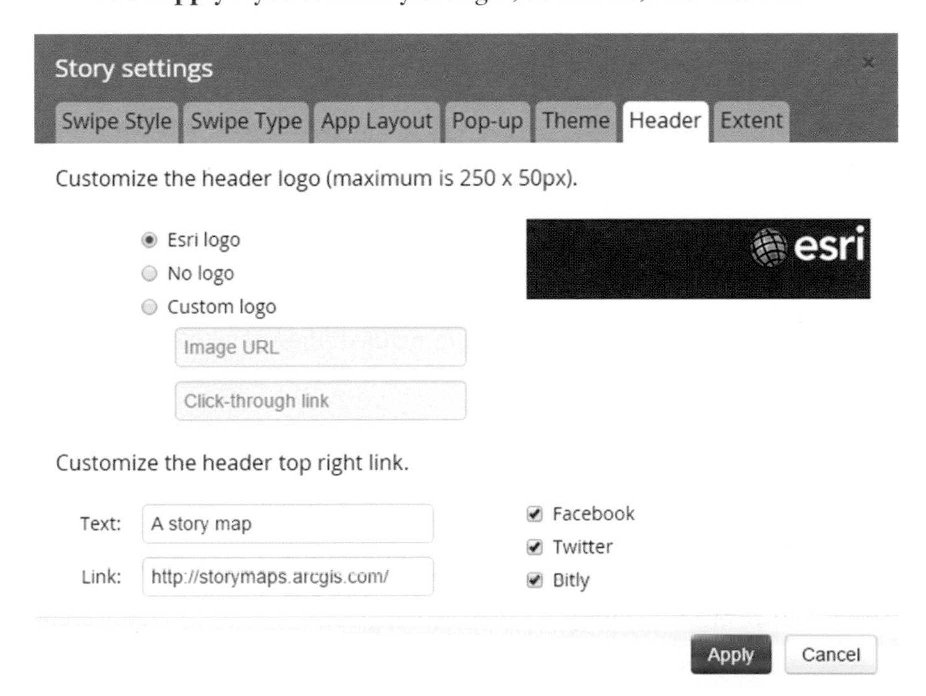

12. Click **Save** in the banner. The **Story saved** window appears. Click **Close**.

13. Click **Share** in the banner.

14. In the Share your story window, click **Share publicly**.

Note the URL in the text box under **Share your story**. This URL is the short version of your app URL. You can share this URL so your audience can access your app.

15. Click **Open**.

Your app will open in a new window or new browser tab. Next, you will explore this app.

16. Click a location of your interest on the map. Examine the contents in the pop-up.

17. Hold and move the vertical bar left and right to compare the income and the home value layers. As the vertical bar crosses the pop-up anchor, note that the content in the pop-up toggles between the two layers.

18. Click each numbered tab of the swipe series to compare the income and home value in each area.

19. If you are in the United States, click the **Locate** button to zoom to your area, and compare the income and home value in your area.

Optionally, you can change the swipe style of your app to the spyglass mode and experiment with this mode by performing the following steps:

20. In your web browser, go back to your app builder mode.

21. Click **Settings** in the banner.

22. In the **Story Settings** window, under the **Swipe Style** tab, select the **Spyglass** layout.

23. Click **Apply**. Note that a message pops up informing you that you need to save and reload the story. Click **OK**.

24. Click **Save** in the banner.

25. Reload the web page by refreshing your web browser.

You story map reloads in spyglass mode.

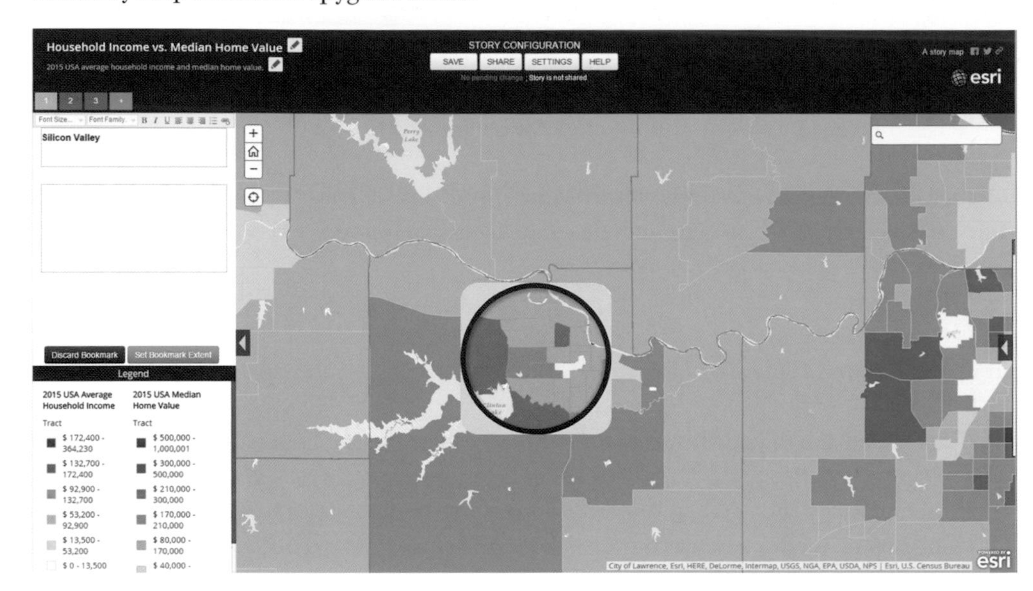

26. **Hold and drag the spyglass on the map to view the home value layer under the household income layer.**

If you want to switch back to the swipe mode, you can click **Settings** and select **Vertical bar** under **Swipe Style**, click **Apply**, save your app, and reload the page.

In this section, you created an app that can compare household income with home value. Imagine you are considering moving to a new area for a new job offer. This app can help you get a sense of how your salary compares to median incomes in the new area and what your home will probably cost. This app can help you decide if you should accept the offer or even provide a solid justification for you to negotiate for a better salary.

4.5 Use Story Map Journal to create your own apps

In this section, you will create an app using Story Map Journal to showcase the web apps and maps you have learned and those you will learn in this book. The app supports five main types of resources: text, videos, images, maps, and URL. You will learn about each of these resource types.

1. **In a web browser, go to the Story Maps website (http://storymaps.arcgis.com/), sign in with your ArcGIS Online account, and click Apps in the top menu.**

2. **Find Story Map Journal, and click Build.**

3. **In the welcome window, select Side Panel, and click Start.**

You will be directed to the **Map Journal Builder**.

4. **To enter your map journal title, type Learn ArcGIS Apps with Getting to Know Web GIS, and click the Continue button ❯.**

The **Add Home Section** window appears. A Story Map Journal app is made up of several sections. The home section will act as the cover page of your story. Each section, including the home section, has a main stage and a side panel.

5. **For Step 1: Main Stage Content, click Video, click YouTube, specify a YouTube video URL (for example http://youtu.be/-N7jtYOvstM), click Check, click Select this video, leave Fill as the choice of position, and click Next.**

ADD HOME SECTION

STEP 1: Main Stage Content

This first section is your Home Section, think of it as the 'cover page' to your story. The title you just defined will be displayed with large fonts. Learn More.

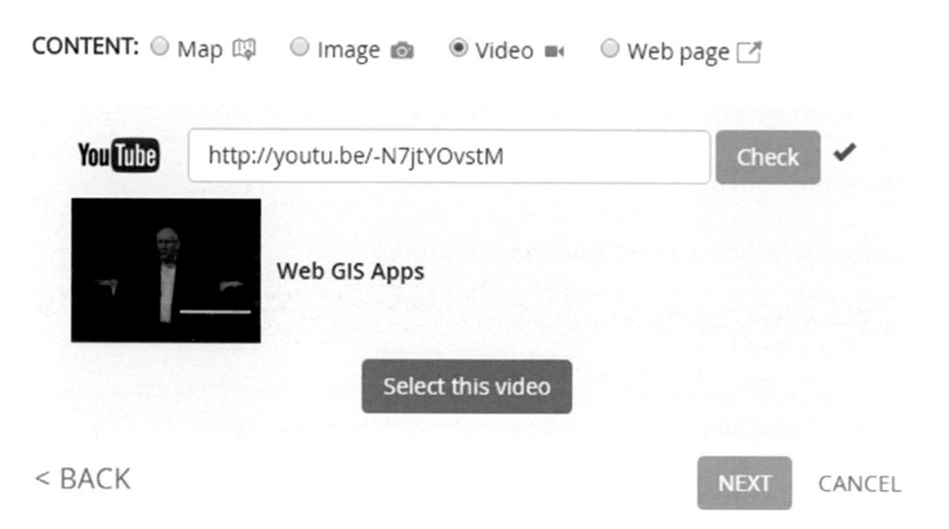

6. For **Step 2: Side Panel Content**, type **I have created several apps and will create more by following the book Getting to Know Web GIS** in the text area.

7. Make **Getting to Know Web GIS** bold and italic.

You can also insert images, videos, and web pages to the side panel. You can even click the source button and write HTML code directly in the text area.

8. Click the **Insert an image, video, or web page** button. Leave **Image** as the selected media, click **URL**, specify **http://bit.ly/1LC1g7u** as the URL, and click **Apply**.

9. Click **Add** to close the **Add Home Section** window. If the **Share Your Story** window appears, close the window.

The home section displays with a video on the main stage and the text and media you specified in the side panel.

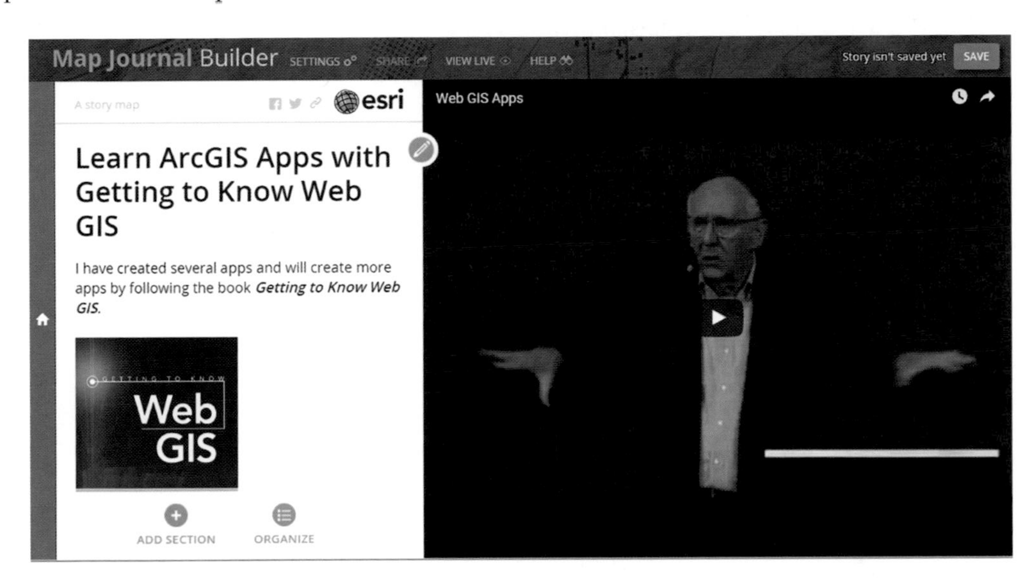

10. Click **Save** (the button at the upper right corner) to save your app.

Save your app after adding each section to avoid losing your work accidentally. Next, you will add several new sections to show the apps you have created.

11. Click the **Plus** button to open the **Add Section** window.

12. For **Step 1: Main Stage Content**, perform the following actions:

- For **Section title**, type **App #1**.
- For **Content**, select **Web page**.
- For **Webpage URL**, type the URL of the app you created in chapter 1 (your URL should use http if the URL of your current page uses http, or use https if your current page URL uses https). If you don't have a URL, use **http://arcg.is/1ZeHq4R** (if your current page URL uses http) or **https://arcg.is/1WJgjg2** (if your current page URL uses https).
- Click **Configure**, and leave **Fill** as the position choice.
- Click **Next**.

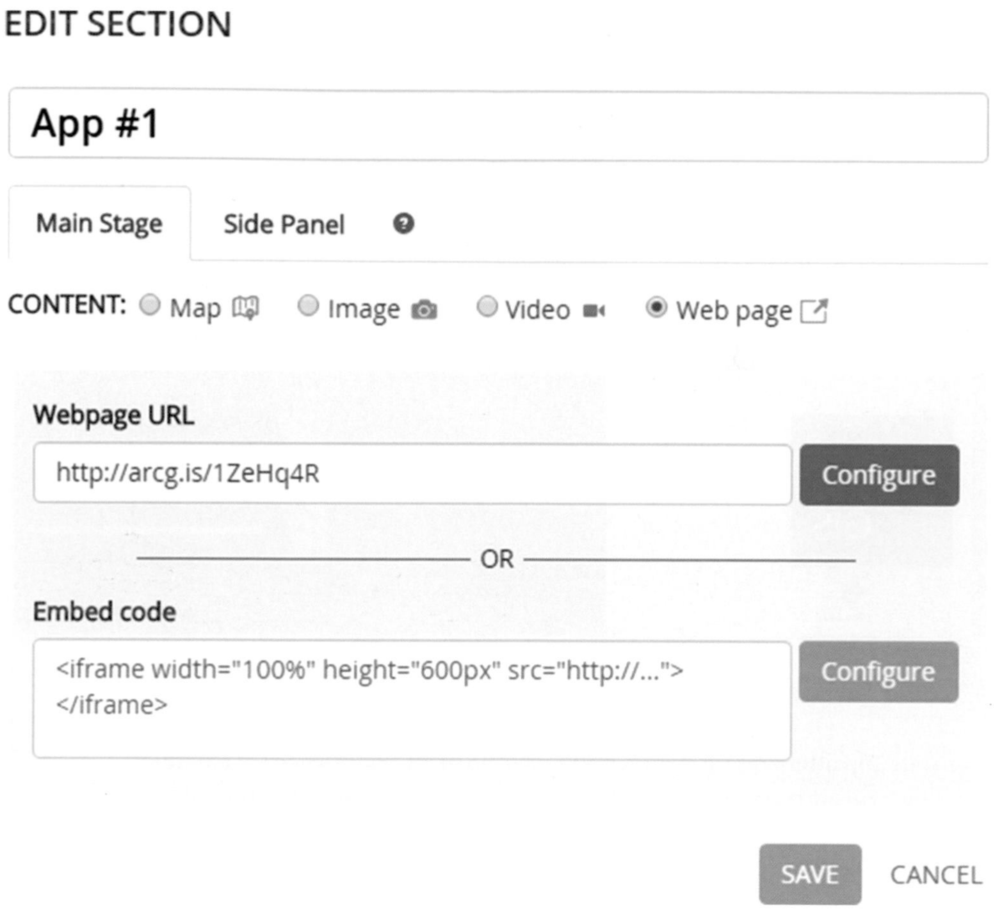

EDIT SECTION

App #1

Main Stage | Side Panel

CONTENT: Map | Image | Video | Web page

Webpage URL

http://arcg.is/1ZeHq4R

Configure

OR

Embed code

<iframe width="100%" height="600px" src="http://...">
</iframe>

Configure

SAVE CANCEL

Note: The webpage URL you specify in this step should use the same http or https protocol as the URL of your current page. Otherwise, the app may not load because some web browsers do not allow mixing http and https contents.

The main stage is configured. The builder leads you to **Step 2: Side Panel Content.**

13. For **Step 2: Side Panel Content**, type **This is the map tour app I created in chapter 1.**

14. Click **Add.**

This section is added. You should see your map tour app appear in the main stage. You can now interact with this live app.

Next, you will add a new section to display a web map type of resource.

15. Click **Add Section.**

16. For **Step 1: Main Stage Content**, perform the following actions:

- For **Section title**, type **App #2**.
- For **Content**, leave **Map** selected.
- For **Map**, choose **Select a map**.

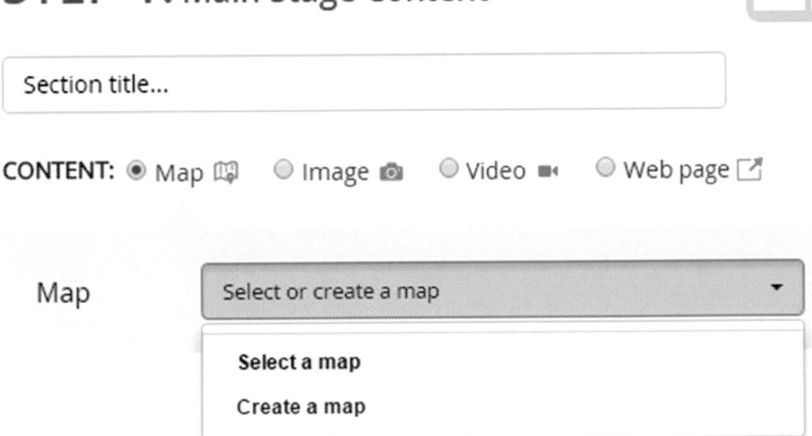

17. In the **Select a map** window, select the web map you created for chapter 2.

If you do not have one, search in **ArcGIS Online** for **Demographics, Population Growth, GTKWebGIS owner:GTKWebGIS**, press **Enter**, and then select the web map that you find.

Next, you will configure how you want the web map to be shown, including the location, content, pop-ups, and extras.

18. **For Location, click Custom configuration.**

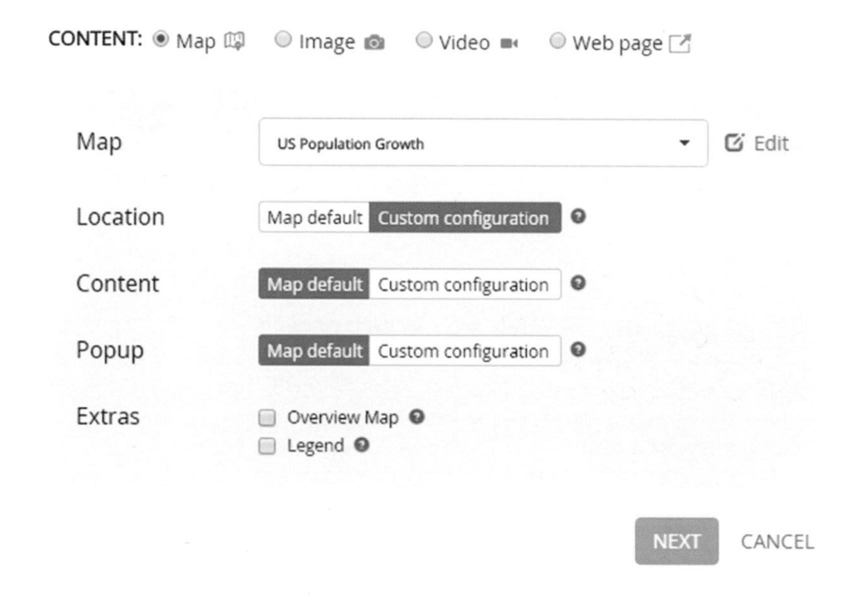

19. **Pan and zoom the map to define a map extent, and then click Save Map Location.**

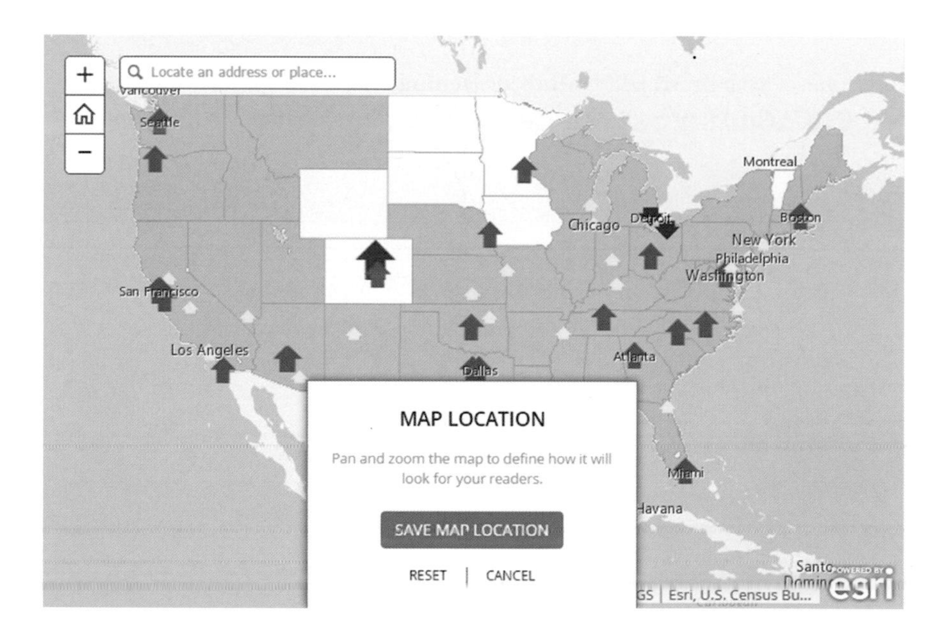

20. For **Content**, click **Custom Configuration.**

21. In the **Map Content** pop-up, turn off the layer on the bottom.

MAP CONTENT

Select which layers will be shown. Map navigation is disabled.

- [x] Top_50_US_Cities
- [x] 2015 USA Unemployment Rate
- [ ] 2015-2020 USA Population Growth Rate

SAVE MAP CONTENT

RESET | CANCEL

22. Click **Save Map Content.**

23. For **Popup**, click **Custom Configuration.** To open the pop-up you want to display on the map, click on a feature, such as a city with a blue downward icon, and click **Save the popup configuration.**

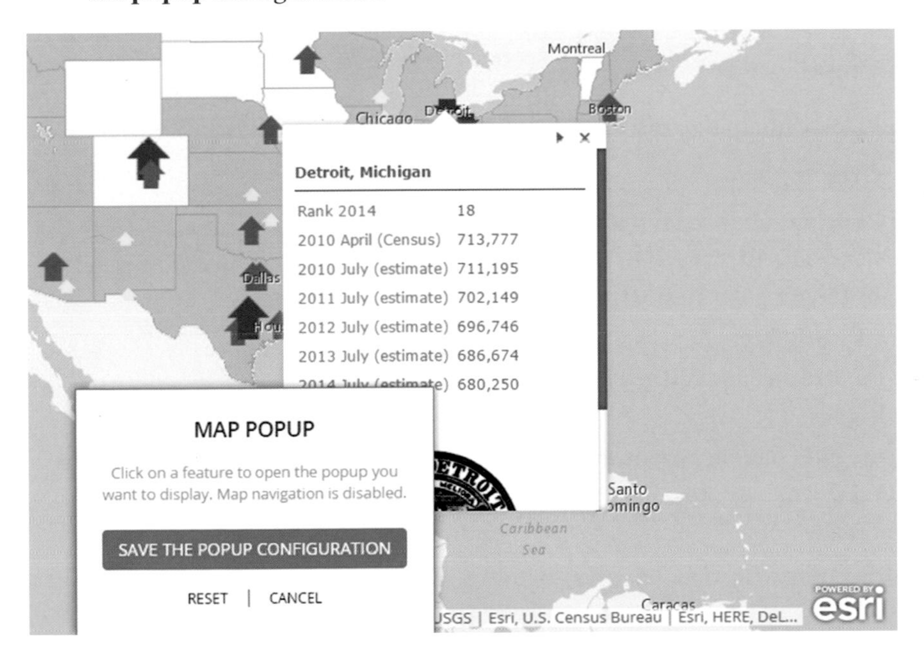

24. For **Extras**, select **Legend**.

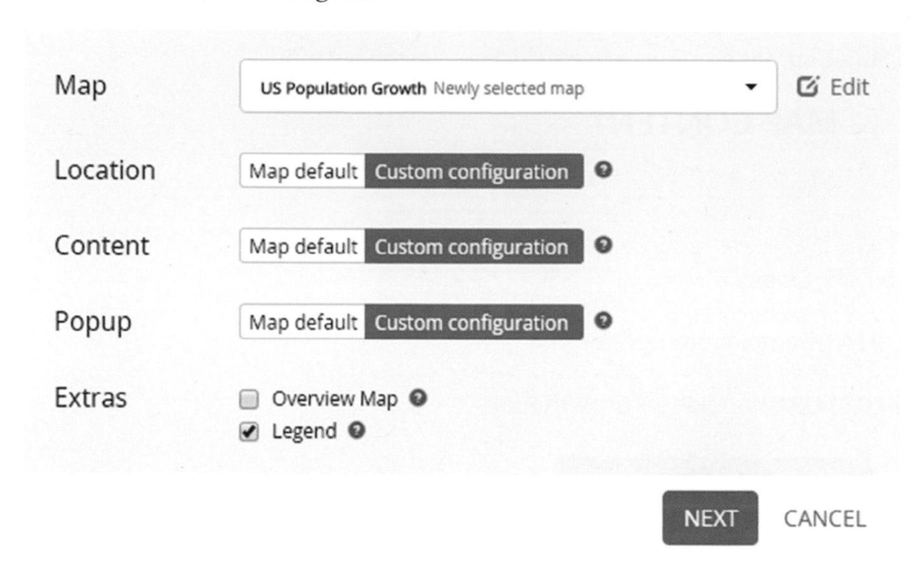

25. Click **Next**.
26. For **Step 2: Side Panel Content**, type **This is the web map I created for app #2** in the text area.
27. Click **Add**.

Section 2 is added. Next, you will add a new section for App #3.

28. Click **Add Section**. In the **Add Section** window, perform the following actions:
    - **For Title, type App #3.**
    - **For Content, choose Web page.**
    - **For Webpage URL, type in the URL of the app you created in chapter 3. If you do not have a URL, use http://arcg.is/1ZgkAtM instead.**
    - **Click Configure.**
    - **For Position, leave Fill selected.**
    - **Click Next.**
    - **For the side panel content, type This web app has a feature layer and can collect data.**
    - **Click Add.**

Next, you will add a new section that uses an image.

29. Click **Add Section**, and perform the following actions:

    - For section **Title**, type **More Apps.**
    - For **Content**, choose **Image.**
    - Click **URL.**
    - For **URL**, type **http://bit.ly/24E6vdd.**
    - Click **Next.**
    - For the **Side Panel Content**, type **I have created other apps and I will learn more apps in other chapters.**
    - Click **Add.**

30. To add a section of a 3D map, perform the following actions:

    - Click **Add Section.**
    - For **Section title**, type **3D Apps.**
    - For **Content**, choose **Web page.**
    - For **Webpage URL**, type **http://arcg.is/1nuTmlV** if the URL in your current page is http, or type **https://arcg.is/1LcQm7K** if the URL in your current page is https.

These URLs are the short versions of an ArcGIS web scene, appended with **&ui=min** in the URLs to simplify the user interface. Refer to the **Questions and Answers** section for information on finding scenes. Refer to the 3D Scenes chapter for how to create scenes. The **&ui=min** eliminates some tools of ArcGIS scene viewer to provide end users with a clean and easy user interface.

- Click **Configure.**
- For **Position**, and leave **Fill** as selected.
- Click **Next.**
- For **Side Panel Content**, type **I will also learn 3D apps and custom apps later in this book.**
- Click **Add.**

The 3D scene of typhoons is added to your story map.

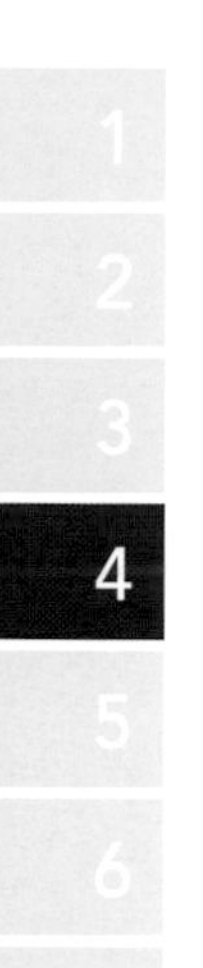

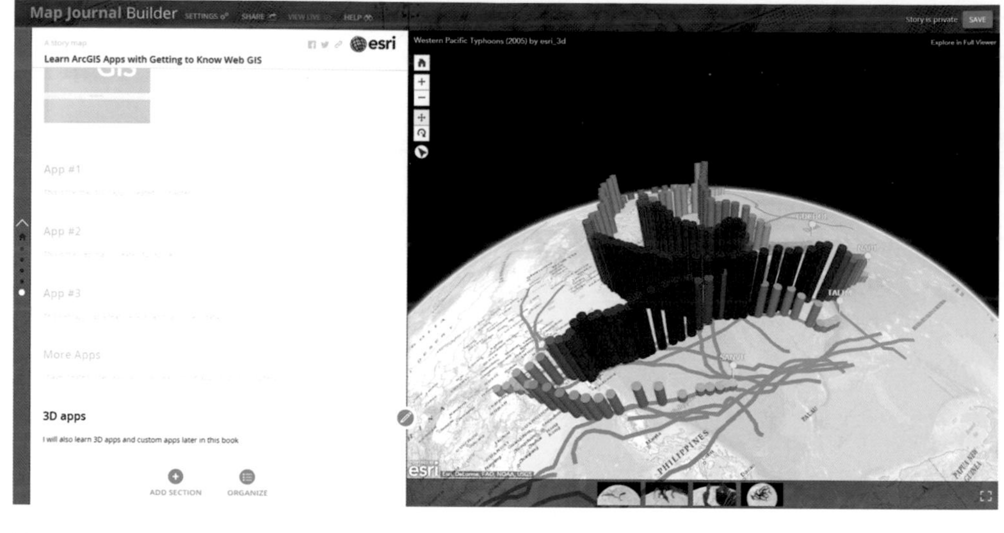

Note: If you see a message saying the scene viewer cannot be opened in your browser or the scene does not show up, it is often because your browser does not support ArcGIS scene viewer. In this case, you need to delete this 3D section.

31. If your browser does not support ArcGIS scene viewer (for example, you received a warning or you do not see the scene displayed), click **Organize**, delete the **3D apps** section, click **Apply**, and skip the next step.

32. If you can see the 3D scene, perform the following actions to navigate the scene:
    - Use your mouse scroll wheel to zoom in and out.
    - Click the **Pan** button ✣, and then drag the map to pan.
    - Click the **Rotate** button, and then drag the map to rotate.
    - Click the slides (or bookmarks) at the bottom to see the different views.

33. Click **Save** to save your app.

The **Share Your Story** window will appear if this is the first time you have saved this app.

34. If the window does not appear, click **Share** in the top menu bar.

35. In the **Share Your Story** window, click **Public**. If you see a warning message saying **There are issues in your story content that will be noticeable to your readers**, click **Make Public** (you may need to scroll down the window to see the button) to share your app and its contents.

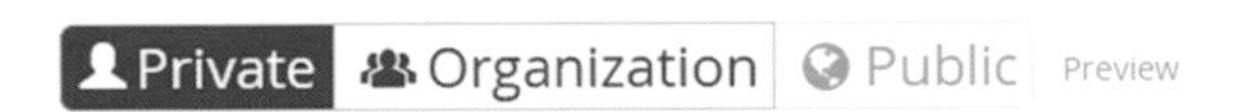
SHARE YOUR STORY
Private
Organization
Public
Preview
There are issues in your story content that will be noticeable to your readers. You can identify and fix these issues below.

We found some issues
IGNORE
Premium content
Your story contains premium content that will consume credits upon making it public. Make your story public and continue, or remove the layers from your story.
MAKE PUBLIC
Status Name
Shared
Section
US Population Growth
3
Layers
Light Gray Canvas
2015-2020 USA Population ...
2015 USA Unemployment R...
Top_50_US_Cities
Manage all your stories
CLOSE

You may see a warning message because this story map contains subscriber content, namely the **2015–2020 USA Population Growth Rate** and **2015 USA Unemployment Rate** layers in the App #2 section. By clicking the **Make Public** button in the step, you allow your end users to see these layers at the cost of your ArcGIS Online credits. If you use different web maps, you may not see this warning message.

36. **Click View story.**

Your app will open in a new window or new tab, and you can explore your story map.

37. **Navigate through your Story Map Journal app from the cover to the end. At each section, interact with the contents on the main stage.**

At each step, you can imagine how you would tell your story.

38. **Share the URL of your app with your audience.**

The URL of the current web page is the URL of your app. You can copy and email the URL to your audience. You also can click the Facebook and Twitter icons to share your app.

QUESTIONS AND ANSWERS

1. **What are the differences between discovering data in ArcGIS Open Data and discovering data in ArcGIS Online?**

   Answer: Open Data focuses on data discovery and download; ArcGIS Online focuses on managing and sharing contents of various formats.

   - All contents in Open Data are downloadable.
   - Open Data contents are mostly authoritative and have more complete metadata.
   - Open Data contents have less duplication. ArcGIS Online contents contain lots of practice data and duplicated data by different publishers.

2. **How do I share my data in ArcGIS Online with ArcGIS Open Data?**

   Answer: The administrator of your ArcGIS Online for Organizations account needs to enable Open Data in the **My Organizations Edit Settings** page. Your administrator also needs to designate a group for Open Data. Data shared with this group will appear in ArcGIS Open Data.

3. **Can I write HTML code directly in story maps?**

   Answer: Yes. While configuring Story Map Journal side panel contents, you can switch the text edit area to an HTML source code editor and write HTML code there. This capability gives developers great freedom in customizing the contents. For example, you can add audio clips to your story, as illustrated.

   Most other Story Maps support HTML code in similar ways. For more details, see links to "Adding audio to your Map Tour" and "Add <audio> to your Map Journal" in the resources section.

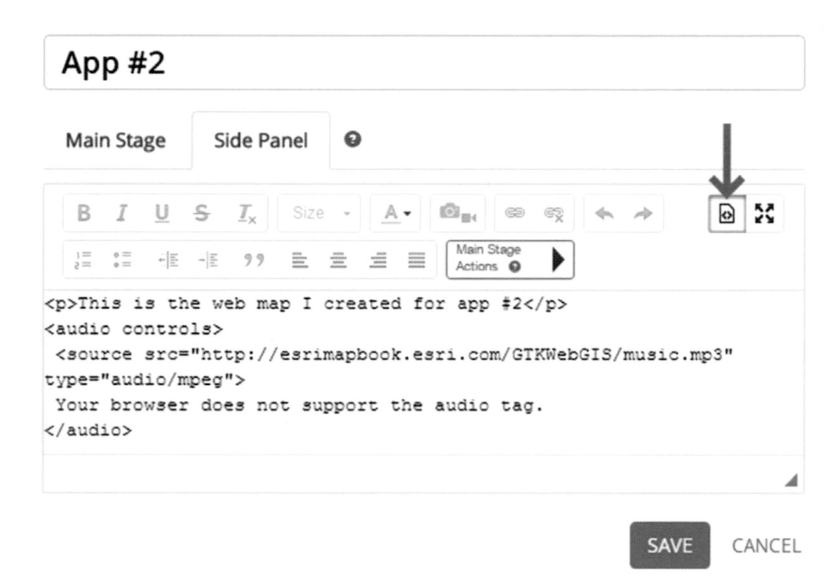

4. **I would like to add a 3D scene to my story map. Where do I find 3D scenes?**

Answer: In a browser, go to ArcGIS Online; do not sign in.

- Click **Gallery** in the top menu, and then click **Scenes** on the left.
- Alternatively, simply search for **scene**, and select **Search for Scenes**.

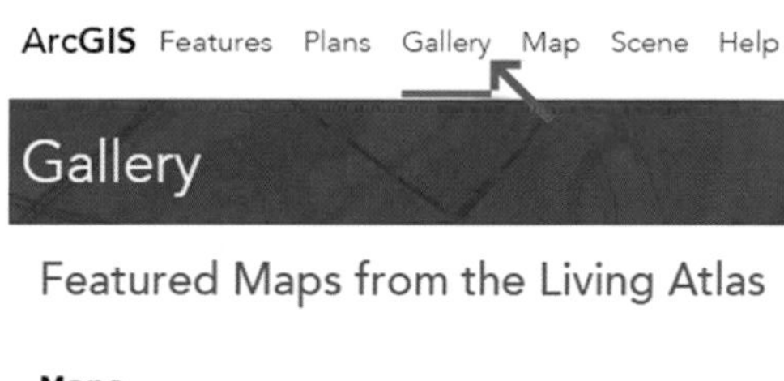

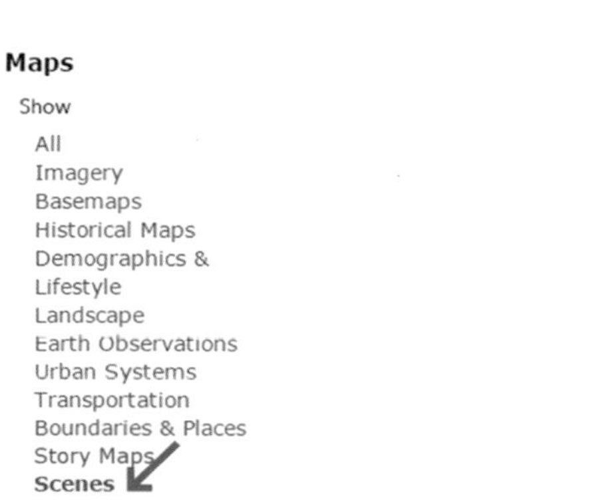

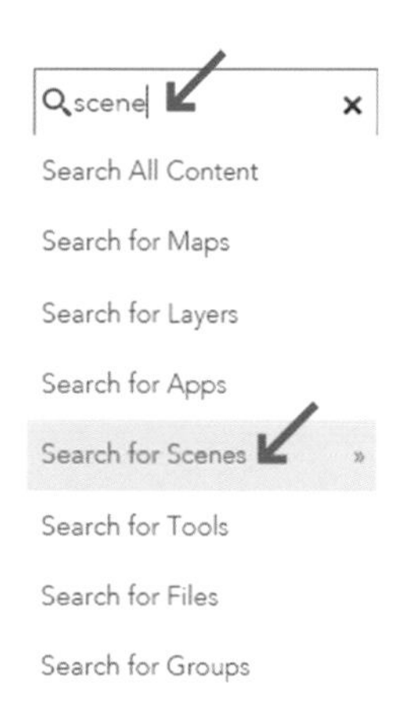

ASSIGNMENT

Assignment 4: Build a story map to tell a story about the following topics:

- Your personal life or passion
- A national or regional economic development strategy
- The spread of a virus
- A historical war or other event
- Spatial patterns of various demographics factors
- Other topics of your choosing

Data:

- Use data from ArcGIS Living Atlas for the World, ArcGIS Open Data, or other sources. You can make up your own data, too.

Requirements: Your story map should have the following characteristics and information:

- A format in map journal, map series, map shortlist, map playlist, or map countdown
- Five types of resources: text, video, picture, map, and web page
- An app created from the **Compare Analysis** app or **Story Map Swipe and Spyglass** app

What to submit: Send an email to your instructor with the subject line web GIS Assignment 4: Your name, and include the following information:

- A list of your data sources, for example, the layer names if you are using Live Atlas and the item URLs if you are using data from Open Data
- The URL to your web app

Resources

ArcGIS product document websites

"ArcGIS Marketplace," https://marketplace.arcgis.com.
"ArcGIS Open Data," http://doc.arcgis.com/en/open-data.
"ArcGIS Solutions," http://solutions.arcgis.com.
"Choose a Configurable App," https://doc.arcgis.com/en/arcgis-online/create-maps/choose-configurable-app.htm.
"Story Maps Gallery," http://storymaps.arcgis.com/en/gallery.

Esri blogs

"Add a Presentation to Your Story Map Journal," by Bern Szukalski, http://blogs.esri.com/esri/arcgis/2015/12/28/add-a-presentation-to-your-story-map-tour.
"Adding Audio to Your Map Tour," by Bern Szukalski, https://medium.com/story-maps-developers-corner/adding-audio-to-your-story-map-map-tours-134a11b85312#.54zav3igs. "Story Maps Developers' Corner," https://medium.com/story-maps-developers-corner.
"Add Your Organization Logo and Links to Story Maps," by Bern Szukalski, http://blogs.esri.com/esri/arcgis/2014/02/04/add-your-logo-and-link-to-your-story-maps.
"Add <audio> to Your Map Journal," by Owen Evans, https://medium.com/story-maps-developers-corner/using-the-html-audio-tag-in-your-story-map-f818d9316252#.xe7z24l46.
"Embedding a Story Map within a Story Map," by Bern Szukalski, http://blogs.esri.com/esri/arcgis/2015/04/21/embed-story-map-in-story-map.
"Enhance Your Story Map Journal Legend," by Bern Szukalski, http://blogs.esri.com/esri/arcgis/2015/12/16/enhance-legend-story-map-journal.
"Using 3D Web Scenes in Story Maps," by Bern Szukalski, http://blogs.esri.com/esri/arcgis/2015/03/20/using-web-scenes-in-story-maps.
"Using Effective Titles and Subtitles in Your Story Maps," by Bern Szukalski, http://blogs.esri.com/esri/arcgis/2015/12/23/effective-titles-and-subtitles-story-maps.
"What's New in Story Maps (November 2015)," by Bern Szukalski, http://blogs.esri.com/esri/arcgis/2015/11/18/whats-new-in-story-maps-november-2015.

Esri tutorial and training websites

"From London to Tokyo" (using the Urban Observatory tool), http://learn.arcgis.com/en/projects/from-london-to-tokyo.
"Get Started with ArcGIS Open Data," http://training.esri.com/gateway/index.cfm?fa=catalog.webCourseDetail&courseid=2827.
"Oso Mudslide—Before and After," http://learn.arcgis.com/en/projects/oso-mudslide-before-and-after.
"Telling Your Story with Esri Story Maps," http://training.esri.com/gateway/index.cfm?fa=catalog.webCourseDetail&courseid=2924.

Esri videos

"ArcGIS Online: Bridging Communities and Data with ArcGIS Open Data," http://video.esri.com/watch/4730/arcgis-online-bridging-communities-and-data-with-arcgis-open-data.

"Getting the Most Out of ArcGIS Web App Templates," http://video.esri.com/watch/4709/getting-the-most-out-of-arcgis-web-app-templates.

"How to Tell Your Story Using Esri's Story Map Apps," http://video.esri.com/watch/4701/how-to-tell-your-story-using-esris-story-map-apps.

Chapter 5
Web AppBuilder for ArcGIS

Have you run into situations where you need more functions than any individual configurable app can provide? Do you wish you could remix the functions of multiple apps? Understanding and using Web AppBuilder for ArcGIS will help you answer these and other questions. Web AppBuilder provides more functionality than any other ArcGIS configurable web apps or templates and is more flexible and configurable. Web AppBuilder comes with more than 30 premade widgets covering functions including mapping, table viewing, charting, querying, routing, geoprocessing, and more. The user community also can create additional widgets. Web AppBuilder allows you to create web apps by selecting, mixing, and configuring widgets in a designer platform—without programming. Web AppBuilder also provides a variety of themes (in other words, styles and layouts) for you to create easy-to-use, friendly, and responsive user interfaces that work for desktop, tablet, and mobile devices.

Learning objectives

- *Know why and when you need Web AppBuilder.*
- *Understand the types of widgets and themes of Web AppBuilder.*
- *Learn the workflow to create web apps using Web AppBuilder.*
- *Configure and use various widgets.*

This chapter in the big picture

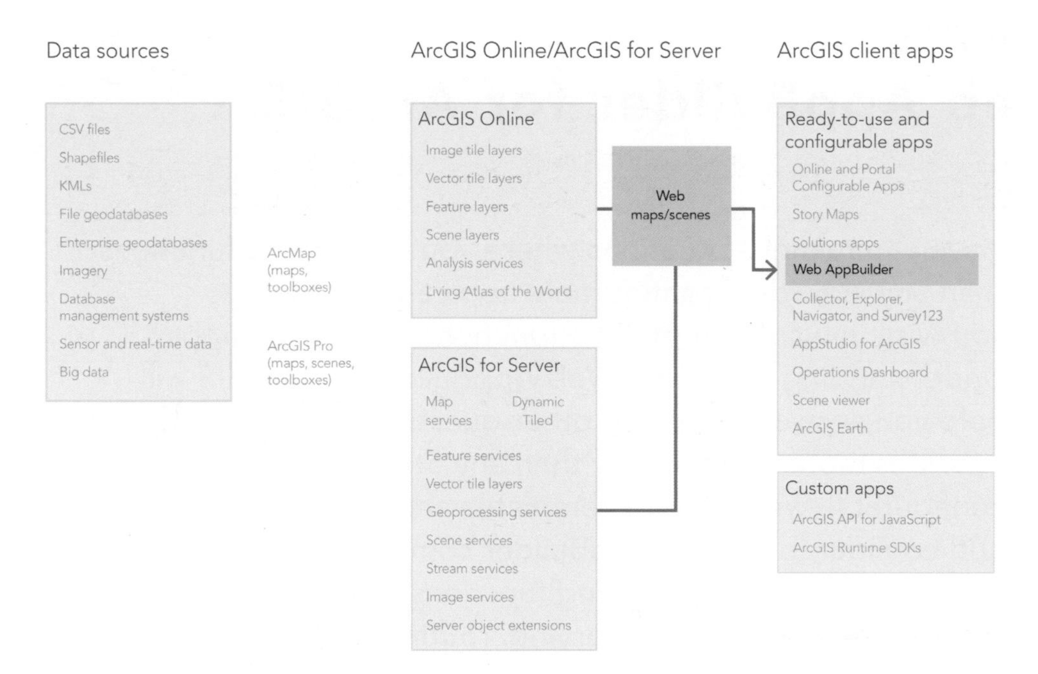

ArcGIS offers many ways to build web apps. The green line in the figure highlights the technology taught in this chapter.

Web AppBuilder is a web client for ArcGIS Online, Portal for ArcGIS, and ArcGIS for Server. You can use Web AppBuilder to create Web GIS apps without programming. The technology is built on HTML5 and ArcGIS API for JavaScript and includes the following key features:

- Creates web apps with pure HTML and JavaScript so that they do not require any plug-ins
- Uses responsive web design technologies, enabling web apps to adapt well to desktops, tablets, and smartphones and work well on screens of any size
- Comes with numerous out-of-the-box widgets, so you can create powerful web apps with rich functionality

- Comes with various themes for you to configure the look and feel of your apps
- Supports the creation of ArcGIS configurable app templates for ArcGIS Online
- Provides an extensible framework for developers to create custom widgets, themes, and apps

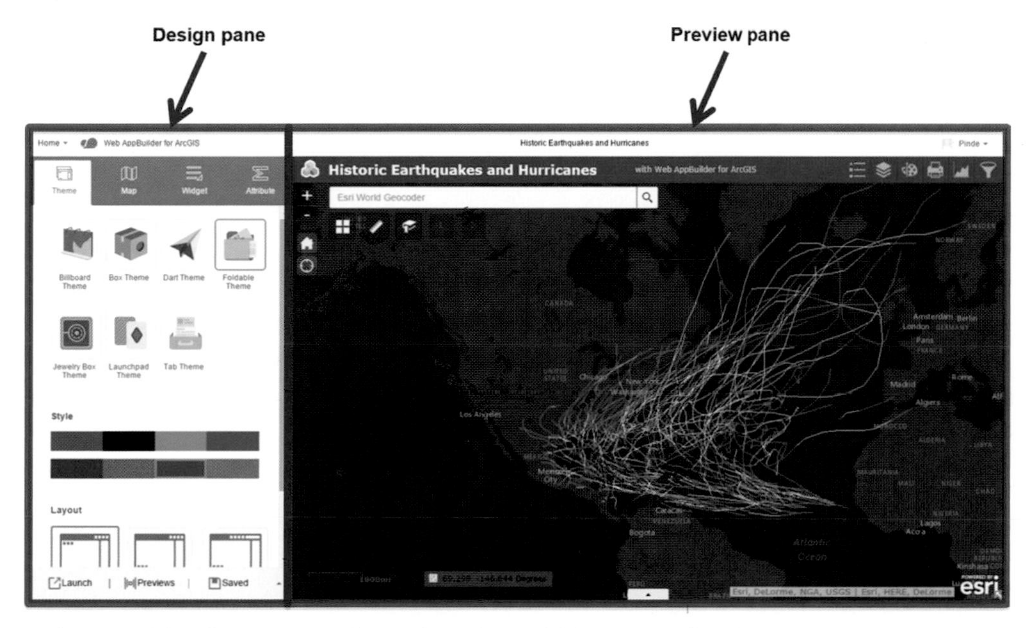

Web AppBuilder allows users to choose from the available user interface themes, web maps, and widgets, in a "what-you-see-is-what-you-get" designer experience. You can immediately see how your app will look as you change the configuration.

Editions of Web AppBuilder for ArcGIS

The Web AppBuilder product family has three editions:

- Embedded in ArcGIS Online
- Embedded in Portal for ArcGIS
- Developer Edition

The first two are embedded editions. The main difference between those two and the third option is that the Developer Edition allows you to create custom widgets and themes, whereas the embedded editions do not. Otherwise, the editions should have similar functionality. For example, they will have the same designer user experience to create web apps, and similar widgets and themes. However, the detailed functionality of the three editions is not equivalent. Typically, new enhancements of the Web AppBuilder are added first to the ArcGIS Online edition, then to the

Developer Edition, and finally to the Portal for ArcGIS edition. Therefore, the editions may differ in themes, widgets, and other aspects during certain periods.

The chapter tutorial is based on the edition embedded in ArcGIS Online, but the skills you will learn apply to all three editions.

Access to Web AppBuilder

For the embedded editions, you can access Web AppBuilder from the map viewer, **Gallery**, or **My Content**.

1. If you opt to use the map viewer, you would click **Share**, click **Create a Web App**, and then click the **Web AppBuilder** tab (see section 5.2).
2. If you start from the **Gallery**, you would go to the **Esri Featured Content** > **Apps** section, and then choose **App Builders**.

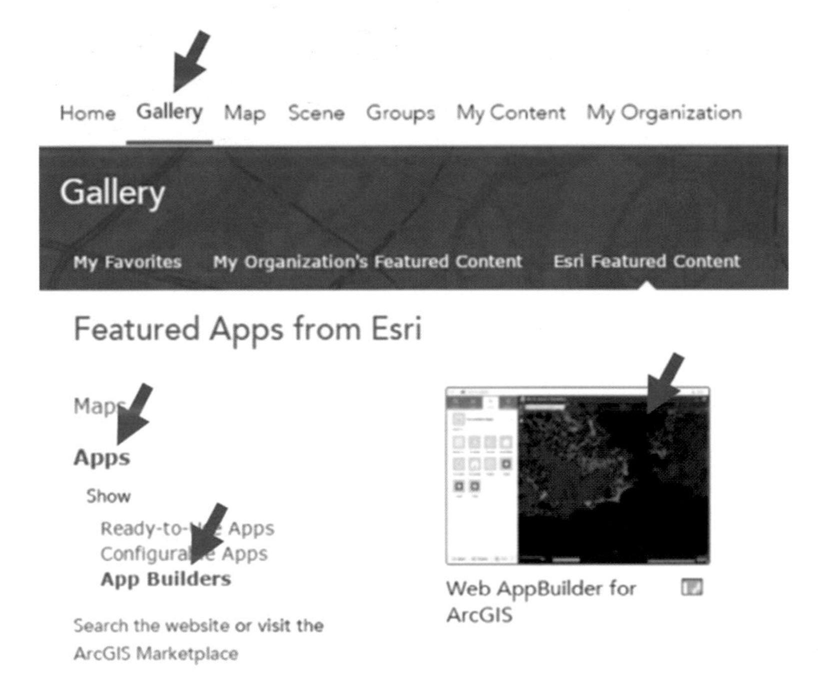

3. For the third option, you would choose **My Content** > **Create** > **App**, and then click **Using the Web AppBuilder**.

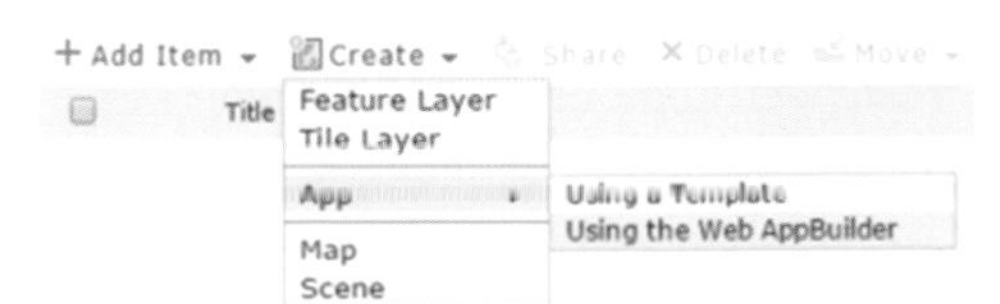

Steps to create a web app using Web AppBuilder

Once you have started the Web AppBuilder, you can create a web app in the following ways:

- **Pick style:** Configure the look and feel of the app by picking a theme. A theme includes a collection of panels, styles, layouts, and preconfigured widgets.
- **Select map:** Select a web map created by you or shared with you.
- **Add widgets:** Widgets give your app functionality, such as print map and query layers. Each theme has its own preconfigured set of widgets. You can hide or show existing widgets and add additional ones.
- **Configure attributes:** Attributes allow you to customize your app banner with a logo, title, hyperlinks, and so on.

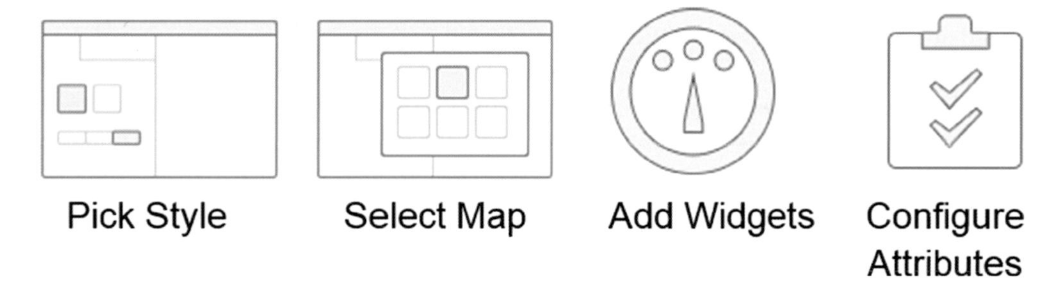

As you configure your web apps interactively, Web AppBuilder generates the configurations in JavaScript Object Notation (JSON) files. Because Web AppBuilder generates the configurations, you do not have to know JSON and the configuration syntax for the widgets themselves.

Web AppBuilder widgets

Web AppBuilder provides functions through widgets. Typically, a widget is a JavaScript/HTML component that encapsulates a set of focused functions. Most widgets have a visual user interface.

Web AppBuilder provides more and more widgets with its new releases. As of this writing, Web AppBuilder provides more than 30 core widgets. You can find additional Web AppBuilder widgets from ArcGIS Solutions widgets (**http://solutions.arcgis.com/shared/help/solutions-webappbuilder-widgets**), which is a set of widgets designed to address specific workflows across industries, and from the user community widgets (**https://geonet.esri.com/groups/web-app-builder-custom-widgets**). You can download and deploy these widgets to your Web AppBuilder or web app.

Types of Web AppBuilder widgets

In general, widgets are categorized as two types: in-panel and off-panel widgets.

- **In-panel widgets:** For example, **Basemap Gallery**, **Bookmark**, and **Chart** widgets are available to the widget controllers and can be added to your application. Each in-panel widget requires user interactions on the panel.
- **Off-panel widgets:** For example, **Attribute Table**, **Coordinate**, and **Time Slider** can be turned on or off but cannot be removed from the application. They can be added to the controller. The off-panel widgets embed in a theme display when the Widget tab is activated.

Based on their relation with map layers, widgets can be categorized into two groups:

- **Data independent widgets:** For example, **Basemap Gallery**, **Measurement**, **Draw**, and **Bookmark** widgets are not related to the operational data layers you have. These widgets need no or little configuration. They are not affected if you switch from one web map to a different web map.
- **Data dependent widgets:** For example, **Query** and **Chart** widgets are related to specific attribute fields of specific layers in your web map. They often require detailed configuration. When you switch from one web map to a different web map, you will need to reconfigure these widgets.

Themes

Web AppBuilder provides a variety of themes for targeted app types:

- **Foldable** and **Tab:** These two themes support all the widget types and can be used for an app with complicated tasks.
- **Billboard:** This theme is designed for apps with simple tasks.
- **Box:** In this theme, which is designed for apps that require a clean look on the map, all the on-screen widgets are turned off by default.
- **Dart:** The widgets in the widget controller display like placeholder widgets. You can have multiple widgets open, and you can move them around. All the on-screen widgets are turned off by default.
- **Jewelry Box:** Designed for apps with a workflow task, Jewelry Box evolved from the Foldable theme with a focused widget on the left of the app.
- **Launchpad:** Designed for users who like Apple Mac style, Launchpad lets you open multiple widgets and move, resize, and minimize widgets.

Web AppBuilder Developer Edition

Web AppBuilder Developer Edition provides a great framework to create new widgets, customize existing widgets, create new themes, and build apps with extended functionality. Custom widgets and apps can be shared for free or sold in ArcGIS Marketplace.

In contrast to the embedded edition, you will need to download the Developer Edition, register the edition with your ArcGIS Online or Portal for ArcGIS, and run the edition on your own computer.

This tutorial

An organization would like to provide a web app that displays data on historic earthquakes and hurricanes to the public.

Data: A web map is provided to you. This web map uses a map service.

Requirements: The web app should have the following capabilities:

- Zoom to the entire United States in its initial view.
- Provide bookmarks so that users can quickly zoom to predefined areas.
- Allow users to print displays as PDFs.
- Allow users to chart and compare selected attributes of selected earthquakes.
- Allow users to query for features based on their attributes.
- Display appropriate logo, title, subtitle, and links in the banner.

System requirements:

- A publisher or administrator account in an ArcGIS Online organization

5.1 Explore the web map

Before you build your app, you will need to familiarize yourself with the web map and the map layers that you will use. Understanding your web map is especially important when you need to use and configure data dependent widgets.

1. **Sign in to your ArcGIS Online Organization account.**

2. **In the search box, type historic disasters GTKWebGIS owner:pinde.webgis, click the search box, and click Search for Maps from the list. On the left of the page, clear the Only Search in your organization checkbox.**

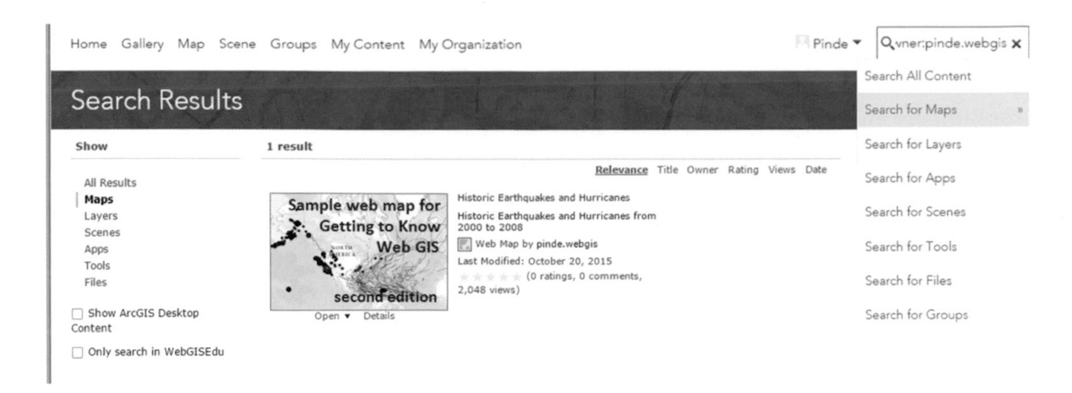

You should see the search result as illustrated (sample web map for *Getting to Know Web GIS*, second edition).

3. **Click the Details link of the web map.**

4. **On the item details page, under the Layers section, notice that this web map has a topographic basemap and an operational layer named Natural Disasters.**

Note: The **Natural Disasters** layer has a URL. The term "MapServer" in this URL indicates that this layer is a map service layer, which can support chart and query widgets (see section 5.4). Other web services types, such as feature layers, also support chart and query. If your operational layer does not have a URL and is not a web service layer, you will not be able to chart or query the layer.

5. **On the item details page, click the thumbnail image to open the web map in the map viewer.**

You should see a time slider appear under the map canvas. The slider indicates that the **Natural Disasters** layer is time-enabled. You can use the **Time** widget of Web AppBuilder to animate the earthquakes and hurricanes to indicate time when you configure your web app. By default, you see only the earthquakes and hurricanes that occurred in the time interval in the time slider.

6. **Click and drag the right thumb of the time slider to the far right.**

With the left thumb of the time slider at far left and the right thumb at far right, you will see all the earthquakes and hurricanes in the layers.

7. **In the Legend pane, click Content, and then click the Natural Disasters layer to expand it and see its sublayers.**

You will see it has an earthquakes layer and a hurricanes layer.

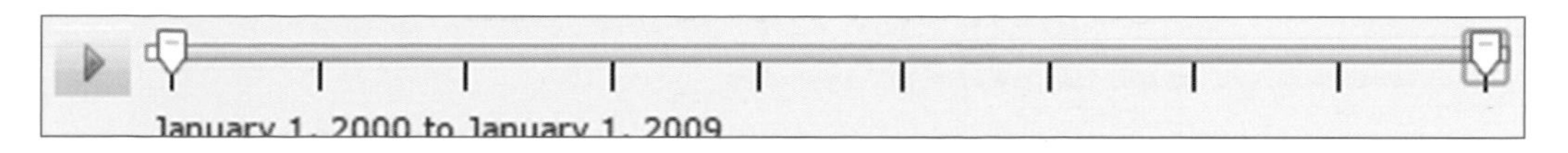

8. **In the map viewer search box, type the name of a hurricane—for example, Katrina—and click the Search button or press Enter.**

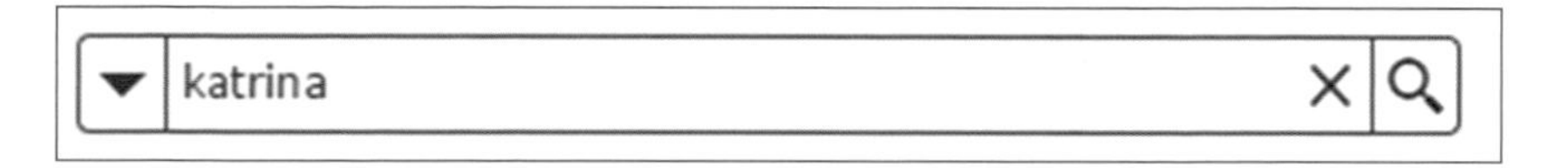

You will find Hurricane Katrina highlighted on the map.

Note: The reason that you can search hurricanes by names in the search box here is that your web map supports feature search. See the **Questions and Answers** section at the end of this chapter for instructions about how to configure a feature search.

9. **In the Contents pane, point to the Earthquakes layer, click the Show Table button, and study the layer attribute fields.**

You will use these fields to configure the chart and query widgets in section 5.4.

If you need to make changes, for example, change the layer styles, enable pop-ups, and configure pop-ups, you can do so now and save it after you are done. Because you are not the owner of this web map, you will need to save it as a new web map if you made any changes.

5.2 Create a web app

1. **Continuing from the previous section, click the Share button.**

2. **In the Share window, click Create a Web App.**

3. **In the Create a New Web App window, click the Web AppBuilder tab.**

Create a New Web App

Configurable Apps | Web AppBuilder

To create a new app with Web AppBuilder, enter a title, tags and summary.

Title: Historic Earthquakes and Hurricanes

Tags: Natural Disasters × Earthquakes × Hurricanes × GTKWebGIS × Sample × v2 ×
Add tag(s)

Summary: Historic Earthquakes and Hurricanes from 2000 to 2008

Save in folder: pinde

BACK | GET STARTED | CANCEL

4. **Specify your app title, tags, and summary, and click Get Started to open Web AppBuilder for ArcGIS.**

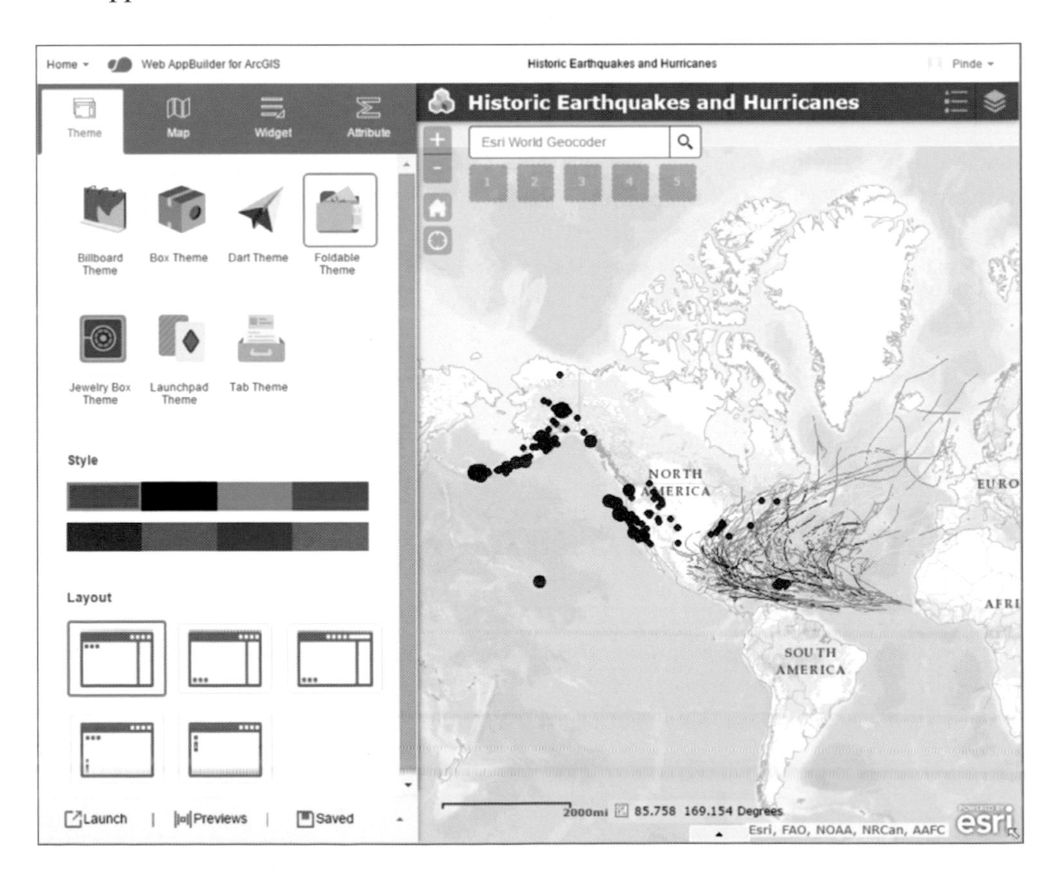

Web AppBuilder has two panes: the design pane on the left and the preview pane on the right. The design pane has four tabs: **Theme**, **Map**, **Widget**, and **Attribute**, which correspond to the four different aspects available to configure your web app.

5. **Click the Theme tab, click Foldable Theme, and choose a color style and layout you like.**

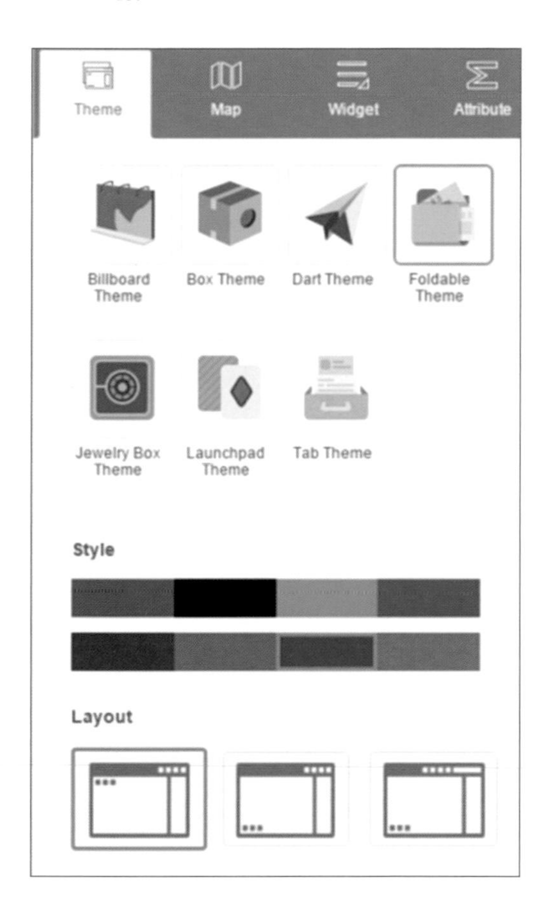

You can experiment using other themes, too. As you make the changes, look to the preview pane on the right, where you can immediately see how the new setting looks.

6. **Click the Map tab**

The **Map** tab allows you to choose the web map to use in the app. You started with a suitable web map to use. Otherwise, you can click **Choose Web Map** to select a different web map.

7. **Pan or zoom the map to cover the earthquakes and hurricanes. Under Set Initial Extent, click Use Current Map View.**

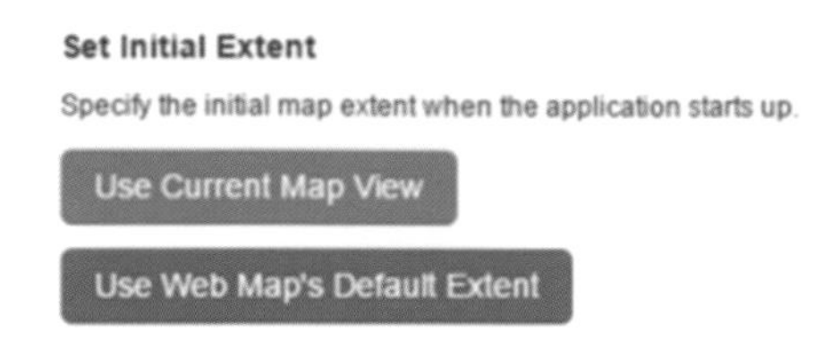

8. **Click the Attribute tab, set the title to Historic Earthquakes and Hurricanes, and in the subtitle, indicate that you designed the app.**

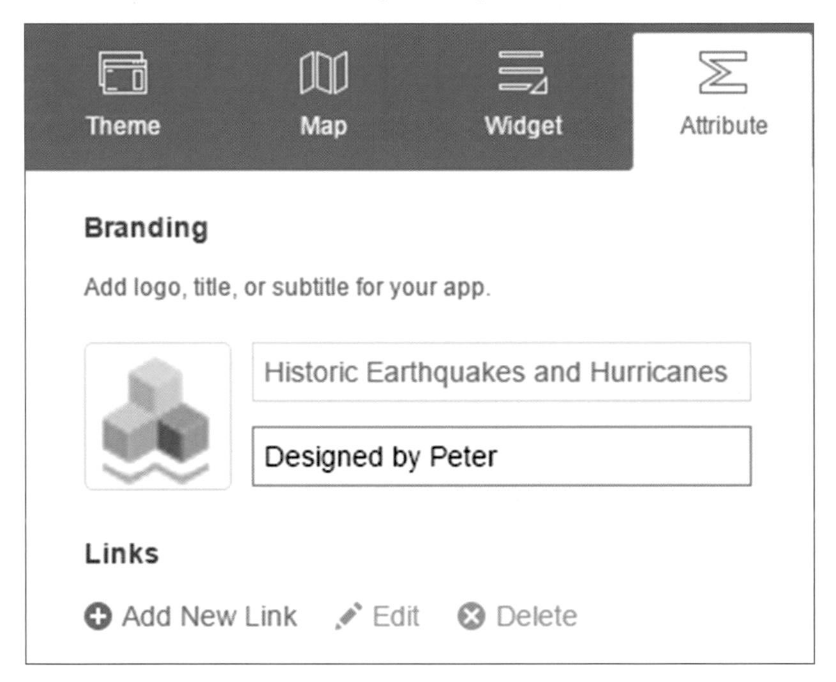

9. **Click the logo icon, and select the image you would like to use—for example, your organization's logo. Optionally, you can click the Add New Link button to add the URL of your organization or your organization's contact page.**

10. **At the bottom of the design pane, click Save.**

As you complete the rest of the steps in this chapter, you should save your configuration frequently so that you do not lose your work accidentally.

11. Explore the default widgets in the preview pane.

Each theme loads with some commonly used default widgets.

- **Click the My Location widget to zoom to your current location. The ability to find your location is especially useful on mobile devices.**
- **As you zoom and pan your map, note the Scalebar widget and Coordinate widget showing your current map scale and your current cursor location.**
- **In the Search widget, search for an address or place name, such as Los Angeles, to zoom your map to that location. Or search for a hurricane, such as Rita.**
- **Click the Default Extent widget to zoom back to the initial map extent you set.**
- **Click the Show Map Overview button in the lower-right corner (or another corner depending on the layout you choose) of your map to bring up the overview window. Click the arrow (now with a reversed direction) to hide the overview window.**
- **Click the Legend button in the preview pane to see the legend.**

You might see the **Legend** button in the upper-right corner or another location, depending on the theme and layout you choose.

- **Click Layer List button .**

Notice that you can expand the **Natural Disasters** layer to see its sublayers. For each layer, you can enable or disable pop-ups, open the attribute table, and view its description.

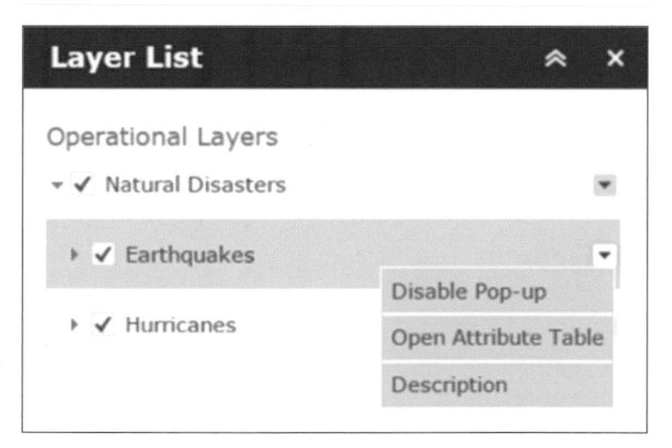

- **Click the arrow at the bottom center of the map to bring up the Attribute Table widget. Click the Earthquakes and Hurricanes tabs to see the table for each layer. The table lists the attributes of the features within the current map extent. Click the Options button to see that you can filter attributes, show/ hide columns, and export the data to the comma-separated value (CSV) file.**

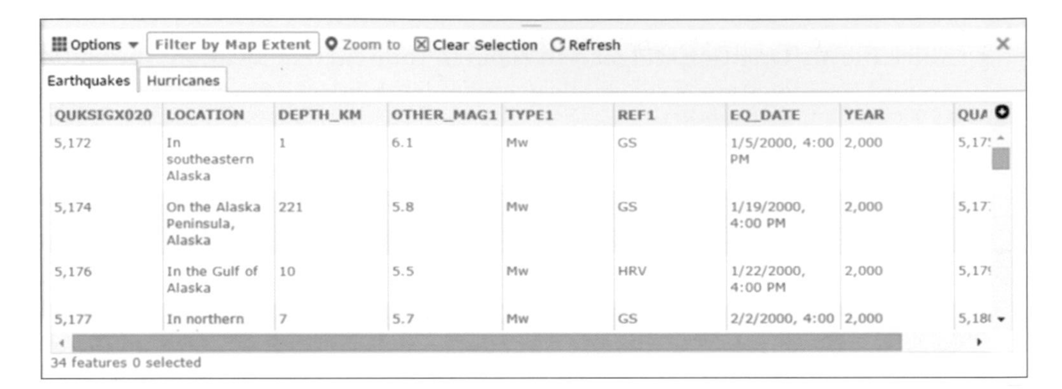

Options
Filter by Map Extent
Zoom to
Clear Selection
Refresh
Earthquakes
Hurricanes
QUKSIGX020 LOCATION DEPTH_KM OTHER_MAG1 TYPE1 REF1 EQ_DATE YEAR
5,172 In southeastern Alaska 1 6.1 Mw GS 1/5/2000, 4:00 PM 2,000
5,174 On the Alaska Peninsula, Alaska 221 5.8 Mw GS 1/19/2000, 4:00 PM 2,000
5,176 In the Gulf of Alaska 10 5.5 Mw HRV 1/22/2000, 4:00 PM 2,000
5,177 In northern 7 5.7 Mw GS 2/2/2000, 4:00 2,000
34 features 0 selected

12. Close the attribute table.

5.3 Configure data-independent widgets

Data-independent widgets often require little or no configuration. For example, **Basemap Gallery**, **Bookmark**, and **Draw** are such widgets.

1. Click the **Widget** tab.

The **Widget** tab shows some widgets that are already added to your app, such as the **Attribute Table**, **Coordinate**, and **Home Button** widgets. Widgets that appear dimmed are turned off but can be turned on.

At the bottom of the list are placeholders for five additional widgets. You will begin by adding several widgets here.

2. **Click the first empty widget button. In the Choose Widget window, click Basemap Gallery, and click OK.**

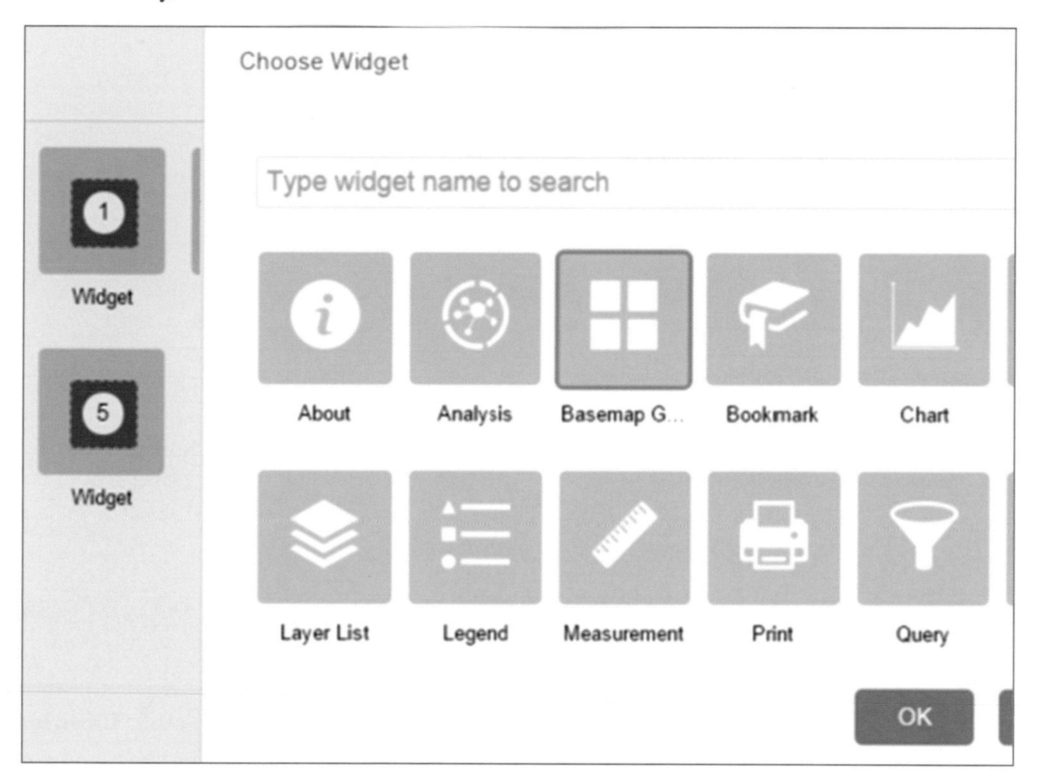

3. **In the Configure Basemap Gallery window, click OK.**

Optionally, you can delete some of the basemap types or add new basemaps. You can also use your own map services as basemaps.

Notice that the **Basemap Gallery** widget has been added to the first widget placeholder in your app.

4. **On the preview pane, click Basemap Gallery.**

You can choose and switch to a different basemap.

Next, you will add the **Bookmark** widget.

5. **Click the current first empty widget button. In the Choose Widget window, click Bookmark, and click OK.**

6. **In the Configure Bookmark window, perform the following tasks:**

- **Click the Click to Add a New Bookmark button.**
- **Specify the title as Western States.**
- **Pan/zoom the map to the western states of the United States.**
- **Optionally, to specify an icon that represents your bookmark, click Thumbnail.**
- **Click OK to add this bookmark.**

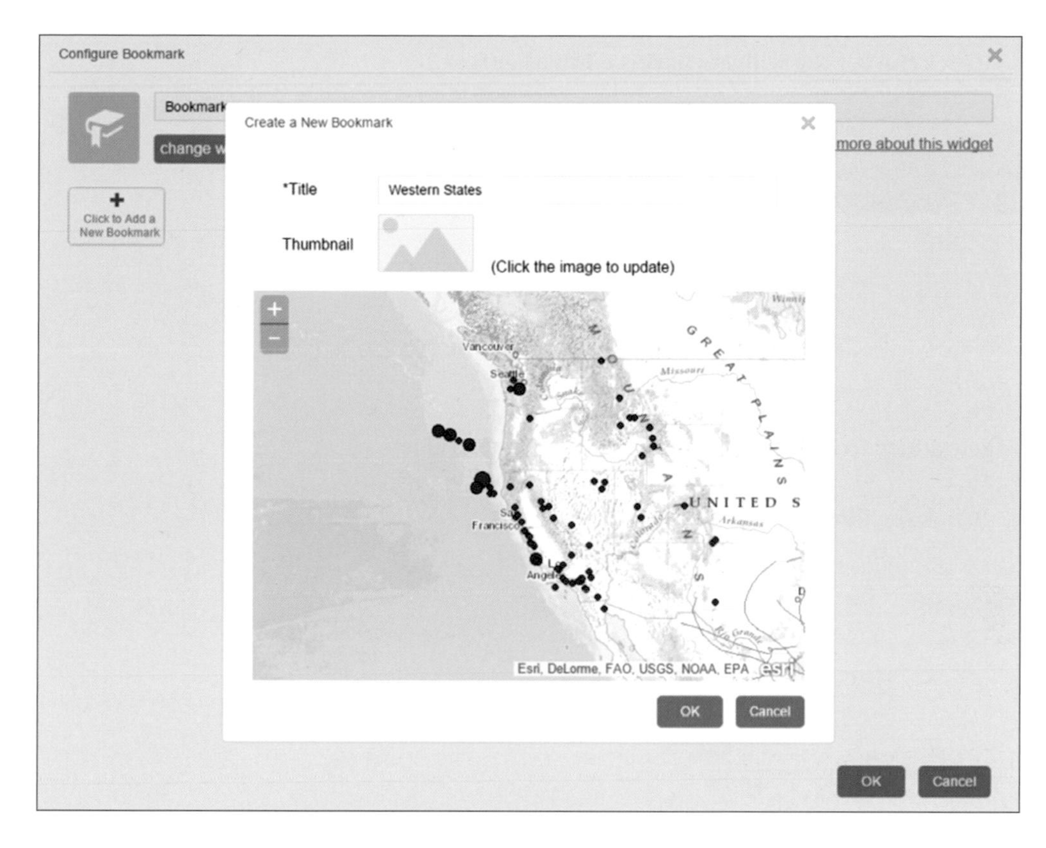

7. **Repeat the preceding step to add another bookmark, such as Southeastern States.**

8. **Click OK to close the Configure Bookmark window.**

The bookmark widget is added to your app and is ready for you to use.

9. **Click the Bookmark widget in the preview pane of your app, and click the bookmarks you defined to see the map extent changes.**

Note: You can add bookmarks in the running mode of the bookmark widget. Such bookmarks live locally in your browser cache, and thus are available only to you. The bookmarks you defined in the configuration mode are contained in the app configuration and are globally available to all users of your app.

Next, you will add additional widgets to the header controller.

10. **Click the Set the widgets in this controller link.**

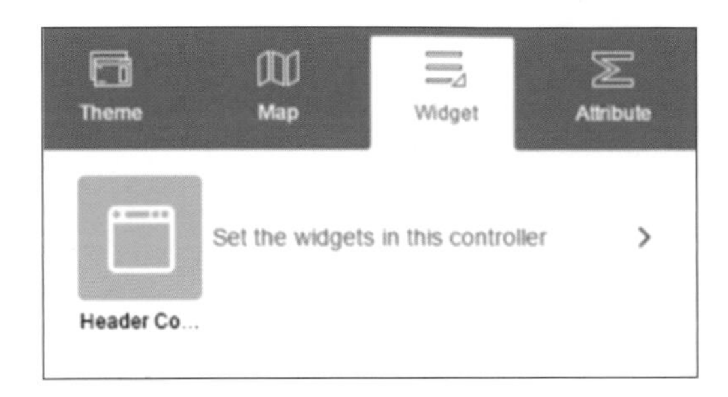

Widgets added to the header controller are in-panel widgets.

11. **Click the Plus button.**

12. **In the Choose Widget window, click Draw, click Print, and click OK.**

You have now added these two widgets to your app.

13. **If you need to change the order of your widgets, click a widget button, drag the button to the desired position, and drop it there.**

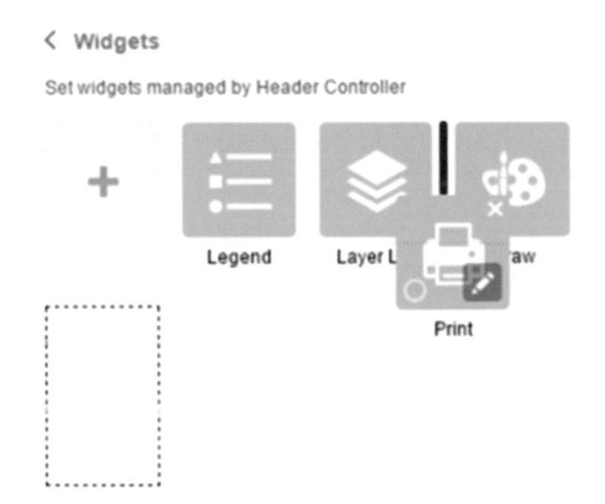

Your two new widgets are data independent and have default configurations. You can use them right way.

14. **In the right corner of the app toolbar, click the Draw button, (or the More button, and then the Draw button). In the Draw window that appears, select a Draw mode.**

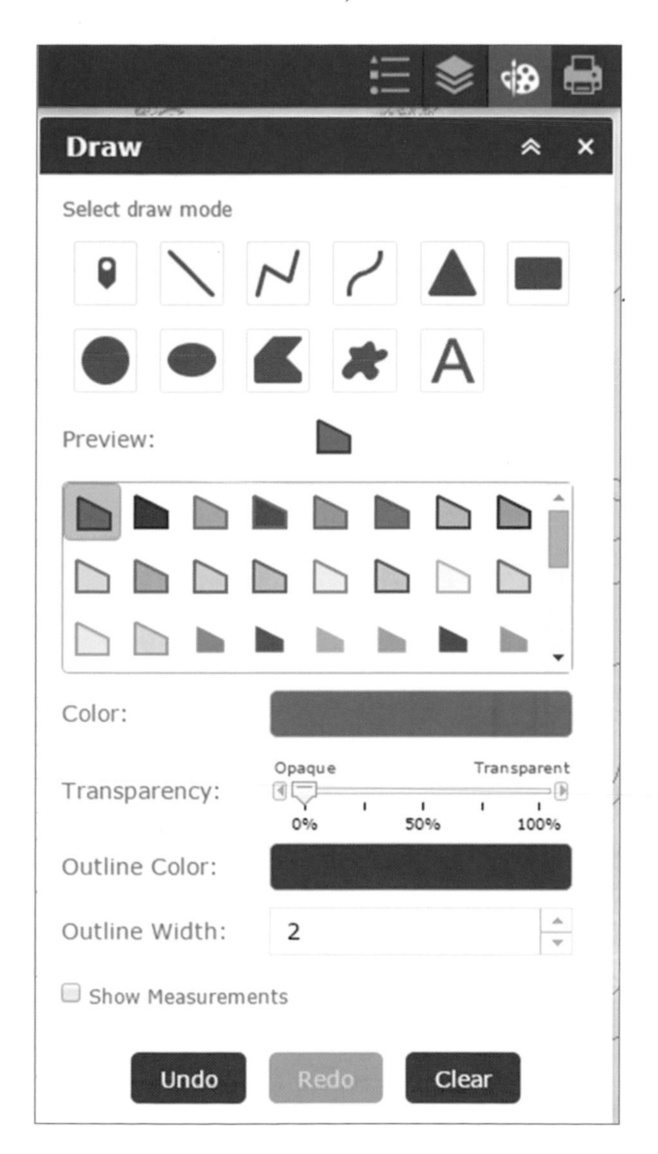

15. **Select a symbol, and experiment with drawing some graphics on your map.**

Data independent widgets may need configuration as well. Next, you will configure the **Print** widget.

16. **Point to the Print widget, and click the pencil.**

17. **In the Configure Print window, notice that the service URL points to the printing geoprocessing service hosted in ArcGIS Online.**

18. **Specify the default title as Historic Earthquakes and Hurricanes, for instance, and click OK.**

Note: If your map has layers from an internal ArcGIS for Server, you will need to change the printing service URL to one that has network access to your internal server. (See the "Questions and answers" section for details.)

19. **On your app toolbar, click the Print widget to test how it works, and in the Print window, click Print to print the current map view, including any drawings, to a PDF.**

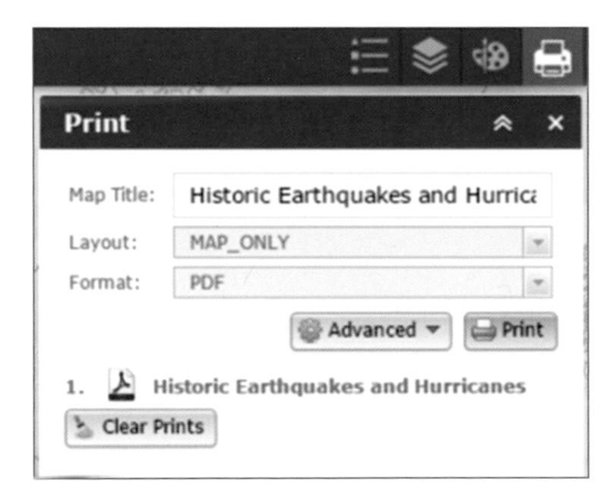

20. **When the printing job is done, click the PDF link, examine the PDF, and then close the print window.**

21. **Click Save to save your web app configuration.**

5.4 Configure data-dependent widgets

In this section, you will enhance your web app by adding the chart and query widgets. You will configure the layer and fields to which these widgets will associate.

1. **As you continue from the previous section, click the Plus button to add additional widgets to the header controller.**

2. **In the Choose widget window, click Chart and Query, and click OK.**

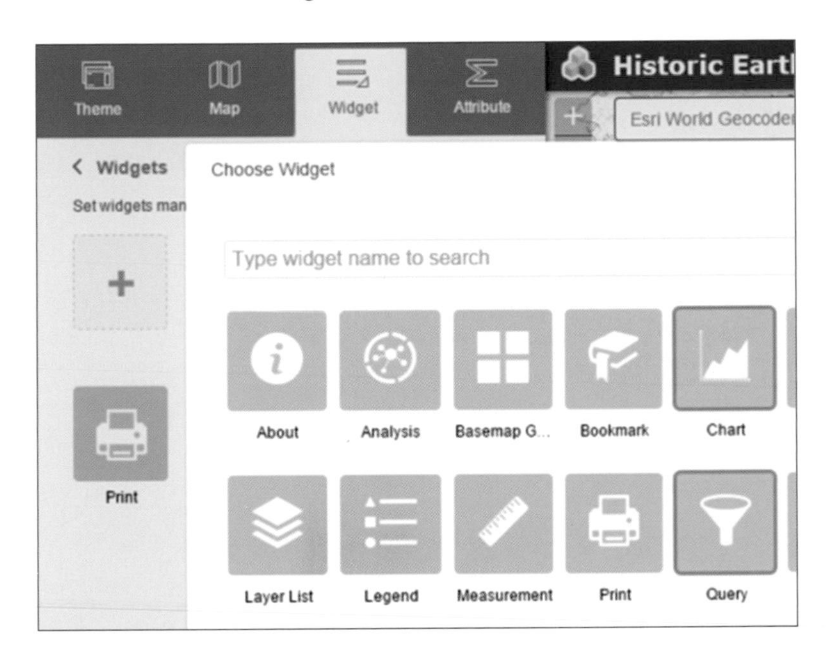

You have added two widgets, but you must add more configurations before you can use them.

3. **Point to the Chart widget, and click the pencil.**

The **Chart** widget allows multiple charts, and each chart can display one or more attribute values of multiple features.

4. **In the Configure Chart window, click Add New.**

Configure Chart

Chart

change widget icon

Chart Tasks

Add New

Name

5. **In the Set Data Source window, choose Select from Map, open the Natural Disasters sublayers, select the Earthquakes layer, and click OK.**

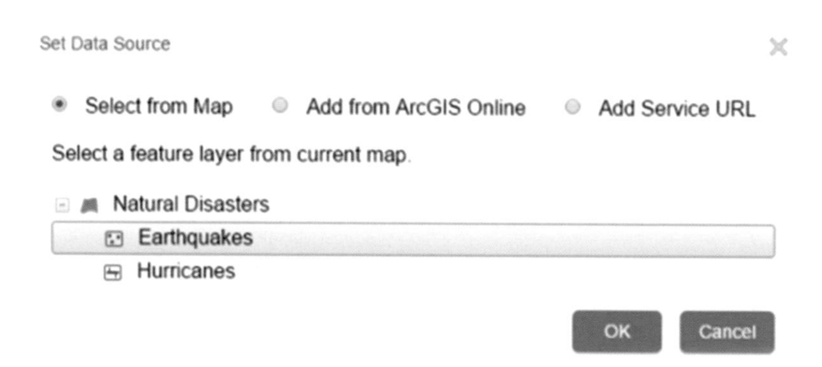

6. **In the Configure Chart window, perform the following tasks to chart earthquake magnitudes:**

   - **For Chart Title, specify Earthquake magnitudes.**
   - **For Value Fields, select Other_Mag1.**
   - **Leave the Category Label as Location.**
   - **For Chart Type, select Column Chart and Line Chart.**

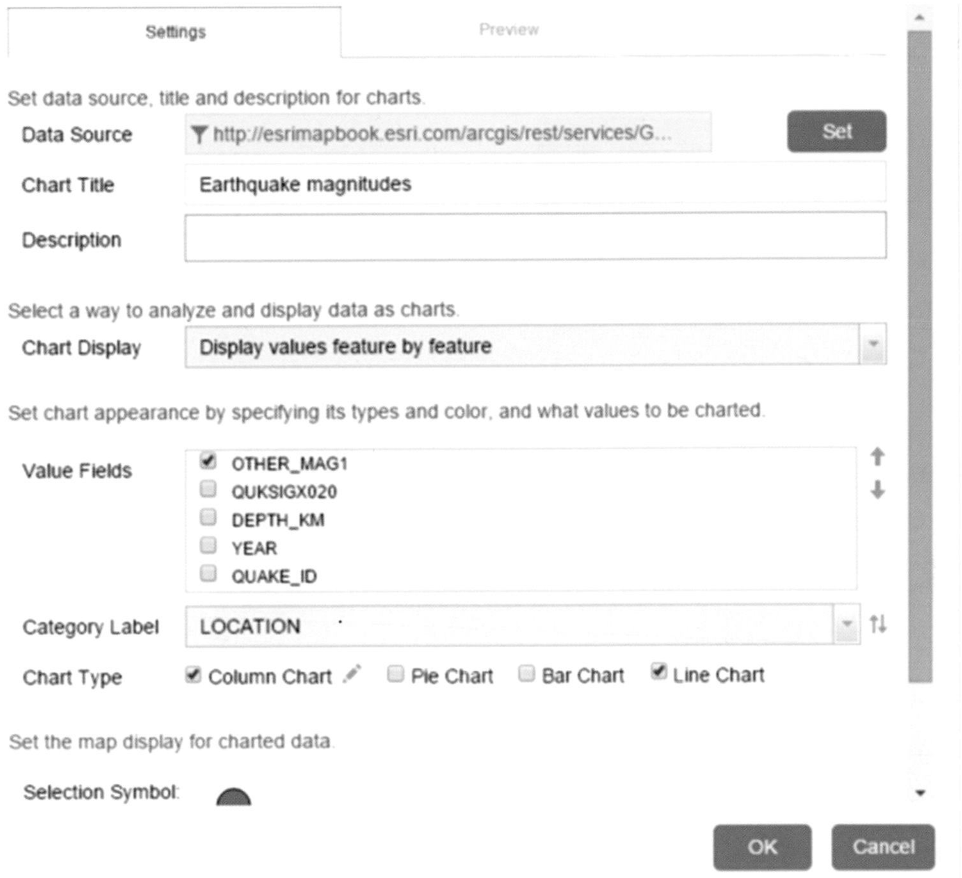

Settings | Preview

Set data source, title and description for charts.
Data Source: http://esrimapbook.esri.com/arcgis/rest/services/G... [Set]
Chart Title: Earthquake magnitudes
Description:

Select a way to analyze and display data as charts.
Chart Display: Display values feature by feature

Set chart appearance by specifying its types and color, and what values to be charted.
Value Fields: OTHER_MAG1, QUKSIGX020, DEPTH_KM, YEAR, QUAKE_ID
Category Label: LOCATION
Chart Type: Column Chart, Pie Chart, Bar Chart, Line Chart

Set the map display for charted data.
Selection Symbol:

OK Cancel

These settings define two charts, which display the magnitude field in a **Column** chart and a **Line** chart. You can also display multiple fields in these charts. The label field helps to identify an earthquake whenever a user points to an earthquake in the chart.

7. **Click OK to close the Configure Chart window.**

Next, you will see how the **Chart** widget works, which also helps you understand what you have configured.

8. **On the toolbar in the preview side of your window, click the Chart button (or the More button and then the Chart button) to open the Chart window.**

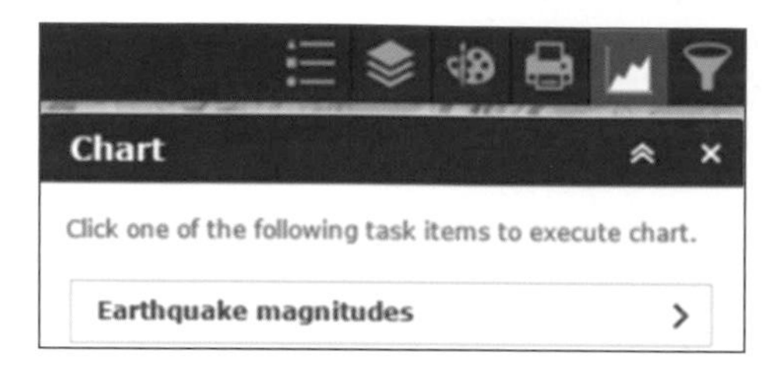

9. **In the Chart window, click Earthquake magnitudes, and select Use spatial filter to limit features.**

10. **Select Only features intersecting a user-defined area, select a shape, draw the shape on the map to select some earthquakes, and click Apply.**

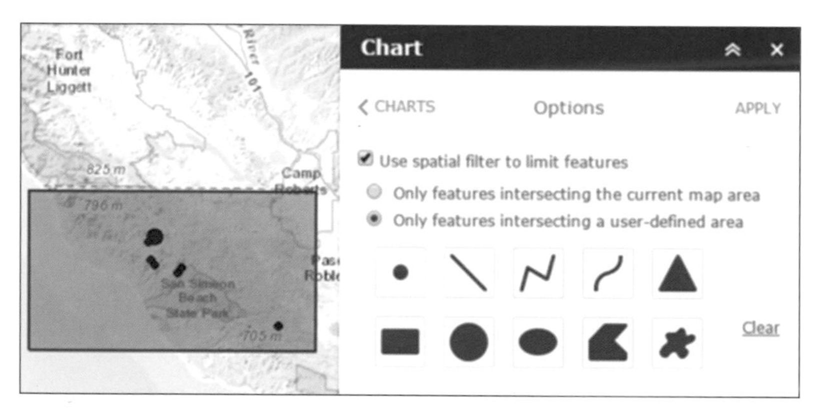

Notice that the **Chart** widget changes to show the results in a column chart.

11. **Perform the following tasks:**

- **Point to a column in the chart. This action displays both the location (that is, the Label field) and the magnitude of the selected earthquake in a pop-up window.**

In the map, the corresponding earthquake is highlighted.

- **Click a column to center the map to the corresponding earthquake.**
- **Click the right-facing arrow ▸ to the right side of the column chart to switch to the second chart.**
- **Close the Chart window.**

Next, you will add a pie chart that can group earthquakes by years.

12. In the design pane, point to **Chart**, and click the pencil.

13. In the **Configure Chart** window, click **Add New**.

14. In the **Set Data Source** window, choose **Select from Map**, open the **Natural Disasters** sublayers, select the **Earthquakes** layer, and click **OK**.

15. In the Configure Chart window, perform the following tasks:

    - For **Chart Title**, specify **Earthquakes by years**.
    - For **Chart Display**, choose **Display feature counts by category**.
    - For **Category Field**, select **Year**.
    - For **Chart Type**, select **Pie Chart**.
    - Click **OK** to close the **Configure Chart** window.

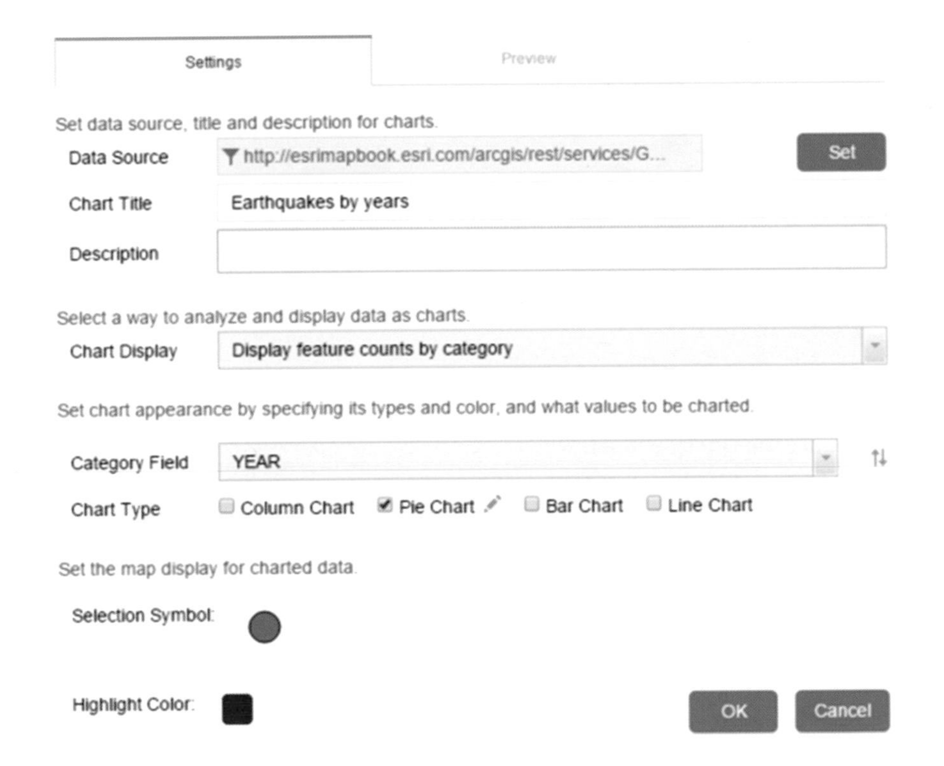

Next, you will explore how the new chart works.

16. On the toolbar of your app, click the **Chart** button.

17. **In the Chart window, click Earthquakes by years. Click Apply to see the pie chart. Point to a slice of the pie to see the year, count of earthquakes, and percentage of the total earthquake count.**

Pointing to a particular slice highlights that year's earthquakes on the map.

18. **Close the Chart window.**

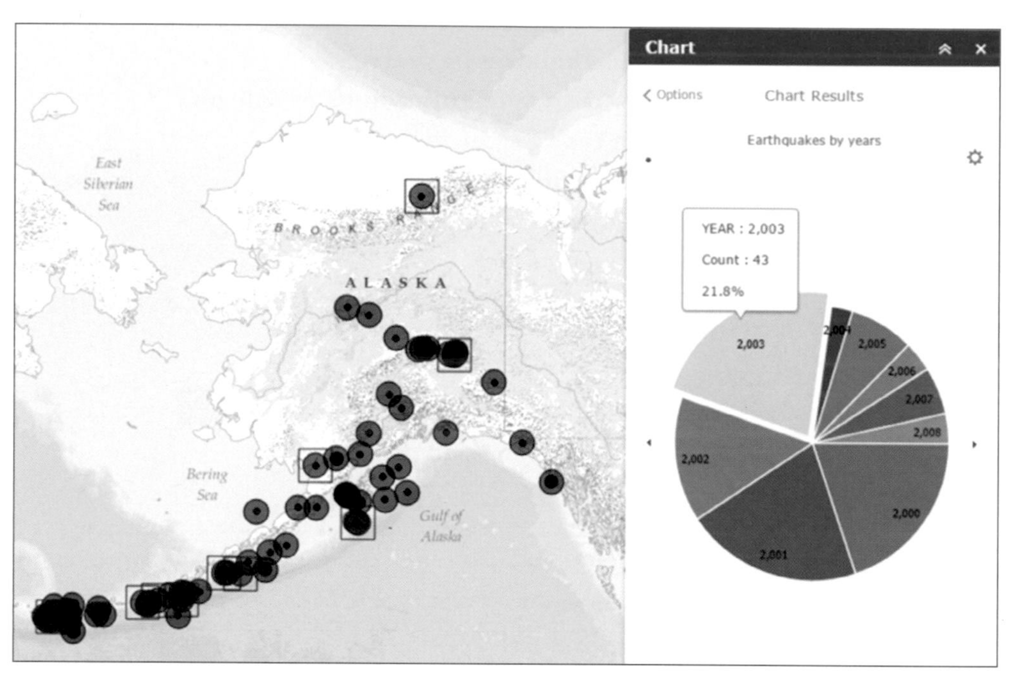

Next, you will configure the **Query** widget.

19. **In the design pane, point to Query, and click the pencil.**

20. **In the Configure Query window, click Add New.**

21. **In the Set Data Source window, choose Select from Map, expand the Natural Disasters map service, click the Earthquakes sublayer, and click OK.**

22. **In the Configure Query window, perform the following tasks:**

- **Click Add a filter expression.**
- **In the first expression, click OTHER_MAG1 (Number).**
- **In the first field, click the arrow, and click Is at least as the operator.**

- **Type 4 as the default value, select the Ask for values check box, and change the prompt to Magnitude is at least and the hint to 4.**

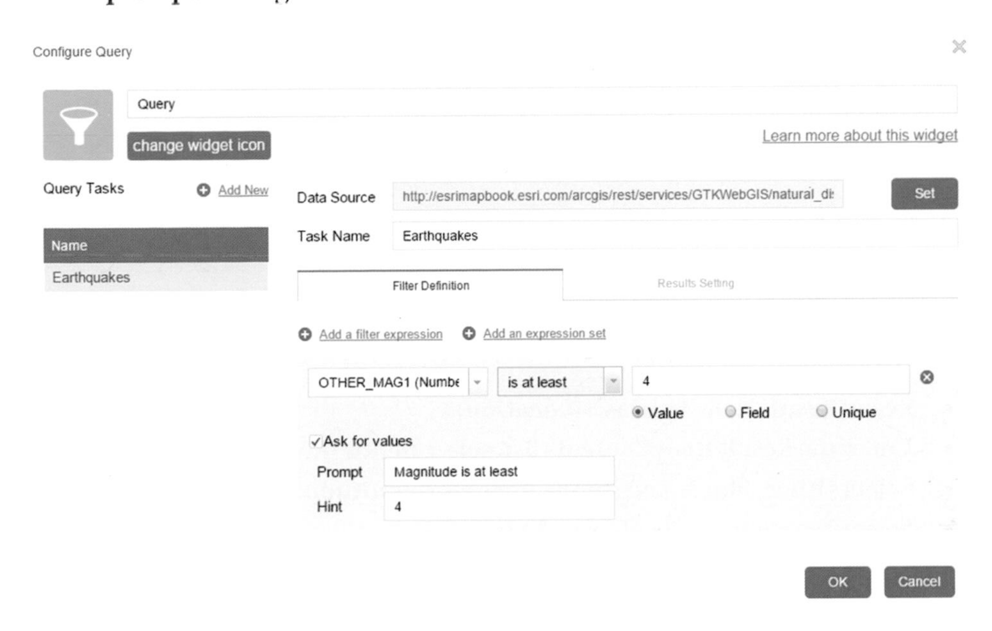

Next, you will add two new filter expressions.

23. Click **Add a filter expression** again. In the second expression (scroll down the configuration window if you don't see the second expression), click **Location** and **contains**, and type **Alaska** as the default value.

24. Select the **Ask for values** check box, and change the prompt to **Location contains** and the hint to **Alaska**.

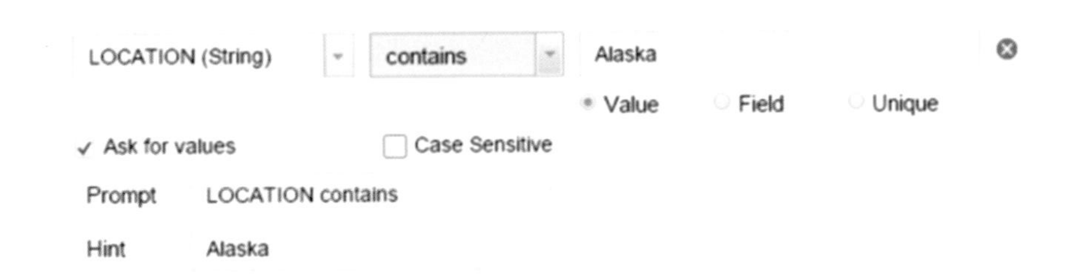

25. Click the **Add a filter expression** button again. In the new expression, click **Year (Number)** and **is**, and select **Unique**, which will retrieve the unique values of the **Year** field and populate the values in a list.

26. Click the arrow, click **2,000** in the dropdown, and select **Ask for values**.

27. **Change the prompt to Year is and the hint to 2000.**

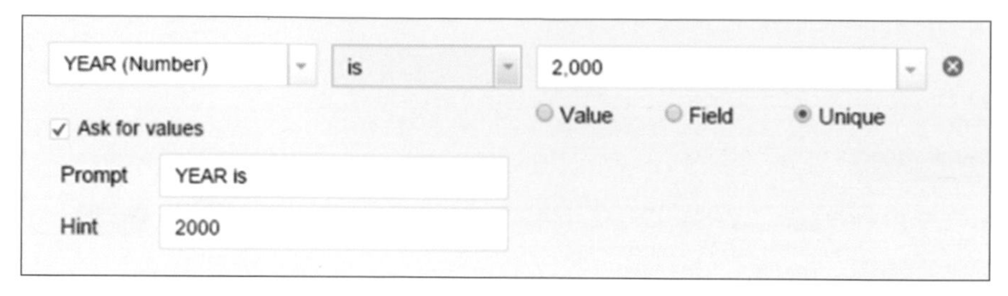

Next, you will configure how you would like to see the query results.

28. **Click the Results Setting tab.**

- **Leave Result Item Title as ${Location}.**
- **Under the Result Item Contents list, select Depth_KM, and set the alias as Depth.**
- **Select Other_Mag1, and set the alias as Magnitude.**
- **Select Year, and set its alias as Year.**

Optionally, you can click **Set symbol for query results** to define how you would like to symbolize the result.

- **Click OK.**

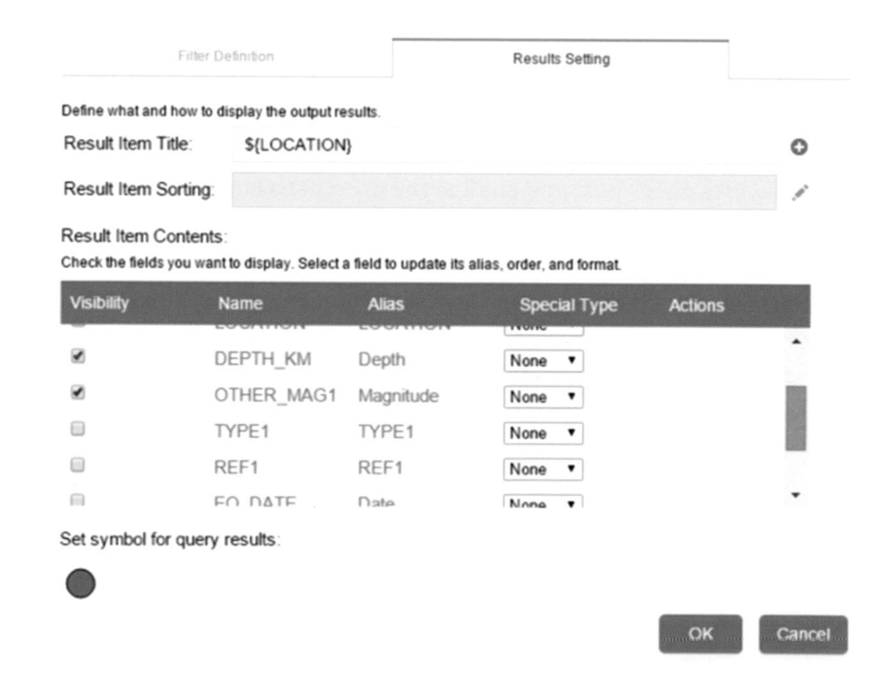

Next, you will try this widget to help you understand the configuration controls.

29. **On the app toolbar, click the Query button, and click Earthquakes, which has the query you defined.**

You should see the search expression you configured.

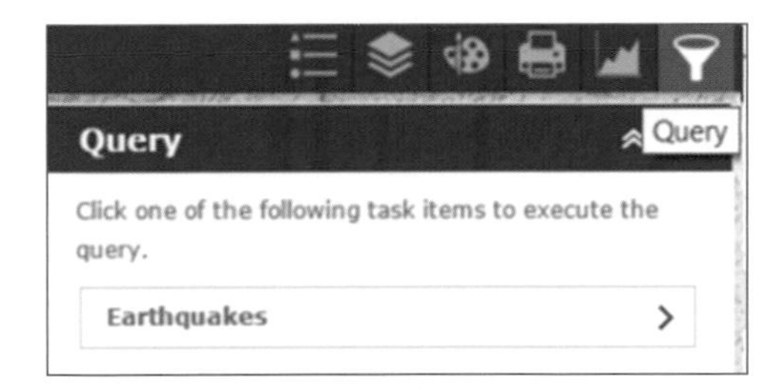

30. **Leave the values as the defaults, or change the values, and click Apply.**

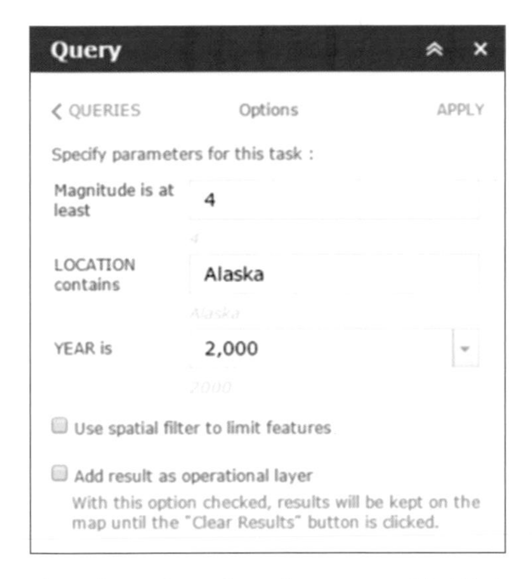

The selected earthquakes appear highlighted on the map and are listed in the **Query** window.

If you select **Add result as operational layer**, the query result will be added as a layer in the **Layer List** widget. In the preview pane, you can open its attribute table in the **Attribute Table** widget.

31. **Click an earthquake in the Results list. The map centers on the earthquake, and a pop-up appears, displaying the attributes you selected when you configured the widget.**

32. **Click Save in the design pane to save your configuration.**

5.5 Preview and share your app

In the previous steps, you saw the effects of your configurations in the preview pane. You can further preview your app in various mobile devices and in its own browser window.

1. **At the bottom of the design pane, click Previews.**

You will see a list of popular mobile devices.

2. **Choose the type of device you wish to preview, or specify a custom screen resolution.**

3. **In the preview window, try each of the widget buttons and see how your app works on mobile devices.**

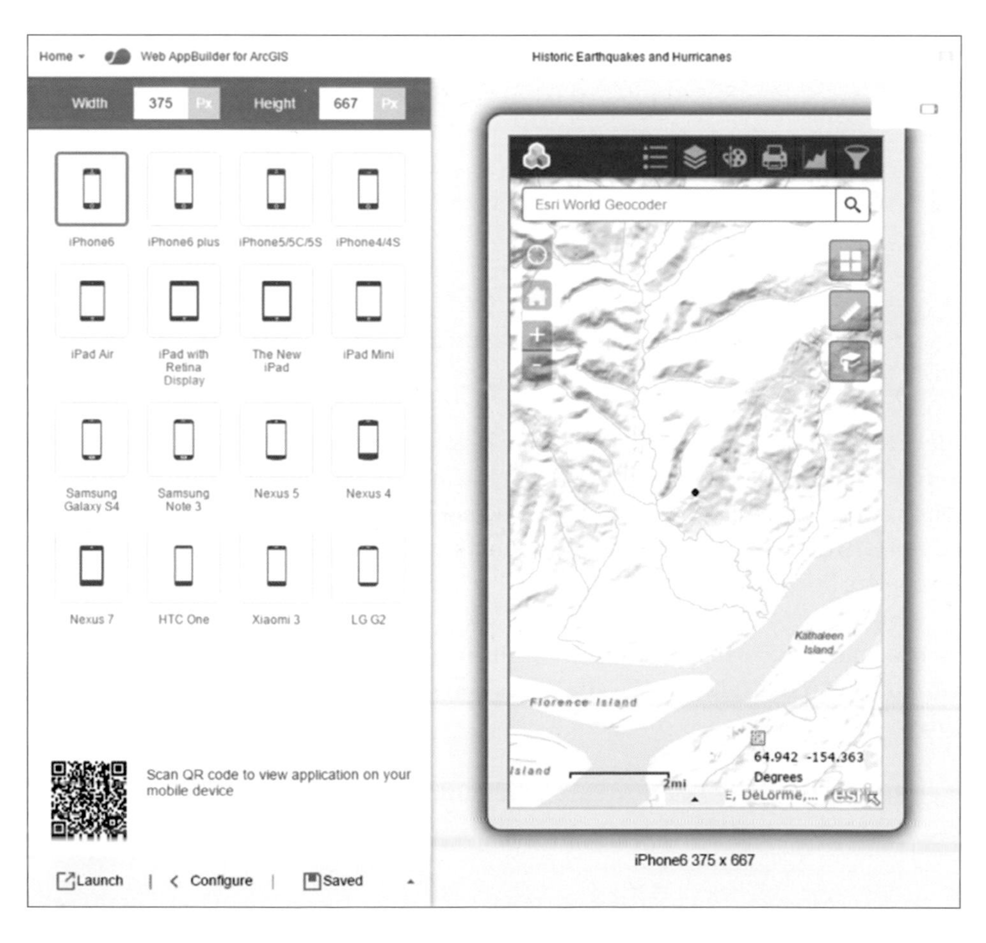

4. **In the upper-right corner of your browser, click the Phone Orientation icon (▯ or ▭) to change the orientation of the device. Try your app to see how it behaves in the new orientation.**

If you have a smart mobile device with a QR scanner app, you can scan the QR code to view your app on your mobile device directly.

5. **Click < Configure in the design pane to return to the configuration mode.**

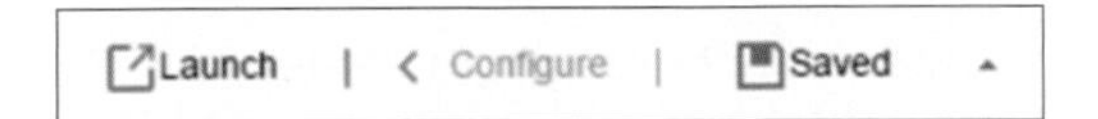

6. **Click Launch to view your app in the full web browser.**

Your app will display in a new web browser or a new tab. The URL is your app URL that you will share with your audience and your instructor.

7. **Go to your ArcGIS Online or Portal for ArcGIS content list, and find and select the app.**

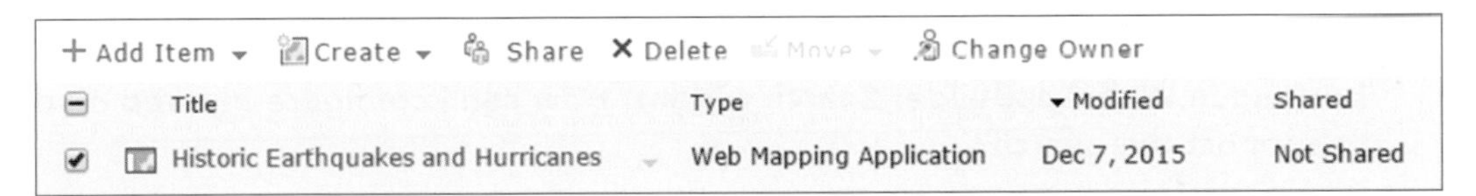

If you need to configure your app further, click the arrow after the app title, and choose **Edit Application**.

8. **Click the Share button to share the app with Everyone, and click OK.**

In this tutorial, you created a web app that provides a set of functions without programming. The widgets you configured are among the widgets most commonly used. You also can try other widgets and explore how they work.

QUESTIONS AND ANSWERS

1. **My Print widget does not work. Instead, an error message appears, saying, "Error. Try again." Why?**

   Answer: A common reason for this error message is that your web map contains a map service layer from an internal ArcGIS for Server, which is inside your network firewall. The default printing service configured in the Print widget comes from ArcGIS Online or Portal for ArcGIS. ArcGIS Online printing service sits outside your network firewall and thus cannot see your internal server. As a result, the printing service cannot ask your map service to generate a map.

   To fix this problem, replace the default printing service URL with an internal printing service URL. Your ArcGIS for Server comes with a built-in printing service, which you can find in the **Utilities** folder of your ArcGIS for Server Services Directory. Some organizations choose to stop this printing service. You can ask your ArcGIS for Server administrator to start the service using ArcGIS Server Manager (navigate to **http://your_ArcGIS_Server_name/arcgis/manager).**

2. **In the tutorial, I searched for hurricanes directly in the place/address search box and in Web AppBuilder Search widget. How can I configure my web map to support this search?**

   Answer: This capability is called "feature search," which allows users to locate features in the same search box as address and place name search. For example, enabling search on your parcel layer would allow users to find specific parcels simply by entering a parcel ID in the search box. For your users, this way to locate features is consistent with the way they locate an address or place name.

   To configure feature search, you can go to the item details page of your web map and click **Edit**. In **Application Settings** and under **Find Location**, you can enable **By Layer**, select the **Layer** and **Field** you want to allow your users to search, and choose a condition for comparison. You can also configure the **Hint text**, which will appear in the search text box and tell users what they can search for.

Application Settings | Select the tools and capabilities to enable in applications that access this web map

- [ ] Routing
- [ ] Measure Tool
- [ ] Basemap Selector
- [x] Find Locations [-]

Hint text: Place, address, or hurricane name

- [x] By Layer

Layer	Field	Condition
Hurricanes	Name	Contains

ADD LAYER

- [x] By Address

3. Can Web AppBuilder for ArcGIS work directly with web services?

Answer: Yes and no.

Web AppBuilder is not designed to work directly with web services when specifying a web app's map content. This workflow (adding map services directly into the map) is not supported and is not recommended. Instead, the Web AppBuilder works with web maps, which can encapsulate web services. Authoring a web map is easy and empowers non-GIS experts to make their own maps. Web AppBuilder is designed to extend the reach of non-GIS experts so they can easily create a custom web app for their web maps. These apps can run on desktops, smart phones, and tablets. This design strengthens the concept of extending the power of GIS and spatial mapping throughout an organization.

However, Web AppBuilder does work directly with web services via its many widgets. You can configure widgets to work directly with your web services by specifying their REST URLs when you set the properties of a widget. For example, you can specify your own basemap services for the basemap gallery widget, a custom geoprocessing service for the geoprocessing widget, or your own layer as the data source for the query and chart widgets.

ASSIGNMENTS

Assignment 5: Build a web app using Web AppBuilder for ArcGIS.

Data: No data has been provided. Instead, create a web map using the feature layer you published before, or find a web map in ArcGIS Online.

Requirements:

- The initial map extent should zoom to your study area.
- The app should have the following capabilities:
  - Provide bookmarks that allow users to quickly zoom to predefined areas.
  - Let users print their maps as PDFs.
  - Enable users to chart selected features with selected attributes.
  - Allow users to query for features based on multiple attributes.
  - Display the appropriate logo, title, subtitle, and links in the banner.

What to submit: Send an email to your instructor with the subject line **Web GIS Assignment 5: Your name**, and include your web app URL and screen captures of your web app, showing that it meets the preceding requirements.

Resources

Online document and blog

"Clarifying 3 Questions about Web AppBuilder for ArcGIS," by Derek Law, http://blogs.esri.com/esri/arcgis/2015/04/06/clarifying-3-questions-about-web-appbuilder-for-arcgis.

Live sites showcase, http://www.arcgis.com/apps/MapAndAppGallery/index.html?appid=1e3085af6e1a48c8908fa624bdfef768.

Web AppBuilder for ArcGIS online help document site, http://doc.arcgis.com/en/web-appbuilder.

Web AppBuilder widgets reference, http://doc.arcgis.com/en/web-appbuilder/create-apps/widget-overview.htm.

Online training and tutorials

"Get Started with Web AppBuilder for ArcGIS," http://training.esri.com/gateway/index.cfm?fa=catalog.webCourseDetail&courseid=2887.

"OSO Mudslide—Before and After," http://learn.arcgis.com/en/projects/oso-mudslide-before-and-after.

Online Videos

"Introduction to the ArcGIS Web AppBuilder: JavaScript Apps Made Easy," by Jianxia Song and Derek Law, http://video.esri.com/watch/4706/introduction-to-the-arcgis-webapp-builder-javascript-apps-made-easy.

Chapter 6
Publishing map services with ArcGIS for Server

In previous chapters, you learned how to create hosted layers and web apps using ArcGIS Online, which runs on a public cloud. When you must build on-premises (locally hosted) or on-premises/cloud hybrid web GIS applications, you will need ArcGIS for Server. ArcGIS for Server is an important component for implementing on-premises and hybrid web GIS. In addition, ArcGIS for Server supports many important types of GIS services that ArcGIS Online cannot yet publish. These service types include dynamic map services, which are commonly used to serve your operational layers. This chapter introduces ArcGIS for Server. Using this technology, you will also learn how to publish map services, enable time on your map services, and create web apps with time animation.

Learning objectives

- *Understand on-premises and hybrid web GIS implementation.*
- *Know the types of GIS web services.*
- *Understand the workflow to publish map services using ArcGIS for Server.*
- *Explore and validate GIS services using ArcGIS Services Directory.*
- *Create web apps with time animation.*
- *Manage and use GIS services with Portal for ArcGIS and ArcGIS for Server federation.*

This chapter in the big picture

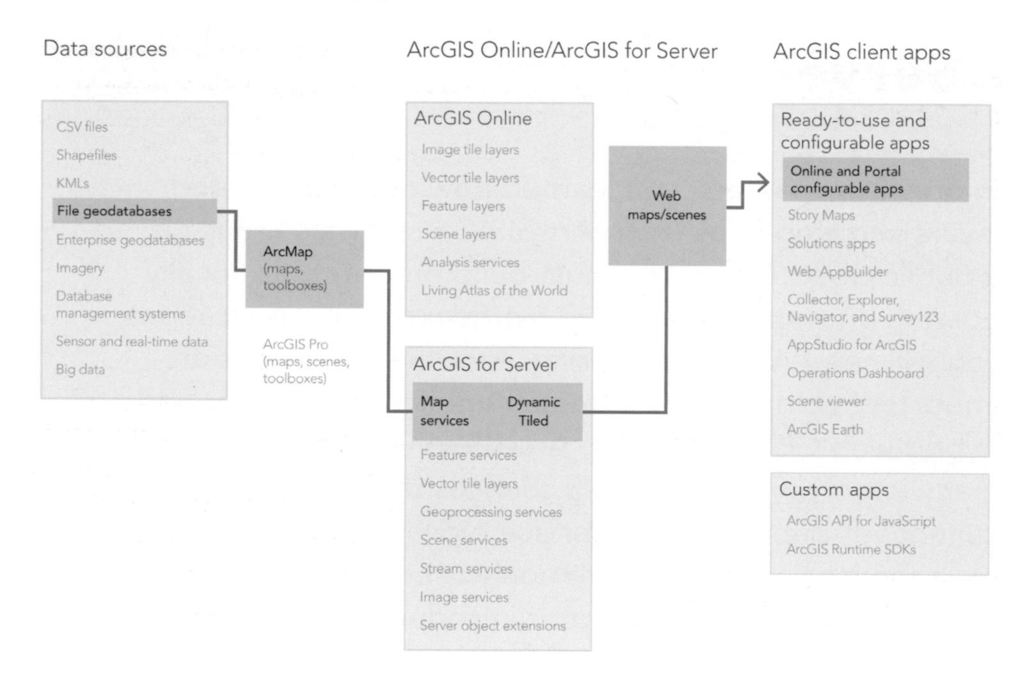

ArcGIS offers many ways to build web apps. The green line in the figure highlights the technology that this chapter teaches.

Why learn ArcGIS for Server?

ArcGIS for Server is an important enterprise component of today's web GIS platform. You can configure ArcGIS for Server as a scalable, standalone GIS server for managing and distributing GIS services. You should understand and learn how to use this product for the following reasons:

- The need for on premises web GIS deployment
- The need for cloud and on-premises hybrid web GIS deployment
- The need to publish feature services, tiled map services, and scene services
- The need to publish these additional following types of GIS services:
  - Dynamic map services
  - Geoprocessing services
  - Geocoding services
  - Stream services (with GeoEvent extension)
  - Image services (with Image extension)

In previous chapters, you learned ArcGIS Online. These services make ArcGIS Online an easy and excellent entry point to web GIS. However, some organizations cannot use a cloud-based implementation. These organizations may not have Internet connections, or their connection may be prohibited or unreliable. Government or industry regulations can prevent organizations from using cloud services or storage. These situations require an on-premises deployment.

ArcGIS for Server provides a web GIS solution that you can install behind your firewall on your infrastructure and still meet your organization's security policies. Server supports single sign-on via Integrated Windows Authentication (IWA), thus allowing greater convenience for enterprise users. And this setup does not require you to spend ArcGIS Online credits to publish services and perform many other operations.

Hybrid deployment combines cloud and on-premises deployment methods. Using this combination, ArcGIS for Server provides the behind-the-firewall data storage, sharing, processing, and security enforcement capabilities, while ArcGIS Online provides cloud-based sharing, dissemination, collaboration functions, and contents for the Living Atlas of the World.

Users can easily access the ArcGIS Online cloud because Esri sets up and maintains the platform for you. However, for an on-premises system, your organization is responsible for setting up and maintaining the system. For this reason, you may encounter challenges related to network, bandwidth, and the SSL (Secure Socket Layer) certificate. Working through these challenges is often a part of any on-premises system.

	Portals	GIS servers	Ready-to-use content
Cloud deployment	ArcGIS Online (portal)	ArcGIS Online (hosted Servers)	ArcGIS Online services
On-premises deployment	Portal for ArcGIS	ArcGIS for Server	Data Appliance for ArcGIS

Web GIS deployment patterns: The cloud deployment row represents all portal GIS servers and ready-to-use contents in the public cloud. The on-premises deployment row represents all components that are within an organization's internal infrastructure. The shaded combination represents one form of a hybrid deployment pattern.

Federated ArcGIS for Server and Portal for ArcGIS

ArcGIS for Server mainly serves GIS services while Portal for ArcGIS serves as the center for managing and accessing these services. You can install the two technologies separately or together (federated). A federated installation provides many of the following advantages:

- ArcGIS for Server extends Portal for ArcGIS by supporting additional types of GIS services that Portal for ArcGIS cannot host by itself.
- ArcGIS for Server services are automatically registered with Portal for ArcGIS, making your organization's GIS services easier to discover and use.
- Portal for ArcGIS serves as a friendly interface for users to access ArcGIS for Server for exploring, creating, and sharing maps and apps.
- Portal for ArcGIS also provides administrative tools for securing critical content, controlling user roles and capabilities, and monitoring and reporting usage statistics.

Types of GIS web services

A GIS web service takes your GIS resource, such as a map, tool, or geodatabase, and makes the resource available and reusable in a wider range of web GIS client apps. ArcGIS offers the following flexible approaches for sharing web services:

- ArcGIS Online
- ArcGIS for Server
- Portal for ArcGIS
- Portal for ArcGIS and ArcGIS for Server

Many types of GIS web services support the full spectrum of web GIS capabilities. The table lists the main types and what products to use to share each type.

Table 6.1 **Main types of geospatial web services and servers that can publish each type**

Service Type	Capabilities	AGOL	AGS	PA	PA+AGS
Dynamic map service	Generates maps on the fly, and returns maps in an image format, such as JPG, PNG, or GIF.		x		x
Tiled map service	Maps tiles are generated in advance. Returns maps in an image format, such as JPG, PNG, or GIF.	x	x	x	x
Feature service	Allows web clients to retrieve vector features and update feature data on the server.	x	x	x	x
Image service	Allows clients to access raw raster data. ArcGIS image service additionally supports rapid map algebra calculation.		x		x
Locator or geocode service	Can convert addresses and place names into X, Y locations.		x		x
Geoprocessing service	Shares server's workflow and analysis functions with web clients.		x		x
Network analysis service	Provides routing, tracing, and allocating functions based on geospatial networks, such as street and pipelines.		x		x
Geometry service	Supports geometric calculations such as buffering, simplifying, calculating areas and lengths, and projecting.		x		x
Stream service	A new type of ArcGIS for Server service that emphasizes low latency, real-time data dissemination for client-server data flows.		x		x
Scene service	Allows web clients to request maps in 3D.	x			x

Dynamic map services and tiled map services

Map services are the most commonly used type of GIS services. Map services allow clients to request maps for a specific geographic extent, and the maps are returned in an image format, such as JPG, PNG, or GIF. Map services can be dynamic or cached.

- For dynamic map services, the server generates a map as the request is received. Dynamic map services are typically used to serve maps whose data is constantly changing or whose data size is not too large to significantly slow service performance.
- Cached or tiled map services fulfill client requests with pre-created tiles from a cache. This type of service is typically used for basemaps or maps with content that is relatively static and changes little over time. Cached map services can require a huge amount of disk space to store. However, once created, these services can significantly improve performance, availability, scalability, and the user experience of your web apps.

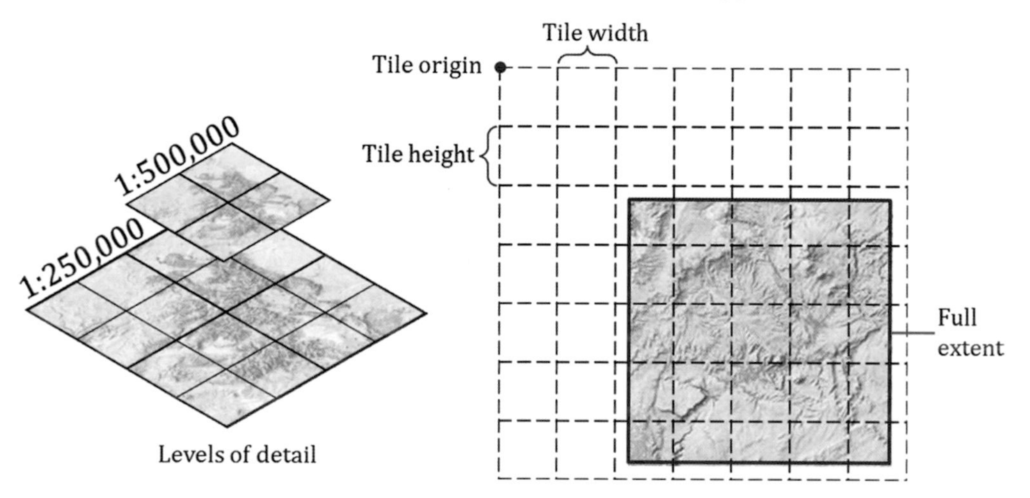

Map caching (left) generates a set of map tile images in advance at predetermined scale levels. The map caching scheme (right) includes the number of scale levels, the scale at each level, tile dimension, tile origin, tiling area, and image format.

Map image layers versus feature layers

You can display maps using the map image layer approach or the feature layer approach. The following figure compares these two approaches.

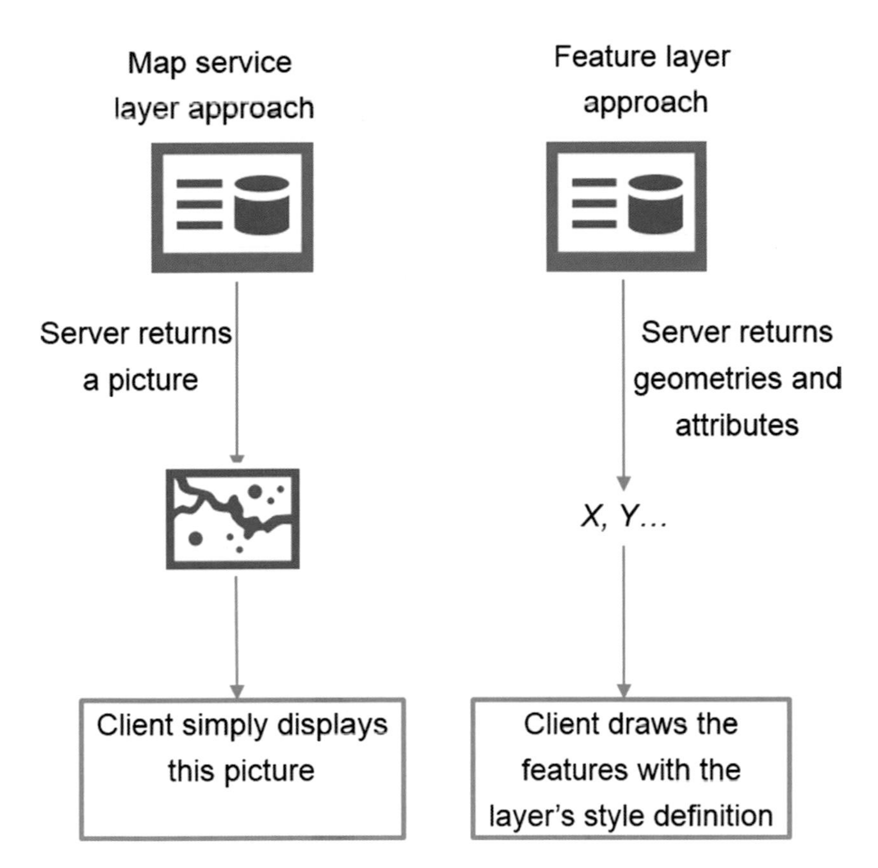
Map service
layer approach
Feature layer
approach
Server returns
a picture
Server returns
geometries and
attributes
X, Y...
Client simply displays
this picture
Client draws the
features with the
layer's style definition

Comparison of a map image layer and a feature layer. A map image layer returns a picture to the client, whereas a feature layer returns the geometries and attributes to the client.

When you add an entire map service to a web map, you are using the map service layer (or map image layer) approach. With this approach, the server draws the map and returns a picture to the client. The client just needs to display the image. This approach exerts less load on the browser side and more load on the server side. This approach has the disadvantage of being less responsive than the feature layer approach. The client side has no GIS data, so clients must route all mouse-click queries to the server. As a result, the client must wait for the server to respond. Yet this approach has the advantage of allowing you to map a lot of data without having to transfer all of that data to the client side.

When you add a layer of a feature service or layer of a map service to a web map, you are using the feature layer approach. With this approach, the server returns the vector data of the features, and the client draws the map. This approach exerts more load on the browser side and less load on the server side. If this approach is not used properly, you risk overloading your end user's web clients with too much data for the clients to draw efficiently. However, this approach can be more responsive because of the GIS data on the web client side. As a result, the client can respond to mouse clicks and queries directly without having to wait for the server.

Notes:

- Hosted feature layers and feature layers for displaying maps may sound confusing. Hosted feature layers are actually feature services hosted on ArcGIS Online or Portal for ArcGIS. Feature layers can mean feature services in the context of hosted feature layers and also can mean vector layers in the context of display maps.
- Both feature services and map services can return vector data and thus support the feature layer approach to display maps. But layers from map services are read only, whereas layers from feature services may support write privileges.

Workflow to publish map services

This section describes the typical workflow to publish map services.

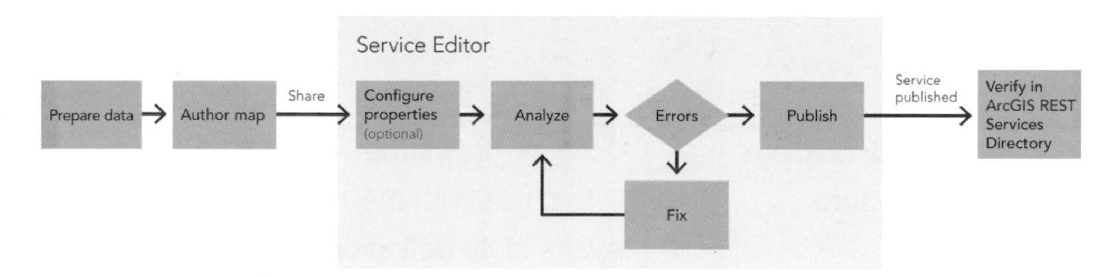

Workflow to publish map services.

- Prepare the data using ArcGIS for Desktop. The following practices are recommended to increase map service performance:
    - For most situations, project your data into Web Mercator so that your map aligns with the popular basemaps without having to be projected dynamically (on the fly).
    - Create indexes on the attribute fields that your users will query in your app.
    - Turn off or delete attribute fields that are irrelevant to your app objectives.
- Author the map document in ArcGIS for Desktop. Add data layers to your map. Configure layer symbols and other properties.
    - Remove unused layers before sharing your document as a service.
    - Do not use complex symbols, such as 3D symbols, that are not supported in map services or that will slow down your map service performance.
- Share your map as a service.
    - Use the **Service Editor** in ArcGIS to analyze your map and review for any errors, warnings, and messages that may result. Once the errors are fixed, you can publish your map service. The publishing process will generate a service definition (SD) file from your map document and upload the file to ArcGIS for Server to create the map service.

- Verify your map service.
  - After you publish the service, use the ArcGIS Services Directory, ArcGIS for Server Manager, or Portal for ArcGIS to verify whether the service works correctly. These products allow you to browse and discover the services, read their metadata, and preview or test them.

Make your data accessible to ArcGIS for Server

This chapter uses ArcGIS for Server to publish and host map services. ArcGIS for Server needs access to your data to serve your service—for example, to generate maps, extract features, and respond to queries. You can perform three tasks to make your data available to ArcGIS for Server:

- Store your data where the data is visible to ArcGIS for Server. The data location can be file based or database based.
- Grant the ArcGIS for Server account permissions to your data folders or databases.
- If your data size is large, register your data folder or database with the server. This registration creates a data store that gives ArcGIS for Server a location with verified access to your data folder or database. The registration also helps ArcGIS for Server adjust the data path or data connection in your map when you publish services.

Notes:

- In this chapter, you will not create an actual data store. For purposes of this tutorial, ArcGIS for Server will automatically create a copy of relevant data so that the server always can access the service-related data. This option works if your data size is not too large.
- In this situation, the term "data store" is a generic name and differs from ArcGIS Data Store, which is a product name. ArcGIS Data Store is an optional component of ArcGIS designed to optimize publishing workflows to Portal for ArcGIS. ArcGIS Data Store allows you to publish hosted feature layers and hosted scene layers to your Portal for ArcGIS.

Time-enabled map services and time series animation

Time is an important dimension of geospatial data. Time-enabled web apps allow you to step through periods of time, revealing the temporal patterns and trends hidden in your data. For example, these web apps offer the following capabilities:

- Display discrete events, such as crimes, accidents, and diseases over time.
- Visualize the value change in stationary objects, such as air quality sensors and weather stations.
- Map the progression of a wildfire or flood.

You can easily use time-enabled map services to animate the geographic features in your web apps. To meet the data requirement for time-enabled services, each feature must have a single date attribute field or beginning and ending date fields. The date fields can be date type or other data types, as long as you can express their values in "sortable" formats, such as YYYY or YYYYMMDD string or number format.

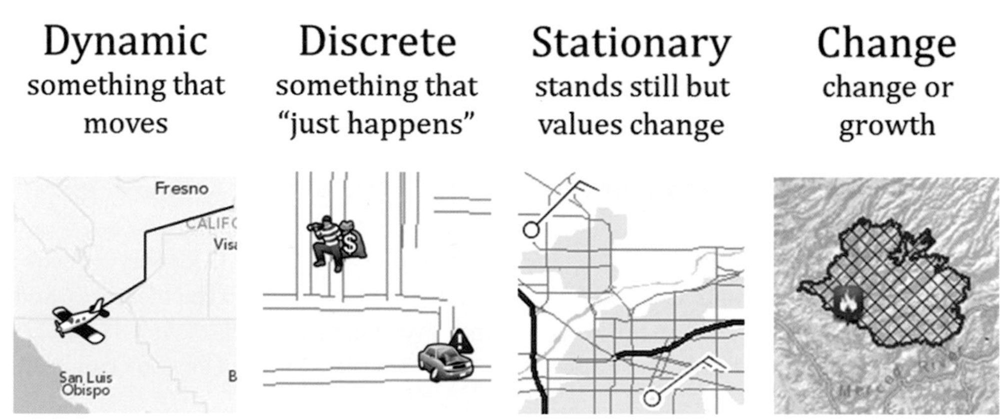

Examples of temporal data.

This tutorial

In this tutorial, you will create a web app to display the spatial and temporal patterns of historic hurricanes and earthquakes in the US region using map animation.

Data: A file geodatabase containing the following two layers covering the US region between 2000 and 2008:

- Major earthquakes with magnitudes greater than 6.6
- Major hurricanes (this layer contains 4,419 lines—too many for the CSV/shapefile upload approach that you learned in chapter 2)

Requirements: Your web app should display the following capabilities:

- All the hurricanes and earthquakes on one map
- Earthquakes and hurricanes by time (for example, by units of months)
- Meaningful pop-up windows whenever a user clicks on an earthquake or hurricane

System requirements:

- ArcMap
- ArcGIS for Server (standard or advanced edition) and an account at the publisher or administrator level; ask your instructor or GIS administrator for the connection information to your ArcGIS for Server
- ArcGIS Online and an account at the publisher or administrator level
- Optionally, Portal for ArcGIS (federated with ArcGIS for Server) with an account at the publisher or administrator level (account should be the same as the ArcGIS for Server account)

6.1 Connect to your GIS server

1. **In ArcMap, click the catalog icon to open the Catalog window.**

For your convenience, click the **Pin** button to turn off the **Auto Hide** mode so that the **Catalog** window stays open.

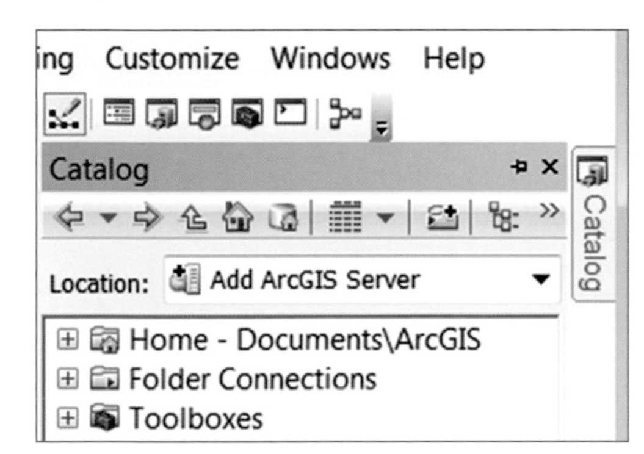

2. **In the Catalog tree, expand the GIS Servers node and double-click Add ArcGIS Server.**

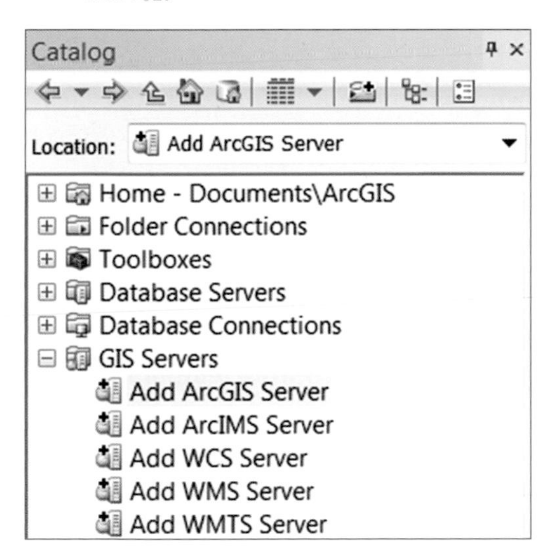

3. In the **Add ArcGIS for Server** window, select **Administer GIS Server** or **Publish GIS Services**, depending on the type of account provided to you, and then click **Next**.

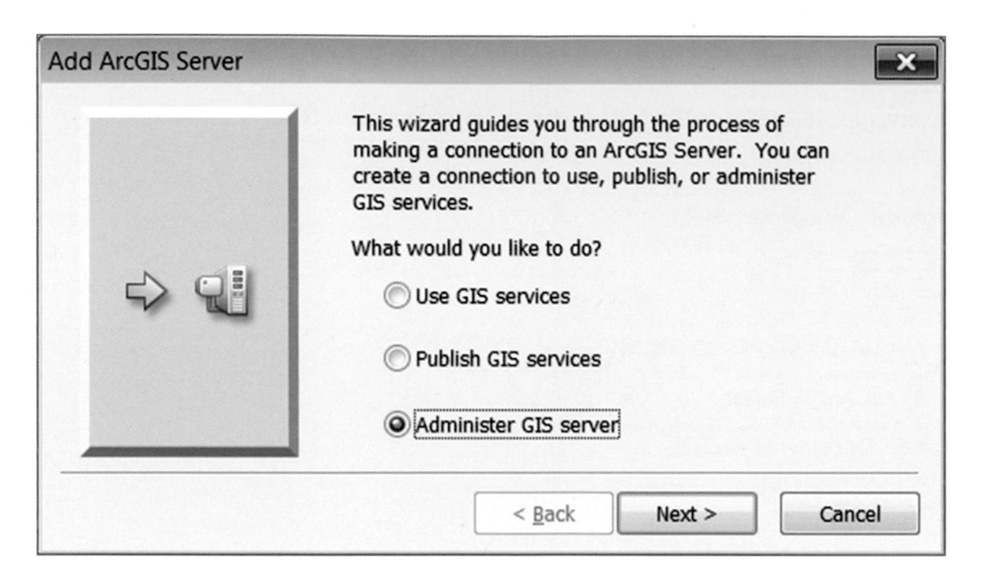

A publisher can publish, delete, start, and stop services. An administrator can additionally edit server properties, such as register data locations.

4. In the **General** window, specify the following server connection information (ask your instructor or GIS administrator if you need assistance):

5. For **Server URL**, type the URL for ArcGIS for Server, which is usually **http://computer_name:6080/arcgis** or **http://computer_name/arcgis.** The **http** can be **https** depending your server configuration.

   - For **Server Type**, click the drop-down box, and select **ArcGIS for Server.**
   - For **Staging Folder**, leave it as is. The folder is used to store temporary files, such as SD files.
   - Enter the user name and password of an administrator or publisher account to the ArcGIS for Server with the URL that you specified earlier. The default user name and password are **siteadmin/siteadmin.**

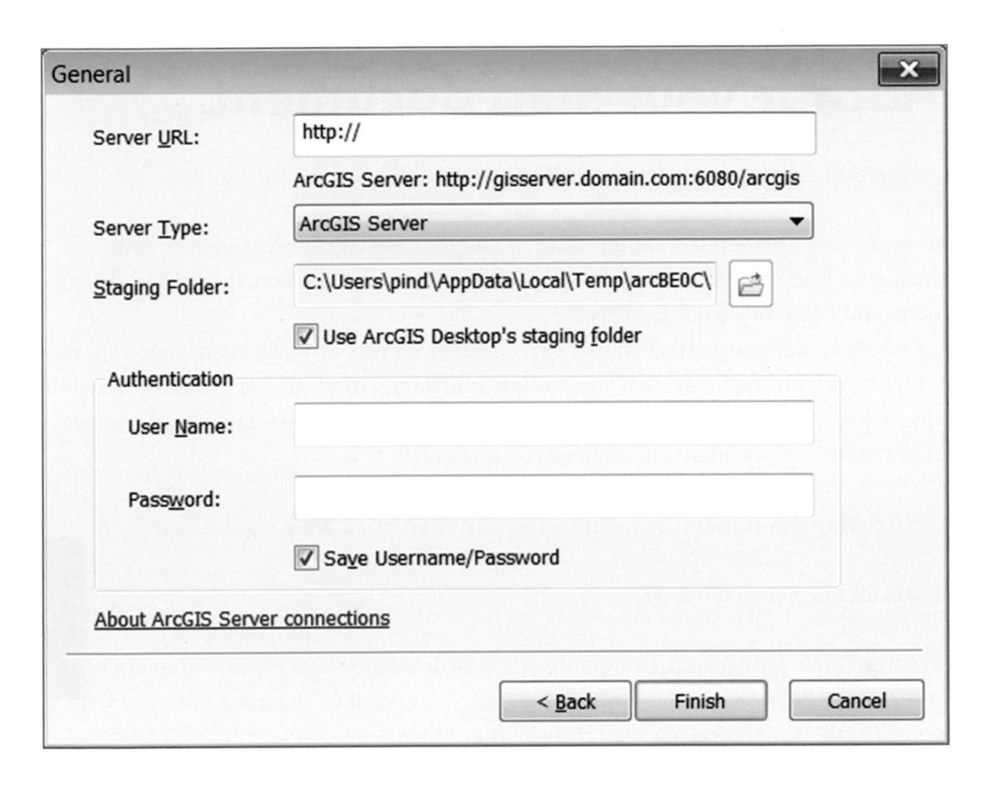

6. Click **Finish.**

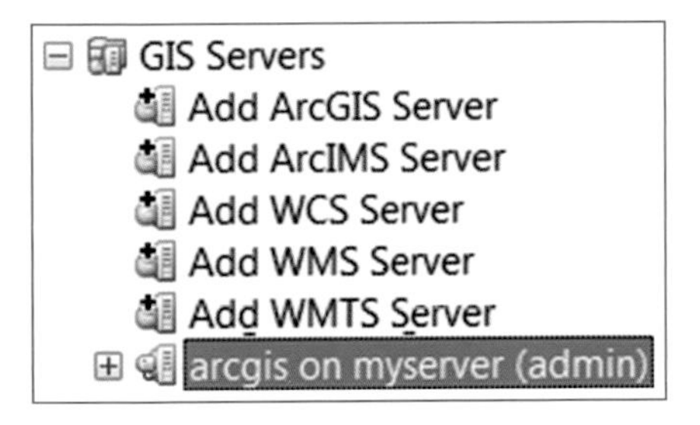

Your connection will appear in the **GIS Servers** node in the **Catalog** tree. Now you have made a connection to ArcGIS for Server that you will use to publish services.

6.2 Author your map document

In this section, you will create a map document and use it to publish a map service.

Notes:

- If you have not downloaded the sample data, navigate to **http://esripress.esri.com/bookresources**; find the title *Getting to Know Web GIS,* second edition; download the sample data; and extract the files to **C:\EsriPress**.
- This book does not address map design or cartography, so you will quickly design a simple map that will work for the tutorial. If you already know how to author maps and you would like to skip these map design steps, follow steps 1 through 3, open the **natural_disasters_key.mxd** file in the **Answers** folder, and proceed to section 6.3.

1. **On the ArcMap main menu bar, click File, and then click New.**

2. **Click Blank Map, and click OK.**

This step creates a new blank map. By default, the coordinate system of your map document and your map service are the same as the first layer that you added to the map. Starting with a blank map prevents you from being affected by previous work.

3. **In the ArcMap Catalog window, click the Connect to Folder button, click Computer > Local Disk C: > EsriPress.**

4. **Select the GTKWebGIS folder, and click OK.**

A folder connection to this book's sample data is created, which allows you quick access.

5. **In the Catalog window, look for Folder Connections. Click the folder C:\EsriPress\GTKWebGIS, find Chapter6\Data.gdb, and expand the Data.gdb file. Two feature classes appear.**

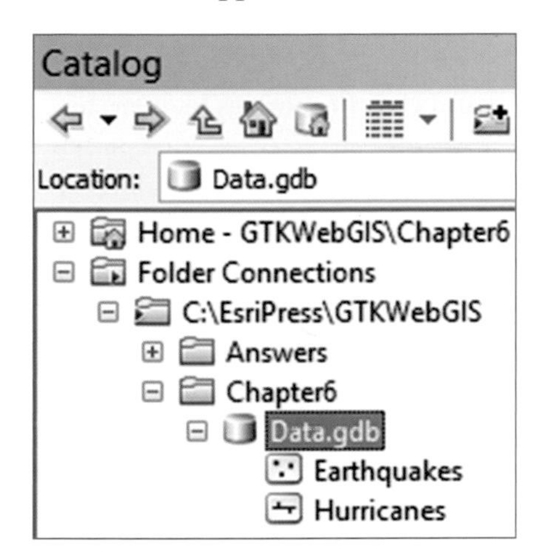

6. **Drag the two layers (Earthquakes and Hurricanes) to your map. They display with the default symbols.**

These two data layers are in the Web Mercator coordinate system. Because they are the first layers added to your map, the coordinate system of your map document is also Web Mercator.

7. **Click Save. Save your map document as C:\EsriPress\GTKWebGIS\Chapter6\natural_disasters.mxd. Click Save to close the Save As window.**

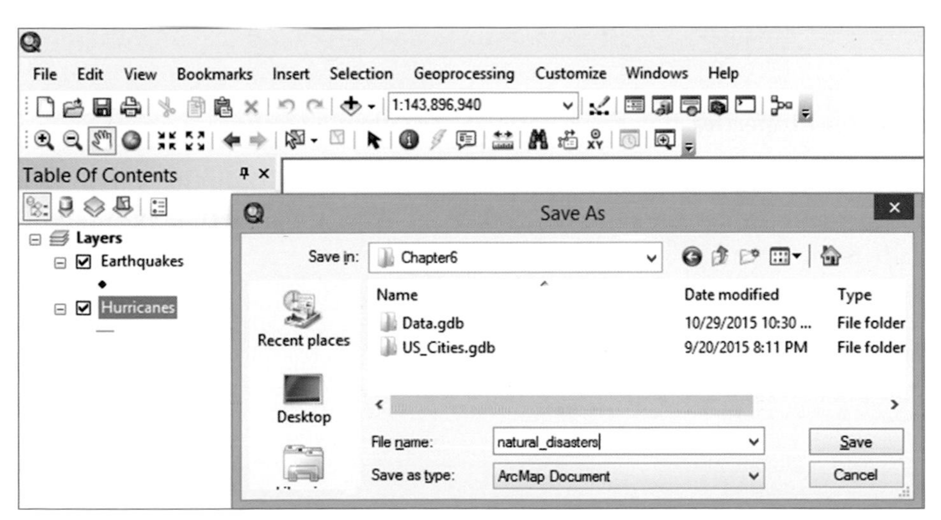

You have now created your map document, and you can publish it as a service. You can enhance the layers' symbols by following steps 8 through 13 to make your map service more intuitive.

8. **In the Table of Contents (TOC), double-click the Earthquakes layer to open the Layer Properties dialog box.**

Alternatively, you can right-click the **Earthquakes** layer and click **Properties** to open the **Layer Properties** dialog box.

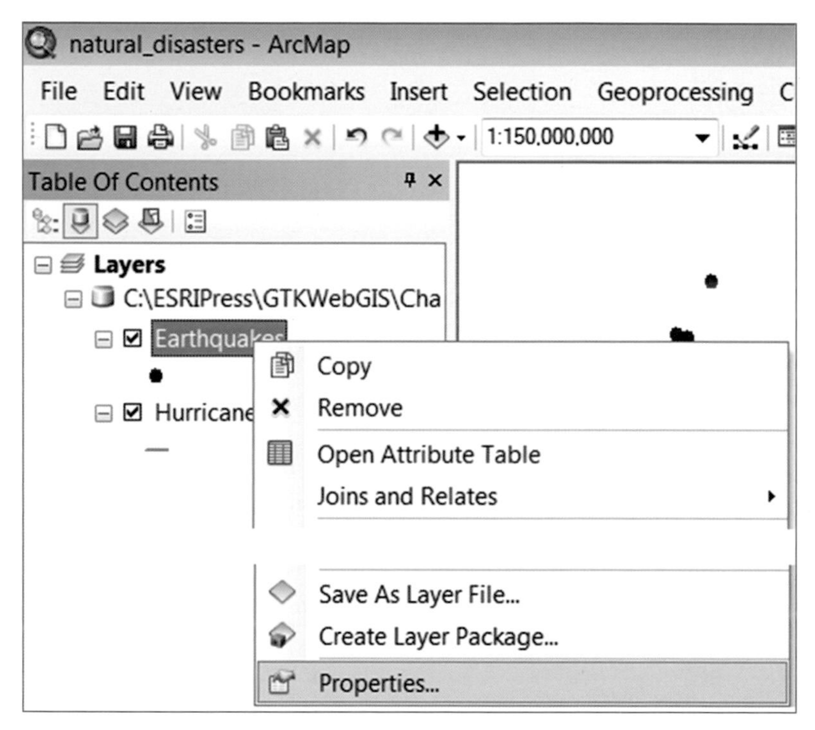

9. **In the Layer Properties dialog box, perform the following tasks:**

   - **Click the Symbology tab, and then click Quantities > Graduated symbols.**
   - **In the Fields group, for Value, click OTHER_MAG1, the earthquake magnitude field.**

10. **In the Classification group, for Classes, set 3. Click Classify, choose the Natural Breaks (Jenks) method, and click OK.**

- **Set Symbol Size from 4 to 12.**
- **Click the Template button to choose an appropriate symbol, such as a black circle filled with red.**

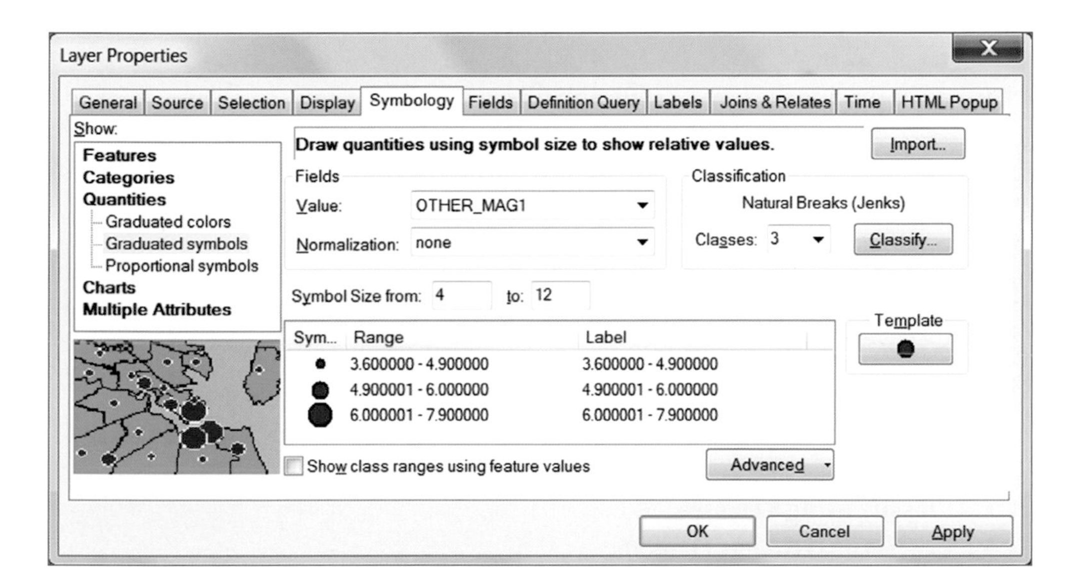

11. **Click the range of the first class break and enter 6. Click the label, and set that as 3.6–6.**

Symbol	Range	Label
●	6	3.600000 - 6.000000
●	6.000001 - 6.000000	6.000001 - 6.000000
●	6.000001 - 7.900000	6.000001 - 7.900000

This setting makes the class breaks and labels more readable.

12. **Click on the range of the second break, and set it as 7. Click on the labels of the second and third breaks, and set them as 6–7 and 7–7.9.**

Symbol	Range	Label
●	3.600000 - 6.000000	3.6 - 6
●	6.000001 - 7.000000	6 - 7
●	7.000001 - 7.900000	7 - 7.9

13. **Click OK to close the Layer Properties dialog box.**

Performing these tasks changes the symbols of the **Earthquakes** layer. The new symbols should appear on the map. Next, change the symbols of the hurricanes.

14. **In the TOC, double-click the Hurricanes layer.**

15. **In the Layer Properties dialog box, perform the following tasks:**

- **Click the Symbology tab, and click Quantities > Graduated colors.**
- **In the Fields group, for Value, click wmo_wind, the hurricane wind-speed field.**
- **Classify using 3 classes.**
- **Use Natural Breaks (Jenks).**
- **Use the green-to-red color ramp.**
- **Change the label of the first class to 10–45.**
- **Change the label of the second class to 45–80.**
- **Change the label of the third class to 80–160.**
- **Click OK to close the Layer Properties dialog box.**

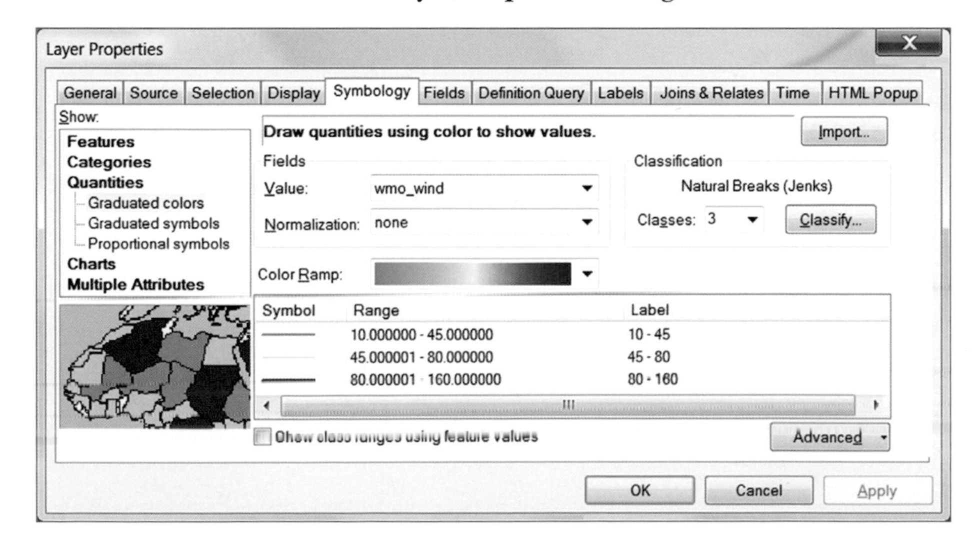

The symbols of the **Hurricanes** layer are updated on the map.

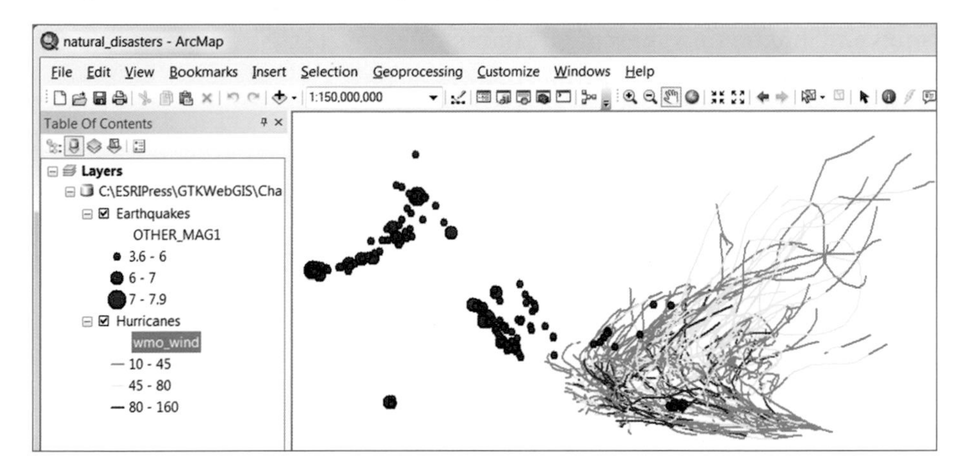

16. **On the ArcMap Standard toolbar, click Save > to save your map document.**

Now you have created a map document with simple but meaningful symbols.

6.3 Enable time on your map layers

You do not have to enable time in your map services. However, if you have date fields or appropriate integer or string fields, enabling time can provide an appealing way to visualize your data.

1. **In the TOC, double-click the Earthquakes layer.**

The **Layer Properties** dialog box appears.

2. **In the Layer Properties dialog box, perform the following tasks:**
    - **Click the Time tab.**
    - **Select the check box for Enable time on this layer.**
    - **Leave Layer Time as Each feature has a single time field.**
    - **For Time Field, select Eq_Date, a date field that stores the time when each earthquake occurred.**

- **Leave Field Format as <Date/Time>.**
- **Click Calculate, which finds the layer time extent (in other words, the minimum and maximum values of Eq_Date).**
- **Click OK to close the Layer Properties dialog box.**

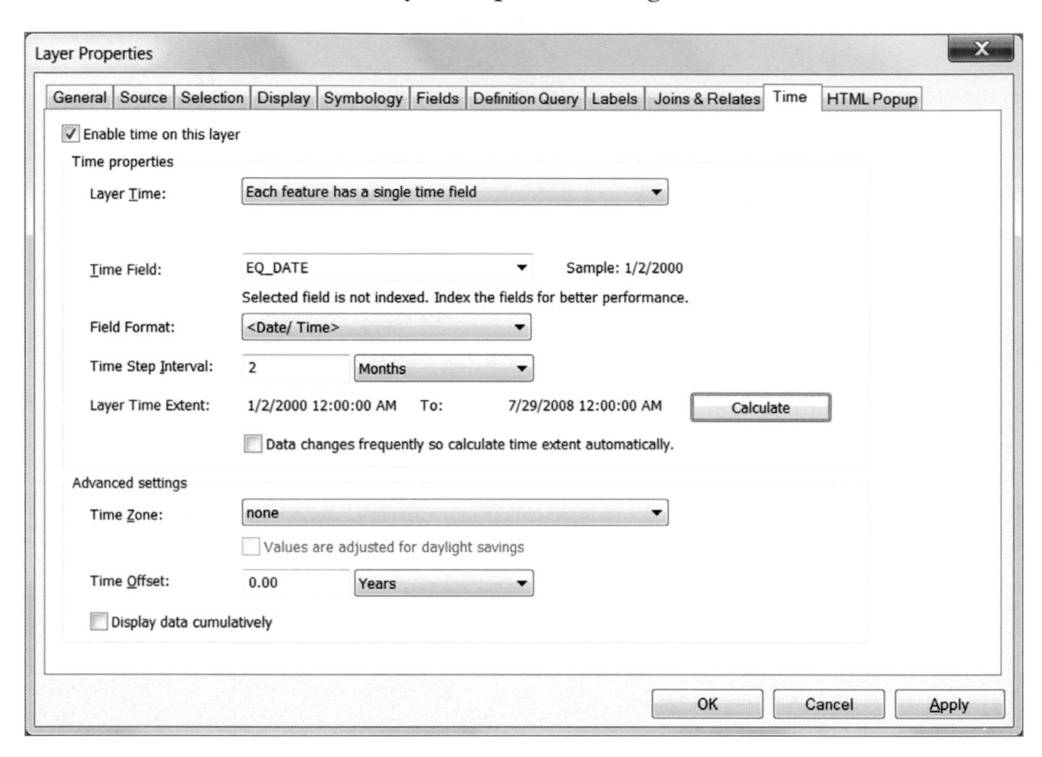

Do not worry about the time step interval because you can configure it later in your web map. Not all data layers have a date field. You can use other fields, such as the **Year** integer field in the **Earthquakes** layer, to enable time on this layer, too.

You have now enabled time on the **Earthquakes** layer. Next, you will enable time on the **Hurricanes** layer.

3. In the **TOC**, double-click the **Hurricanes** layer to open the **Layer Properties** dialog box.

4. In the **Layer Properties** dialog box, perform the following tasks:

- **Click the Time tab.**
- **Select the check box for Enable time on this layer.**
- **Leave Layer Time as Each feature has a single time field.**

- **For Time Field, select H_Date, which is a date field that stores the date of each track of hurricane paths.**
- **Leave Field Format as <Date/Time>.**
- **Click Calculate to find the layer time extent.**
- **Click OK to close the Layer Properties dialog box.**

5. **On the ArcMap toolbar, click the Time Slider button.**

You will play back the animation effects in ArcMap. At the beginning of or during the animation, your data layers may disappear from the map. You will no longer see all of the data because there are no features in these particular time intervals. Your features may not stand out on a white background. Later, you will add a basemap.

- **In the Time Slider window, click the Options button.**
- **In the Time Slider Options dialog box, specify the time window as 2 months, and leave the time window options as Display data for entire time window. Click OK to close the Time Slider Options dialog box.**
- **Click the Play button to play back the animation.**

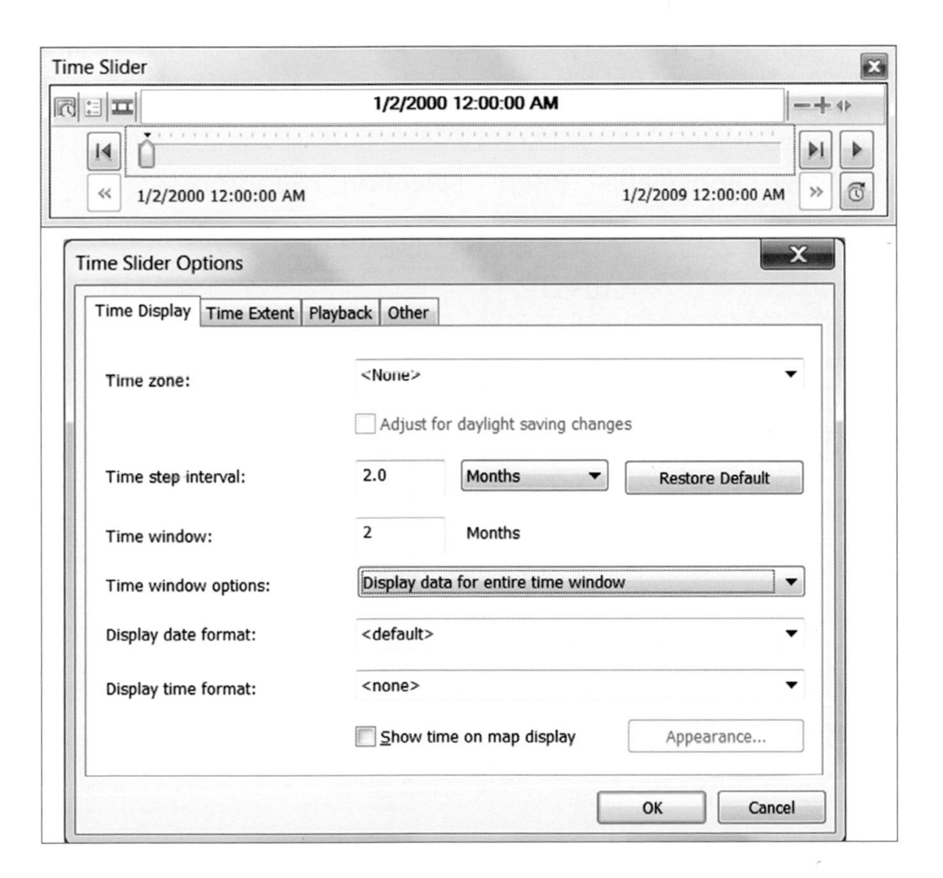

If the playback speed seems too fast, you can change the speed using the **Playback** tab. Do not worry about the **Time Slider** settings. You can configure them later in your web map.

6. **Click Save to save your map document.**

Your map's two layers are now time enabled. This map document can be used to publish not only an ordinary map service but also a time-enabled map service.

You will keep your map open if you are going directly to the next section.

6.4 Publish your map as a service

1. **While you are in ArcMap and have your map document open, from the main menu bar, click File > Share As > Service.**

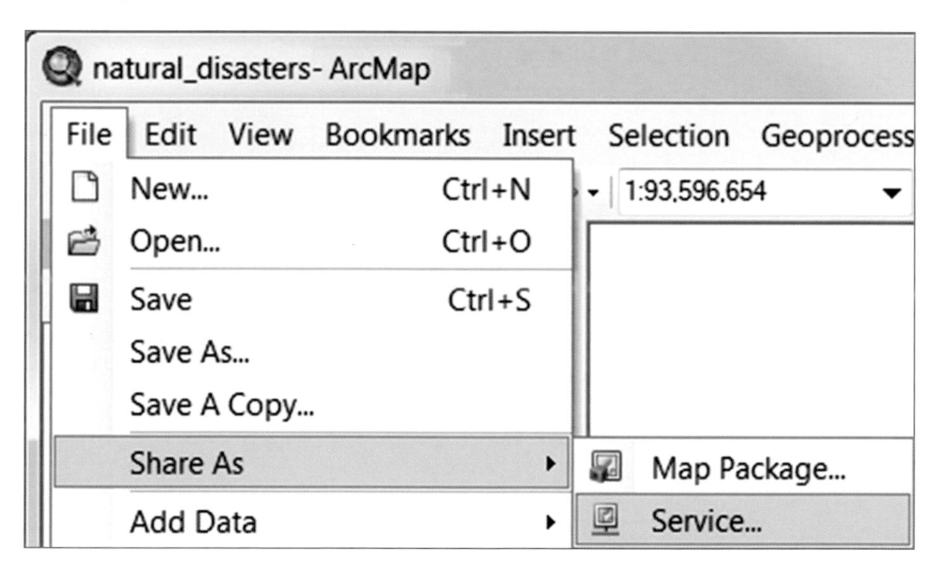

Alternatively, in the **Catalog** window, you can browse to your MXD (map document file extension for ArcMap), right-click it, and click **Share As Service**.

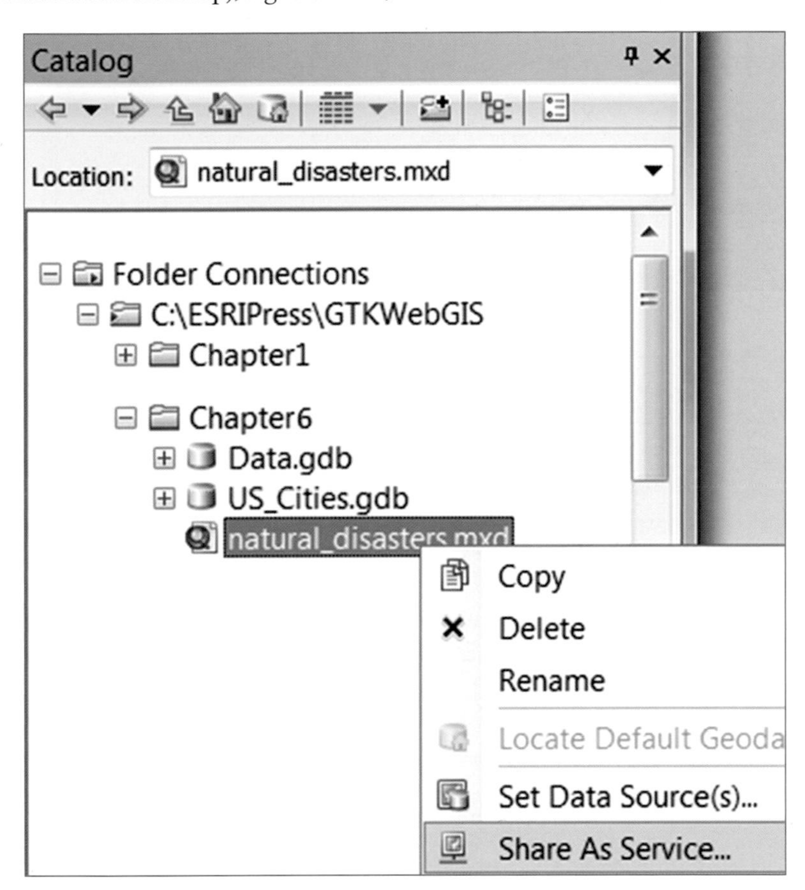

2. In the **Share as Service** dialog box, select **Publish a service** (or **Overwrite an existing service** if you want to overwrite a service you published before), and then click **Next**.

3. **From the Choose a connection list, choose the ArcGIS for Server connection you created in section 6.1. Specify a service name, such as natural_disasters, using only alphanumeric characters and underscores, and click Next.**

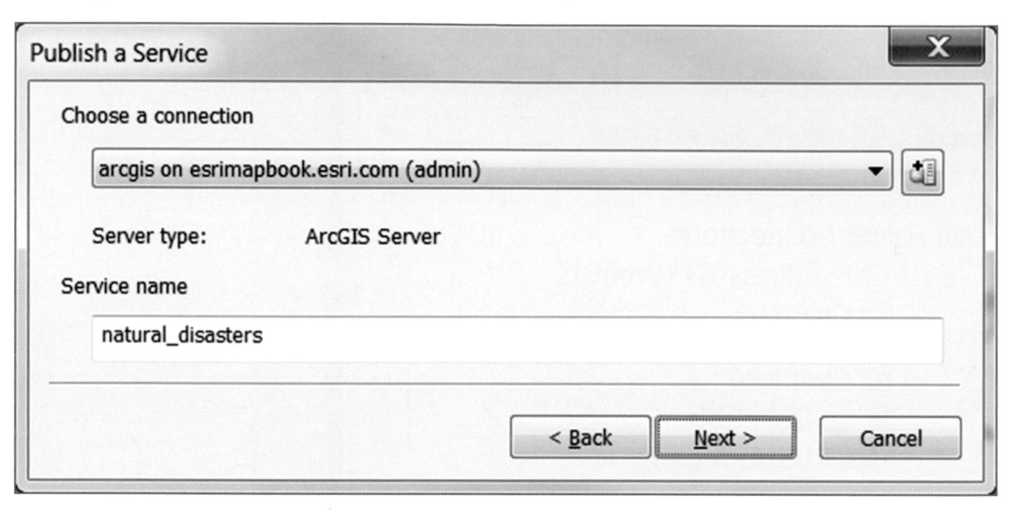

Notes:

The combination of service names and folders should be unique to a server. In a classroom where many students share one server, either add your name as a suffix to your service name or place the service in your own folder.

4. **Select Create new folder, specify a folder name (for example, your name or your project name), and click Continue.**

By default, ArcGIS for Server publishes services to its root folder. Services can be organized into subfolders. You can name your subfolder after your project name, your organization name, or your own name.

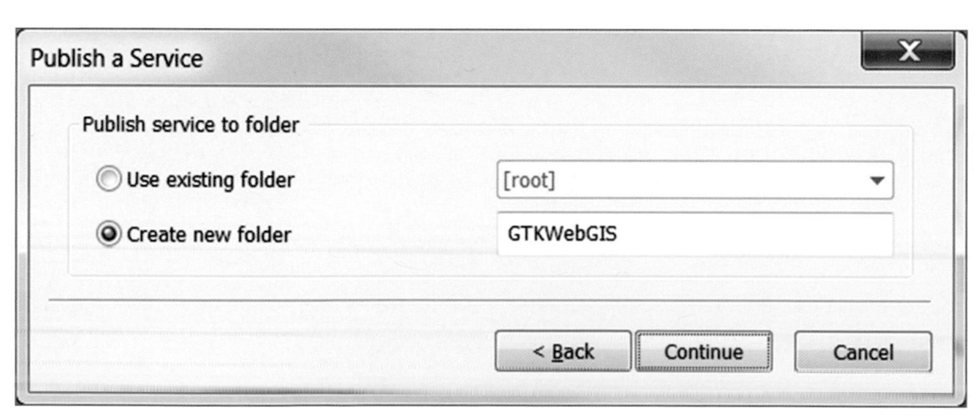

5. **At the top of the Service Editor, click the Analyze button ✓.**

Your map document is analyzed to identify errors or warnings.

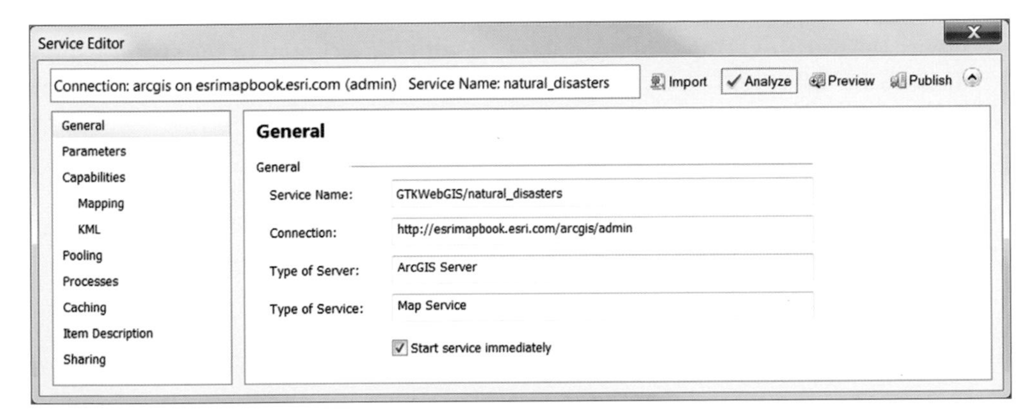

- To create a larger viewing area when configuring your map service, click the **Collapse** button next to the **Publish** button at the top of the **Service Editor**.

6. **The results of your analysis appear in the Prepare window.**

Three types of results are possible. You can right-click a result to get help on how to address a problem.

- **Errors (⊗):** These issues must be fixed before you can publish your map document as a service.
- **Warnings (⚠):** These issues may affect performance, appearance, and data access.
- **Messages (ⓘ):** These messages suggest methods and best practices to optimize your GIS resource for deployment on the server.

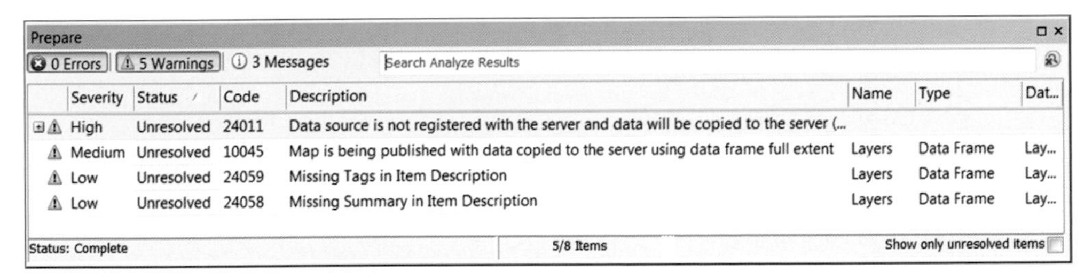

The **Prepare** window should not indicate any errors. However, you will find a few common warnings and messages.

7. Ignore the warning, **Data source is not registered with the server and data will be copied to the server** (see the discussion earlier in this chapter on making your data accessible to ArcGIS for Server).

8. Ignore the warning, **Map is being published with data copied to the server using data frame full extent.** You want the data copied to the server—all data within the full extent of the data frame will copy to the server.

9. For **Missing Tags in Item Description**, double-click the message in the **Prepare** window, and then fill in the **Tags** on the **Item Description** page with the following information: **natural disasters**, **earthquakes**, and **hurricanes.**

10. For **Missing Summary in Item Description**, double-click the message in the Prepare window, and fill in **Summary** on the **Item Description** page with the following information: **This map shows the historical earthquakes and hurricanes from 2000 to 2008.**

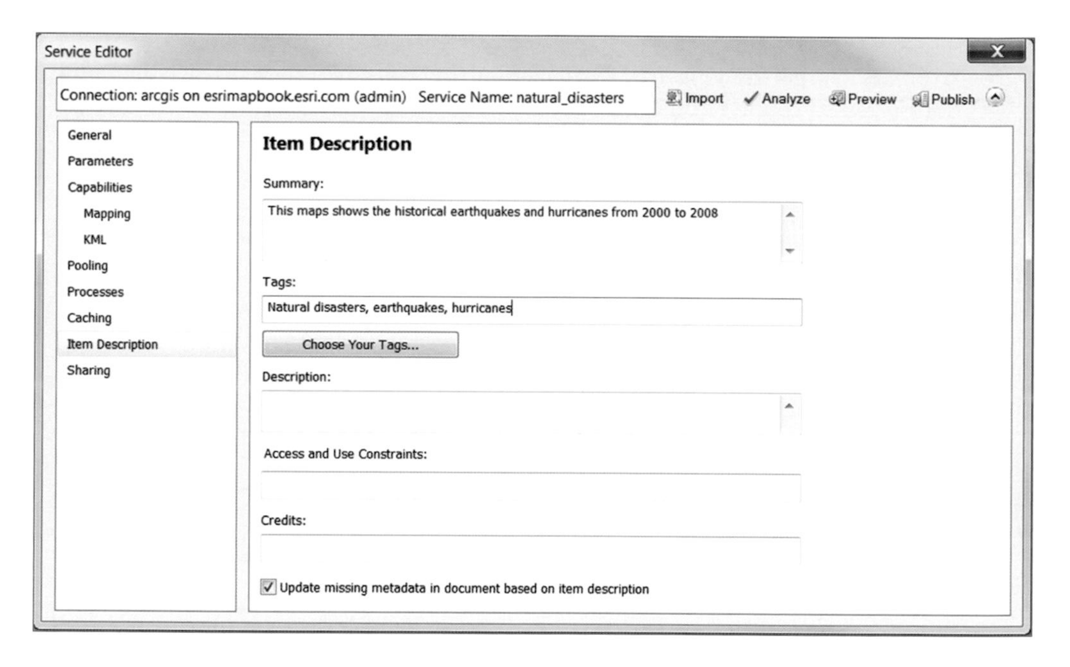

11. Ignore all other warnings.

You can set other properties in the **Service Editor**. For example, you can turn on the following additional capabilities:

- **KML** (Keyhole Markup Language) so that KML clients can use your service
- **OGC WMS** (Open Geospatial Consortium Web Map Service) so that WMS clients can use your service

12. **In the Service Editor, click the Caching link on the left, and study the options to create a cached map service.**

For **Draw this map service**, the selected option is **Dynamically from the data**. This option will create a dynamic map service. You also have the option to select **Using tiles from a cache** to create a cached map service. Notice that the **Estimated Cache Size** is pretty large. The size changes as you change the levels of details.

Note: You will leave the option **Dynamically from the data** selected. Otherwise, you will not be able to animate your map service in the steps later.

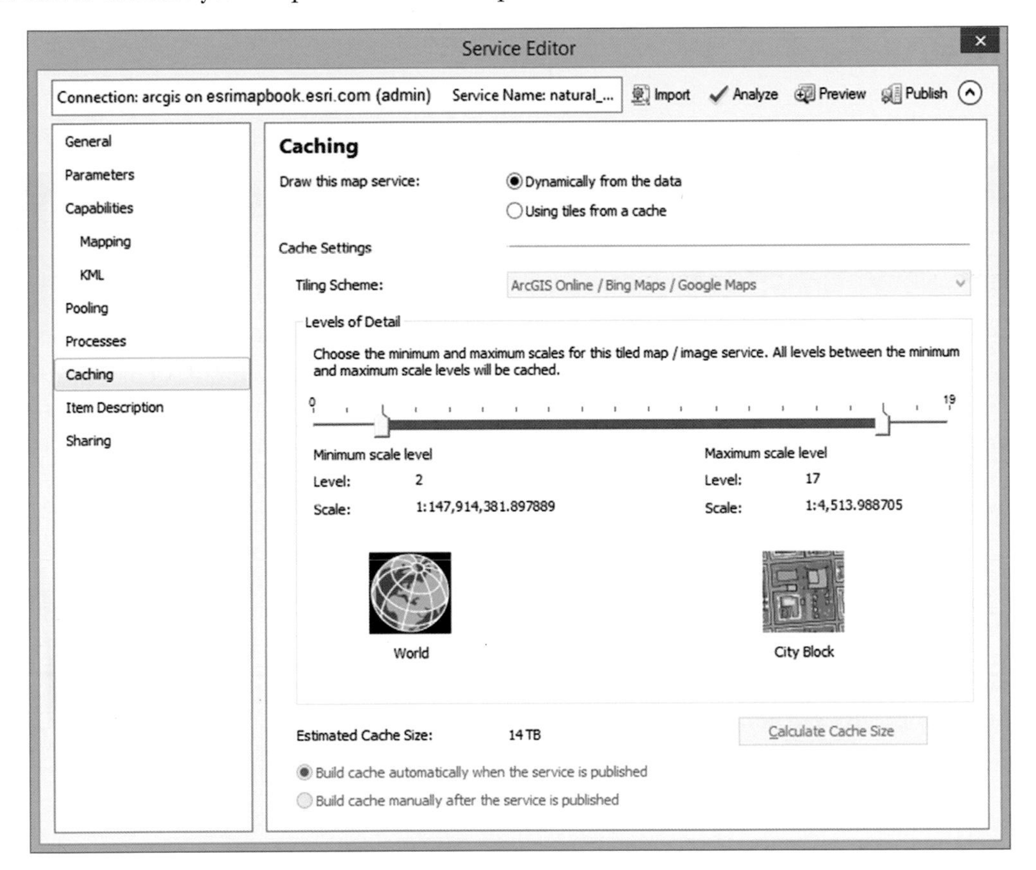

13. **In the upper-right corner of the Service Editor, click Publish.**

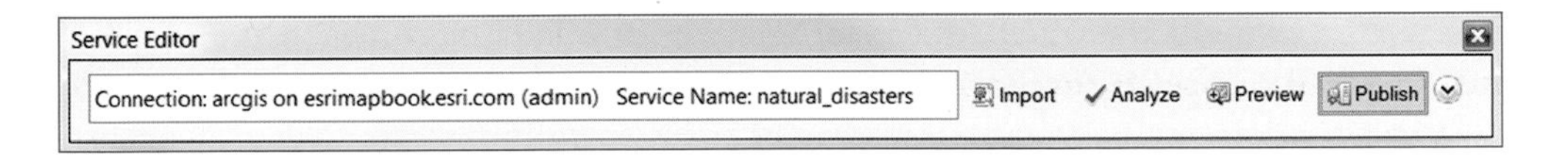

1
2
3
4
5
6
7
8
9
10

14. In the Copying Data to Server dialog box, click OK.

ArcMap creates and copies an SD file with the data to your ArcGIS for Server. The size of your data and your network bandwidth will affect the publishing time.

15. Note the Service Publishing Result message that your map service published successfully. Click OK to close the message.

6.5 Explore your service using Services Directory

Where is my service? What does the service look like? Does the service work correctly? You may have such questions after publishing your map service. You can answer your questions using the ArcGIS for Server Services Directory, ArcGIS for Server Manager, or ArcCatalog. Here, you will use Services Directory.

1. If your ArcGIS for Server is federated with Portal for ArcGIS, the service you published is private by default; follow steps 1 through 4 in section 6.8 to share the service with everyone before continuing to step 2. If your ArcGIS for Server is not federated, continue to step 2.

If you do not know whether your ArcGIS for Server is federated with Portal for ArcGIS, ask your instructor or system administrator.

2. Determine your Services Directory web address.

The URL should have the following pattern: **<http or https>://<server name>:<port_number>/<instance name>/rest/services.** You will use this URL to open the Services Directory for a given GIS server. The <instance name> in the typical ArcGIS for Server installation is **arcgis.** If you are using a computer on which ArcGIS for Server is installed, you can go to **Start** > **All Programs** > **ArcGIS** > **ArcGIS for Server** > **Services Directory**, which will open ArcGIS Services Directory in a web browser. Contact your instructor or system administrator if you are uncertain about the URL. You will write the URL to your Services Directory here:

____

Note: You may get a message saying your connection is not private or is untrusted; this message may appear because your ArcGIS for Server is configured to use https and a self-signed SSL certificate. You may still proceed to the site. If you have questions, you can ask your instructor before proceeding.

3. **Start a web browser, and type the URL that you previously inserted on the blank line to go to your Services Directory.**

Alternatively, if you are using Windows 7 on your server machine, you can go to **Start** > **All Programs** > **ArcGIS** > **ArcGIS for Server** > **Services Directory**. If you are using Windows 8.x, you can go to Apps and find the Services Directory.

The home page of your ArcGIS REST Services Directory appears, listing all the folders and services in the root directory. ArcGIS REST Services Directory is a component of ArcGIS for Server and displays information about the services available at the server.

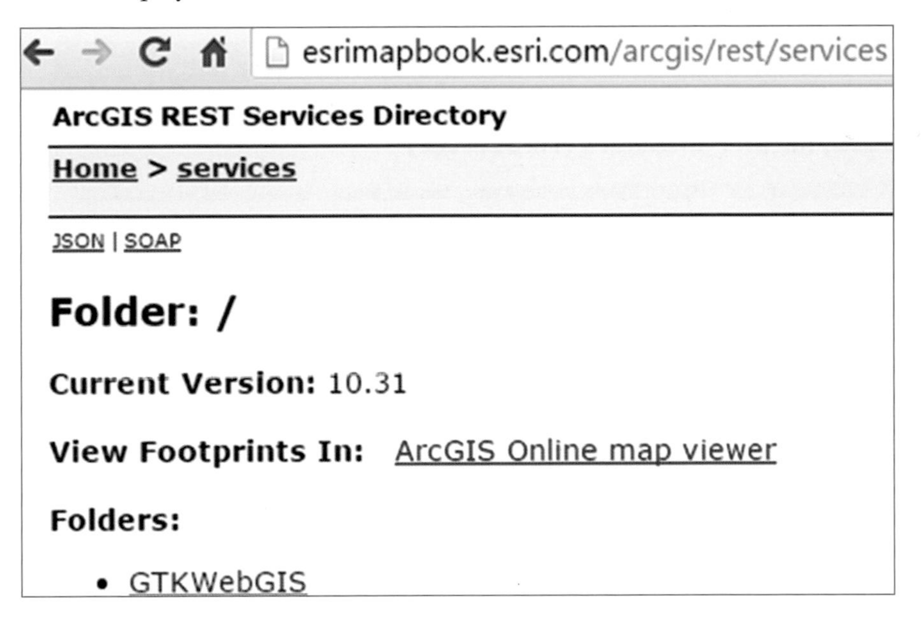

4. **Click the folder name you specified when publishing your service.**

The name of the service you published should appear along with the **MapServer** service type. These results confirm that your service has been published.

Services:

- GTKWebGIS/natural_disasters (MapServer)

5. Click your map service. The metadata page of your map service appears, displaying the following links and information:

- **View In:** You will use these links to preview your map service and test whether your service works. You will try them in the steps following this step.
- **Layers:** This link displays the names of the layers in the map service.
- **Spatial Reference:** The reference **102100** confirms that your map service is in Web Mercator. Another often-used spatial reference is **4326**, which is Geographic Coordinate System in WGS1984.
- **Time Info:** This information confirms that your map service has been time enabled.
- **Supported Operations:** Your web clients can ask your map to perform the following operations: **Export Map, Identify,** and **Find.**

ArcGIS REST Services Directory

Home > services > GTKWebGIS > natural_disasters (MapServer)

JSON | SOAP

GTKWebGIS/natural_disasters (MapServer)

View In: ArcGIS JavaScript ArcGIS Online map viewer Google Earth ArcMap ArcGIS Explorer

View Footprint In: ArcGIS Online map viewer

Service Description:

Map Name: Layers

Legend

All Layers and Tables

Layers:

- Earthquakes (0)
- Hurricanes (1)

Description:

Copyright Text:

Spatial Reference: 102100 (3857)

Time Info:

Time Extent:
[2000/01/02 00:00:00 UTC, 2008/11/14 00:00:00 UTC]
Time Reference: N/A

Units: esriMeters

Supported Image Format Types: PNG32,PNG24,PNG,JPG,DIB,TIFF,EMF,PS,PDF,GIF,SVG,SVGZ,BMP

Supported Operations: Export Map Identify Find Return Updates Generate KML

Note: The URL of this page—**http://esrimapbook.esri.com/arcgis/rest/services/GTKWebGIS/natural_disasters/MapServer**—is the REST (Representational State Transfer), a popular type of web service interface and endpoint to your map service. You will use this service via this URL. For example, you will specify this URL when you add this service to your ArcGIS Online map viewer manually or when you develop custom apps using JavaScript and ArcGIS Runtime SDKs. You will write the address for your map service REST endpoint here and use this URL address later in this tutorial. Remembering the pattern of this URL will help you later.

---

6. **In the View In links, click ArcGIS JavaScript. Your map service displays in a new window or tab. Zoom in and out of your map.**

View In: ArcGIS JavaScript ArcGIS Online map viewer Google Earth ArcMap ArcGIS Explorer

This step confirms that your map service is not only published but also working.

7. **Return to your map service page, look for the View In links, and click ArcGIS Online map viewer.**

Your map service displays in the ArcGIS Online map viewer. If the service overlays correctly on the ArcGIS Online basemap, the coordinate system of your data and map service is correct. Otherwise, the coordinate system may be wrong. Refer to the **Questions and Answers** section at the end of this chapter. When your map service is time enabled, a time slider should appear below the map.

8. **Click the Play button to see the temporal patterns of the earthquakes and hurricanes.**

You will learn how to configure the time slider in the next section.

9. **Go back to the map service page in ArcGIS REST Services Directory, and under Layers, click Earthquakes.**

Layers:

- Earthquakes (0)
- Hurricanes (1)

The metadata page of this layer appears. This page lists the following information:

- **Geometry Type** tells you it is a point layer.
- **Drawing Info** lists the layer's symbology.
- **Time Info** tells you the layer has been time enabled.
- **Fields** are the attributes exposed from this layer and the types of fields.

1
2
3
4
5
6
7
8
9
10

- **Supported Operations** (including **Query**) tell what operations web clients can ask your layer to do.

Note: The URL of this page (for example, **http://myserver.esri.com/arcgis/rest/services/pinde/natural_disasters/MapServer/0**) is the layer's REST endpoint or the layer's URL. This endpoint or URL is essentially your map service URL appended with a number. This number starts with 0 for the first layer, 1 for the second layer, and so on.

You will use this URL when you use this layer with various web clients. Remember this pattern. Write the URL for this layer in your map service here:

____

10. **Click your web browser's back button to return to your map service page in the ArcGIS REST Services Directory.**

Alternatively, in the navigation links at the top of the page, you can click your map server's name.

11. **On your map service page, under Layers, click Hurricanes. Review the layer's metadata page (similar to the page for the Earthquakes layer).**

In this section, you explored your map service using the ArcGIS REST Services Directory and confirmed that your map service is published, running, in the correct coordinate system, and time enabled.

You can do more with the directory. Later, you will learn how the directory can help you understand and explore the REST API (application programming interface) for use with your JavaScript and mobile clients.

6.6 Create a time-enabled web map and web app

1. **In a web browser, navigate to ArcGIS Online (http://www.arcgis.com) or your Portal for ArcGIS, sign in, and go to the map viewer.**

2. **On the map viewer menu bar, click Add, and then click Add Layer from Web.**

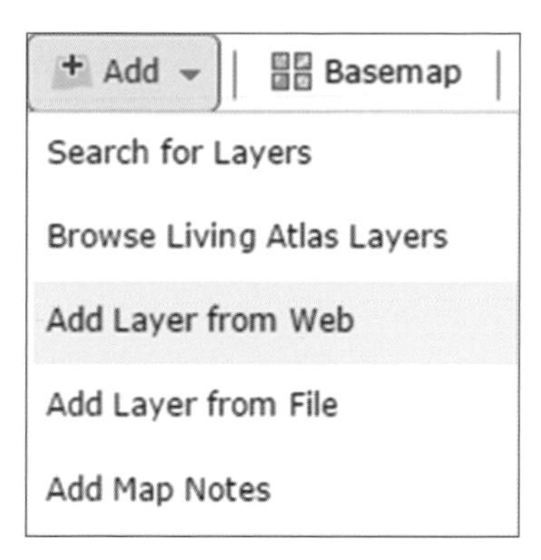

3. **In the Add Layer from Web window, specify the URL of your map service, and then click Add Layer.**

Tip:

- Instead of manually typing the URL, go to your map service page in the **Services Directory**, copy the URL from the address bar, and then paste it in the URL box.

Your map service displays in the ArcGIS Online map viewer. The view you see is similar to the view that appeared when you clicked the **View In ArcGIS Online map viewer** link in the previous section.

4. **In the Contents pane, click the natural disasters layer to expand its sublayers. Point to the Earthquakes layer, click the More Options button, and choose Enable Pop-up.**

Contents
natural disasters
Earthquakes
Hurricanes
Enable Pop-up
Configure Pop-up

This choice enables pop-up windows in the **Earthquakes** layer. If you click an earthquake on the map, the default pop-up window appears.

5. **Configure the pop-up window to enhance the window for the Earthquakes layer using the skills you learned earlier in chapter 2.**

6. **Repeat the preceding two steps to enable and configure the pop-up for the Hurricanes layer.**

7. **On the menu bar, click Save > Save to save your web map.**

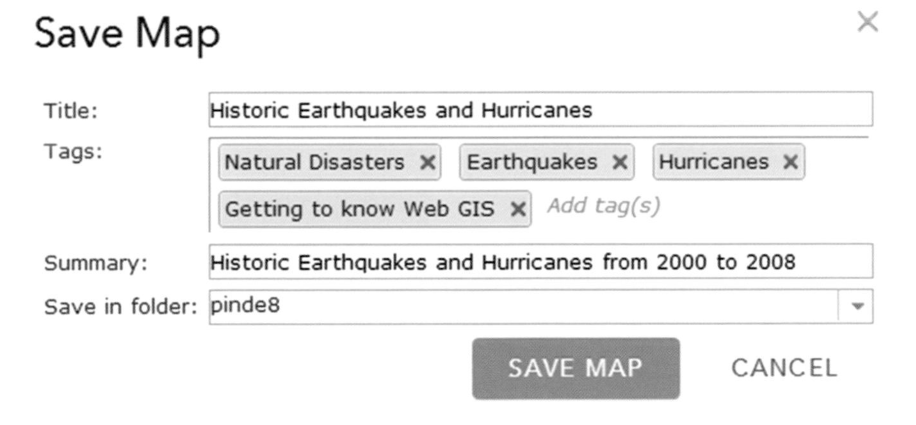

ArcGIS Online recognizes that your map service is time enabled and displays a time slider. The slider allows you to configure the time animation playback speed, time span, time window, and slider labels.

8. **To the right of the slider, click the Configure button.**

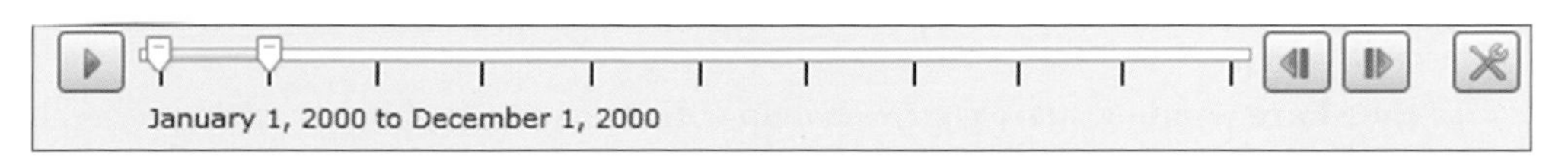

9. **In the Time Settings window, click Show advanced options.**

10. **Under Time Display, set display data in 6 Month intervals. Leave As time passes as only display the data in the current time interval selected. Click OK to close the Time Settings window.**

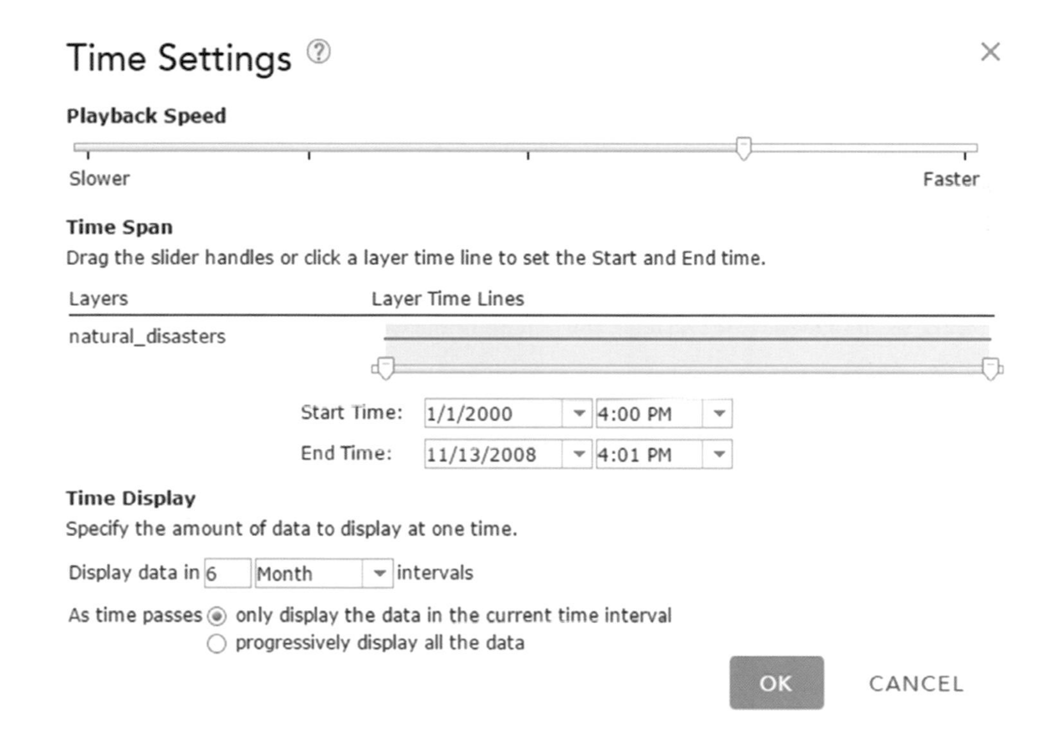

11. **On the Time Slider, click the Play button to test whether the settings play back correctly.**

12. **On the menu bar, click Save to save your web map.**

You have used your map service to create a web map and configure its time properties. Now you are ready to use this map to build a web app. You learned how to create a web app using ArcGIS Online templates in previous chapters. Therefore, this chapter does not contain as much detail. If questions arise, refer to those chapters.

1
2
3
4
5
6
7
8
9
10

13. **While web map is open in the map viewer, on the map viewer menu bar, click the Share button.**

14. **In the Share window, share your web map with everyone (public) or with only certain groups or your organization.**

15. **Click Create a Web App.**

This action brings up the ArcGIS Online configurable apps gallery. You will need to choose an app that supports time animation, such as the **Time Aware** app. You can also create your web app using Web AppBuilder for ArcGIS by using the **Time** widget to animate your map. You learned Web AppBuilder for ArcGIS in another chapter. However, in these steps, you will use the **Time Aware** app.

16. **In the configurable apps gallery, find the Time Aware app. Click Time Aware, and click Create App.**

If you are using an older version of Portal for ArcGIS, the user interface may be different. For example, instead of the **Create App** button, you may need to click **Publish**.

If you want to change the title and logo, you can fill in the web app information on the pane at right.

17. **Click Save and then Done.**

The step you just performed will lead you to the details page of your new web app.

Optionally, you can click the **Edit** button to update the item details or click the **Configure App** button to change your app configuration.

18. **Open your web app by clicking the thumbnail image, or click Open > View Application.**

The URL of this page is the app URL that you can share with your instructor and others so they can use your app.

Note:

- If your server uses https and a self-signed SSL certificate, your users may run into a warning message saying the connection is not private or is untrusted. For a production server (and if the server must use https), you should configure the server with an official SSL certificate.
- If your ArcGIS for Server is installed on a private server within your organization and is accessible only through the internal network, the app you created using your map service will work only inside your organization. Users outside your organization network cannot access your app. See the **Questions and Answers** section for more information.

By default, your web app opens with a time slider. When you click the **Play** button on the time slider, the temporal patterns of these disasters unfold. Hurricanes occur mostly along the Southeast Coast of the United States, mostly in the second half of the year. Thus, the slider reveals a pattern that you would not otherwise easily see.

6.7 Administer your web services

This section is not required for you to create web apps but is useful to learn so you can debug, troubleshoot, and maintain your services.

1. **Perform the following tasks to start, stop, restart, or delete your services in the ArcMap Catalog window:**
    - **Click GIS Servers.**
    - **Click your ArcGIS for Server connection.**
    - **Click your folder (if applicable).**
    - **Find your map service.**
    - **Right-click your service. You see the options to stop, start, restart, or delete your map service.**

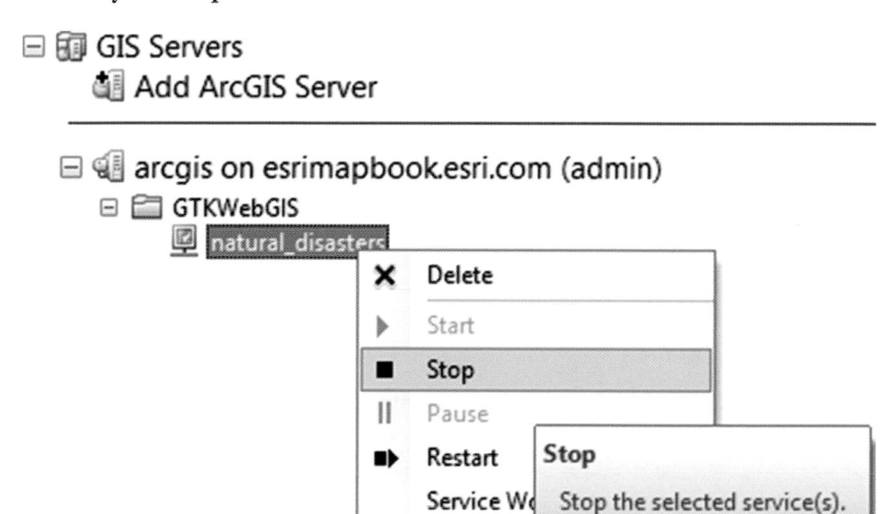

2. Perform the following tasks to start, stop, or delete your services in Server Manager:

   - **Start a web browser, and navigate to your server manager (the URL pattern is http://yourServerName:6080/arcgis/manager, https://yourServerName:6080/arcgis/manager.**
   - **http://yourServerName/arcgis/manager, or https://yourServerName/arcgis/manager).**
   - **Sign in with an ArcGIS for Server administrator or publisher account.**
   - At the top, click **Services.**
   - **On the main menu bar, click the Manage Services button.**
   - **On the left side of the page, locate your folder, and click the folder (if applicable).**
   - **On the right side of the page, find your service, and click the Stop, Start, or Delete button to stop, start, or delete the service.**

3. To examine the log files of your ArcGIS for Server:

   - **While you are logged in to ArcGIS for Server Manager, near the upper-right corner of the page, click the Logs button.**
   - **Filter the logs by log level and age.**
   - **Click the Query button to retrieve and display the logs.**

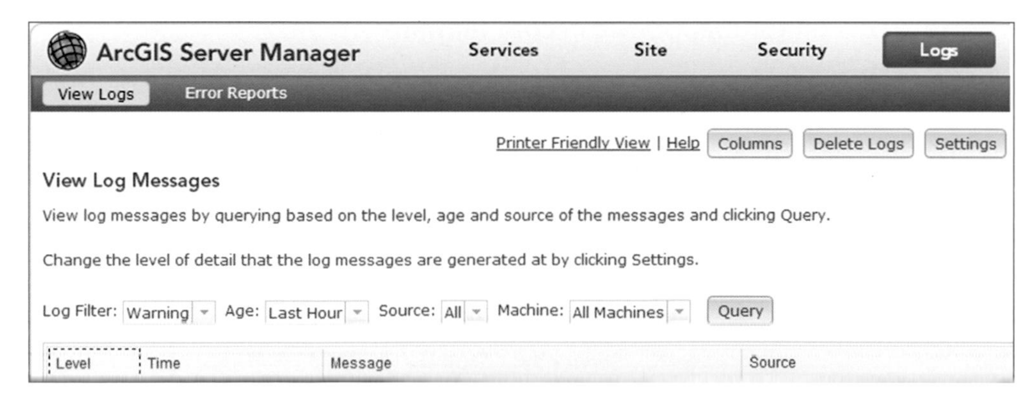

By default, the log level is set at the **Warning** level. If you want a detailed log to help you debug your server problems, you can click the **Settings** button and set the log level to **Fine**, **Verbose**, or **Debug**. Be aware that these detailed log levels will slow down your server performance. If you use a detailed log level, remember afterward to change the level back to **Warning**.

6.8 Manage and use services in Portal for ArcGIS (optional)

This optional section requires a federated Portal for ArcGIS and ArcGIS for Server. You will skip this section if your service is not federated. Otherwise, you can complete the following steps:

1. **Start a web browser, go to your Portal for ArcGIS homepage, and sign in with the same account you used to connect to your ArcGIS for Server in section 6.1.**

2. **Ask your instructor or system administrator if you do not know the URL of your Portal.**

3. **Click My Content, and find the map service you just published in the content list.**

+ Add Item ▾ Create ▾ Share × Delete Move ▾ Change Owner

	Title	Type	▾ Modified	Shared
☐	natural_disasters	Map Image Layer	Oct 27, 2015	Not Shared

4. **Click your map service to see its item details.**

5. **On the item details page, click Share to share the map service with everyone.**

Portal for GIS conveniently manages the access to your services if your ArcGIS for Server and Portal for ArcGIS are federated. Optionally, you can click the **Edit** button to enhance the metadata for your service.

6. **Under the Map Contents section, find the URL of your map service, and click the service.**

Map Contents

natural_disasters

http://esrimapbook.esri.com/arcgis/rest/services/GTKWebGIS/natural_disasters/MapServer

Clicking the service displays the ArcGIS Service Directory page, which you saw in section 6.5.

7. **From this page, return to the item details page of the map service, click the Open button, and choose Add layer to new map.**

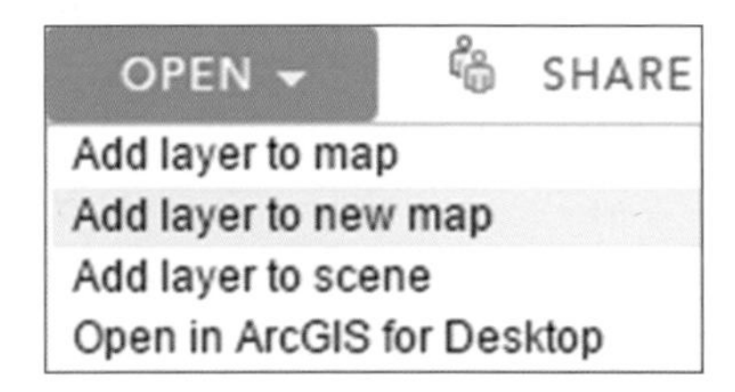

This step leads you to the map viewer, where you can create web maps and web apps. You created your web maps/apps in section 6.6 and can stop right here.

This optional section showed you that you can manage the access to your service and easily create and share web maps and web apps in Portal for ArcGIS.

QUESTIONS AND ANSWERS

1. **Why should I project my data into Web Mercator, and how would I project that data?**

   Answer: Unless your study area lies within the northern or southern polar regions or faces some special requirements, your map service will most likely be displayed on top of other basemaps, such as the basemaps in ArcGIS Online. If your data is not in the northern or southern polar regions and you don't face special requirements, you should project your data to Web Mercator first so that ArcGIS for Server will not have to project your data dynamically (on the fly). Projecting your data to Web Mercator can improve the performance of both the map service and the web app.

In the ArcGIS **Catalog** window, expand **Toolboxes** > **System Toolboxes** > **Data Management Tools.tbx** > **Projections and Transformations** > **Project** if you want to project a vector layer, or **Raster** if you want to project a raster layer. Then double-click **Project** (or **Project Raster**) to open the **Project** dialog box.

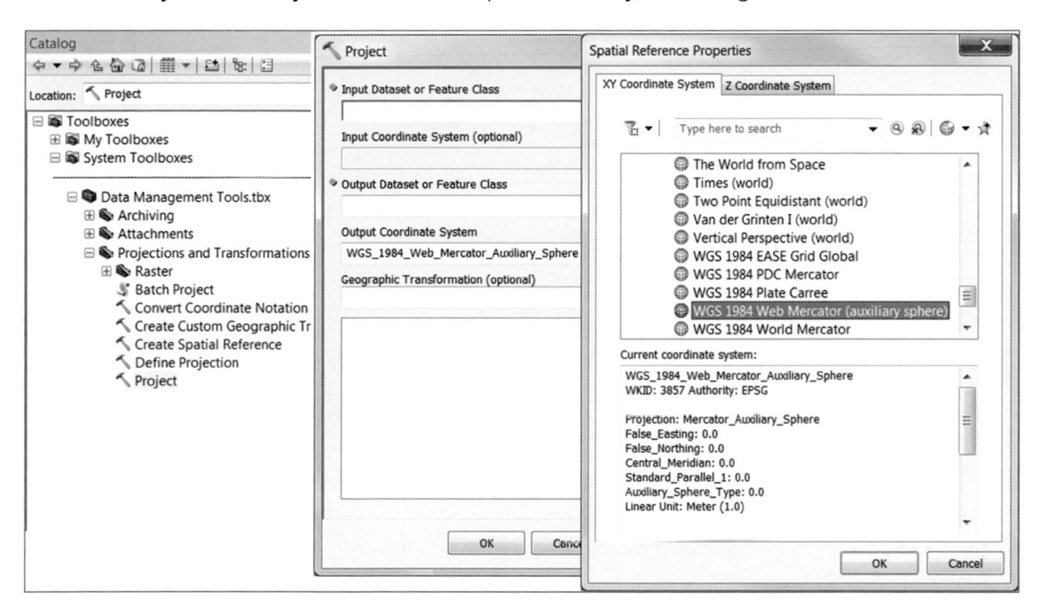

In the **Project** dialog box, specify the input and output feature classes. For **Output Coordinate System**, click **Projected Coordinate Systems** > **World** > **WGS 1984 Web Mercator (auxiliary sphere)**. Most online maps use this Web Mercator coordinate system.

2. **When I display my map in the ArcGIS Online map viewer, the map appears off the western coastline of Africa. This location is incorrect. What is wrong?**

   Answer: You selected the wrong coordinate system for the data layer.

   Your data is almost certainly in latitude, longitude (degrees). Either you are missing coordinate system information or you have defined (not projected) the coordinate system as Web Mercator by mistake.

   To correct this problem, open the ArcGIS Catalog window, find this data layer, right-click the layer, and click **Properties**. In the **Properties** dialog box, click the **XY Coordinate System** tab, click **Geographic Coordinate Systems** > **World** > **WGS 1984**, and click **OK**. You can then correctly project this layer to Web Mercator, create a new map document with this new layer, publish a new service, or overwrite the previous one.

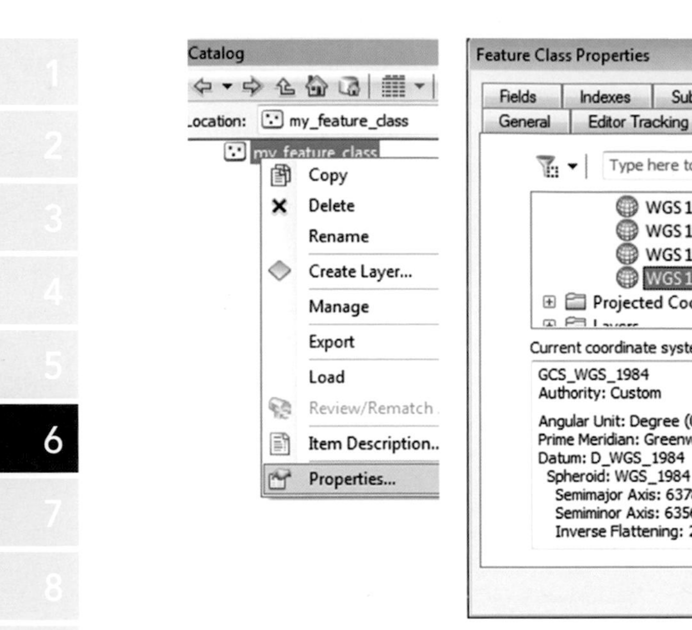

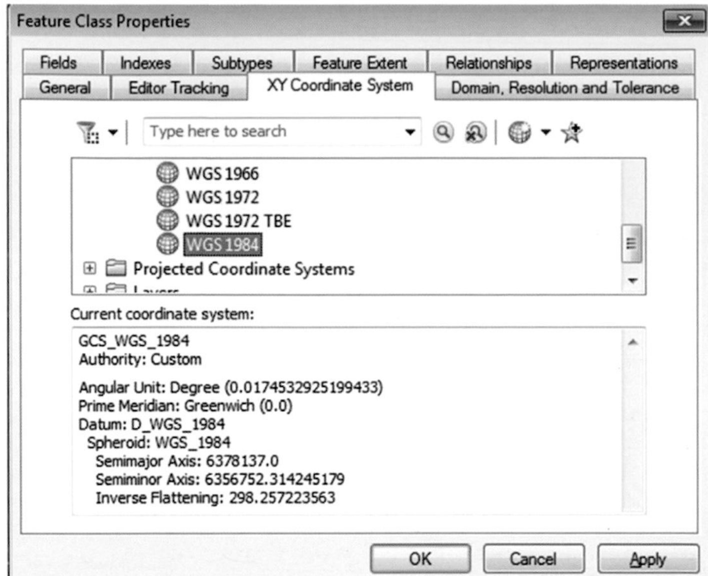

3. **After I published my map service, I updated my map document. Will my map service update automatically?**

   Answer: No.

   During the process of publishing a map service, ArcGIS for Server optimizes your MXD by converting it to an SD file and then storing the file in its own workspace. From then on, your map service uses this SD file and therefore cannot use the changes you made to your MXD file. You will need to publish your MXD again as a new service, or you can overwrite the previous service.

4. **After publishing my map service, I updated my GIS data using my ArcGIS for Desktop. Will my updates appear in my map service automatically?**

   Answer: It depends.

   By default, ArcGIS for Server copies your data and relies on this copy. ArcGIS for Server cannot read any updates you have made to the original copy of your data.

   To make the updates appear in your map service automatically, you must register a data store and configure your ArcGIS for Desktop and ArcGIS for Server to share the same database.

5. **I added my map service to my web map but cannot change the symbols in my Earthquake and Hurricane layers. Can I change the symbols?**

   Answer: When you added the map service to your web map, you used the map image layer approach. In this case, the map draws on the server side. Your browser simply receives a map picture from your GIS server and therefore cannot change the symbols.

   To fix this issue, add individual layers of a map service to ArcGIS Online. In the map viewer, click **Add** > **Add Layer from Web**, and specify the URL for the layer (for example, **http://esrimapbook.esri.com/arcgis/rest/services/GTKWebGIS/natural_disasters/MapServer/0**). The feature layer approach allows you to change the layer symbols.

6. **One of the layers in my map service contains more than a thousand features. I tried to add this individual layer to my web map, but not all the feature layers display. Why?**

   Answer: You added your layer using the feature layer approach. To avoid loading too many features into your browser, ArcGIS for Server returns up to one thousand features by default. You can modify this limit in the service properties of the service, as illustrated in the figure. However, be careful not to return more features than your end users' browsers can handle. Using the dynamic map service approach (in other words, use the map service URL, for example, **http://esrimapbook.esri.com/arcgis/rest/services/GTKWebGIS/natural_disasters/MapServer**, when adding layer from the web) can solve this problem.

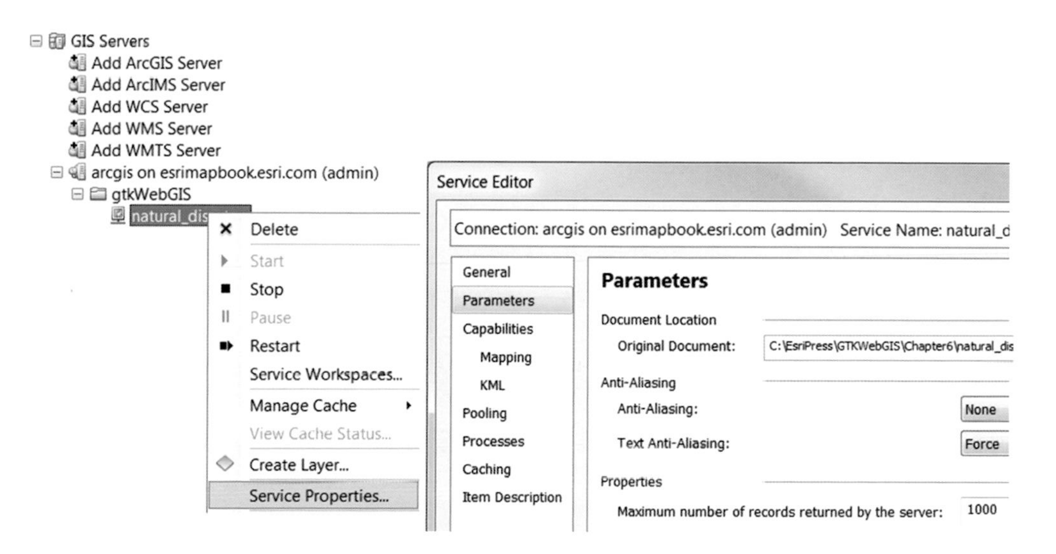

7. **I created a web app using my own map service. However, I can view this web app only when I am using the computers at my university campus. Why?**

   Answer: ArcGIS for Server is installed on a private server at your university and is accessible only through the internal network. For some situations, this limitation means enhanced security. For other situations, this limitation can be inconvenient.

   If you want to make your map service and app accessible off campus, try one of the following solutions:

- Use VPN (Virtual Private Network): If your university allows VPN access, you can use VPN to get into your university network even when you are off campus.
- Use a reverse proxy: A reverse proxy is a type of proxy server that retrieves resources on behalf of a client. Your university system administrator can set this proxy up.
- Make your GIS server a public server and allow users to access the server via HTTP or HTTPS from the external network.

ASSIGNMENTS

Choose one of the following two assignments. The first assignment provides the data for your assignment; the second assignment encourages you to use your own data.

Assignment 6A: Create a web app to animate the population change of main US cities.

Data: C:\EsrIPress\GTKWebGIS\Chapter6\US_Cities.gdb, a file geodatabase containing a US_Cities feature class. This layer presents the historic population of the one hundred most populated US cities between 1790 and 2000.

Requirements:

- Publish a time-enabled map service.
- Create a web app to animate population change in these cities.

What to submit: **Send an email to** your instructor with the subject line **Web GIS Assignment 6A: Your name**, and include the following information:

- The URL to your map service REST endpoint
- The URL to your map service item details page in your Portal for ArcGIS
- The URL to your web app

Tips:

- Attributes do not include date fields. However, you can use the **YEAR** field to enable time in the layer.
- To display the magnitude of population change clearly over time, use proportional symbols on the **Population** attribute field (illustrated in the following figure). Using these symbols allows users to see the how the symbol sizes of these cities increase or decrease over time.
- Configure your web map to display data in 10-year intervals.

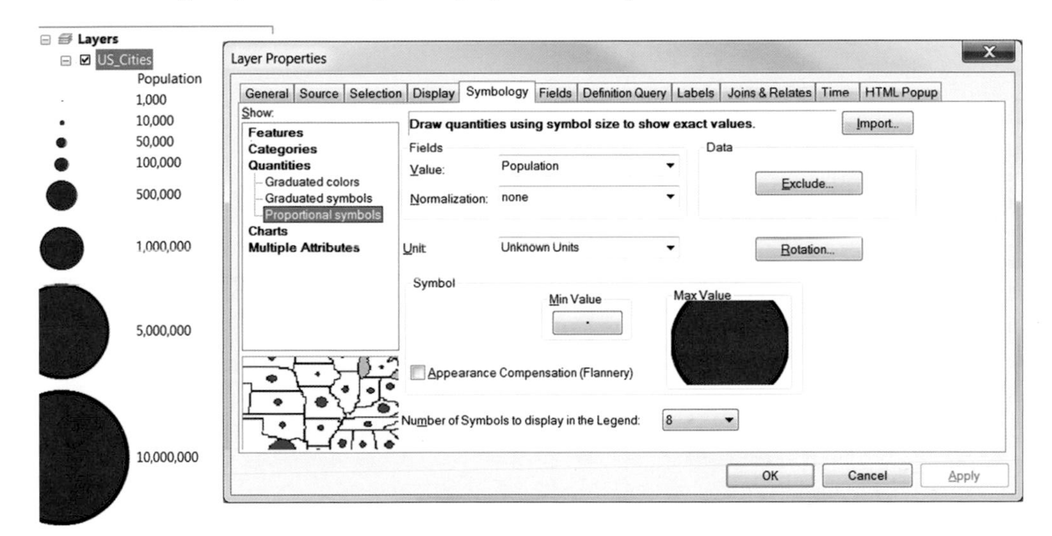

Assignment 6B: Publish a map service, and use the service in a web app.

This assignment intentionally allows you to use your own data.

Data: Use your own data (for example, the data and maps that you used in a real project, still have from another GIS course, or downloaded from another location).

Requirements:

- Publish a map service (this service does not need to be time enabled if no appropriate attribute field enables time).
- Create a web app based on this map service.

What to submit: Send an email to your instructor with the subject line **Web GIS Assignment 6B: Your name**, and include the following information:

- The URL to your map service REST endpoint
- The URL to your web app

Resources

ArcGIS for Server Help document site

"Federate an ArcGIS for Server Site with Your Portal," http://server.arcgis.com/en/portal/latest/administer/windows/federate-an-arcgis-server-site-with-your-portal.htm.

"How to Publish a Service," http://server.arcgis.com/en/server/latest/publish-services/windows/how-to-publish-a-service.htm.

"Serving Time-Aware Layers," http://server.arcgis.com/en/server/latest/publish-services/windows/serving-time-aware-layers.htm.

"What Is a Map Service?" http://server.arcgis.com/en/server/latest/publish-services/windows/what-is-a-map-service.htm.

"What Is Map Caching?" http://server.arcgis.com/en/server/latest/publish-services/windows/what-is-map-caching-.htm.

"What Types of Services Can You Publish?" http://server.arcgis.com/en/server/latest/publish-services/windows/what-types-of-services-can-you-publish.htm.

Esri Training website

"ArcGIS 4: Sharing Content on the Web" (instructor led, fee required), http://training.esri.com/gateway/index.cfm?fa=catalog.courseDetail&CourseID=50130533_10.x.

Esri videos

"ArcGIS for Server: An Introduction," by Derek Law, http://video.esri.com/watch/4696/arcgis-for-server-an-introduction.

"ArcGIS for Server: Publishing and Using Map Services," by Craig Williams and Tanu Hoque, http://video.esri.com/watch/4414/arcgis-for-server-publishing-and-using-map-services.

"Portal for ArcGIS: An Introduction," by Derek Law, http://video.esri.com/watch/4715/portal-for-arcgis-an-introduction, by Derek Law.

Chapter 7
Spatial analytics and geoprocessing services

As a key aspect of GIS, spatial analytics allows users to discover relationships, patterns, and trends in geospatial data. Traditionally, the power of spatial analytics was limited to GIS professionals with access to desktop GIS software. However, ArcGIS Online now provides powerful spatial analytics capabilities and supporting data to a wide spectrum of web users and clients through user-friendly interfaces. ArcGIS for Server allows you to share your models and scripts as geoprocessing services, which support tailored analysis in web apps. ArcGIS Online and ArcGIS for Server spatial analysis capabilities allow GIS professionals and nonprofessionals alike to gain geographic insights and make informed decisions.

Learning objectives

- *Understand ArcGIS Online analysis capabilities.*
- *Know the rich collection of data available from ArcGIS Online.*
- *Create web apps that use ArcGIS Online analysis.*
- *Author and publish geoprocessing services with ArcGIS for Server.*
- *Use geoprocessing services in web apps.*

This chapter in the big picture

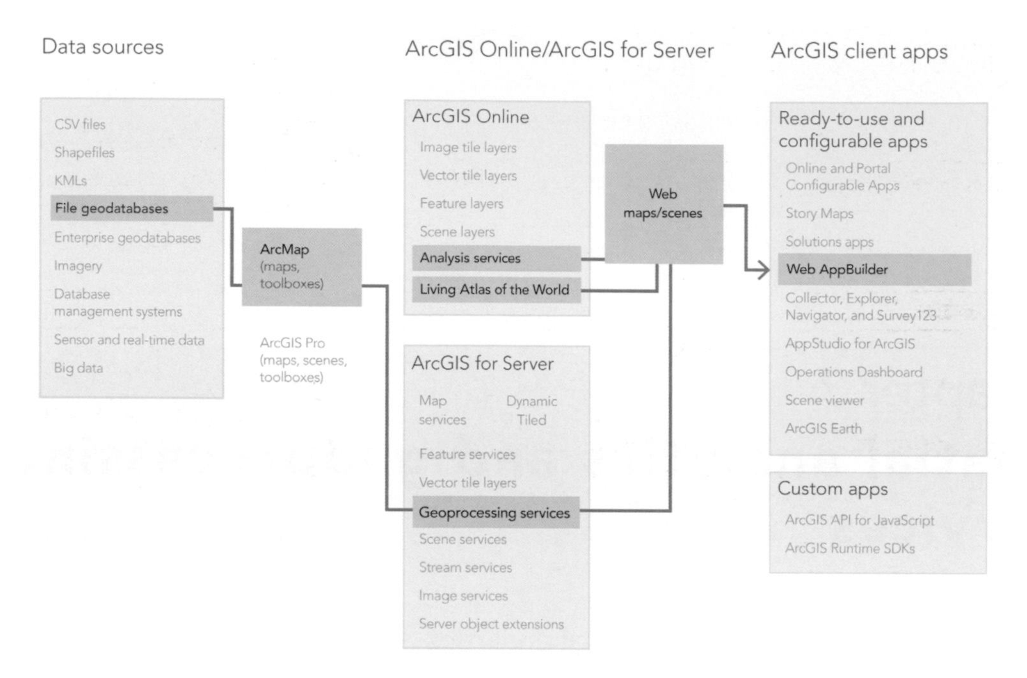

ArcGIS offers many ways to build web applications. The green lines in the figure highlight the technology that this chapter teaches. This chapter introduces two ways to perform geospatial analysis online, one using ArcGIS Online and one using ArcGIS for Server.

Spatial Analysis using ArcGIS Online versus ArcGIS for Server

Spatial analytics holds great practical value for our personal life and for enterprise operations. You most likely use spatial analysis every day, for example, by finding the optimal route to work or home using the apps on your smartphone. Your banks, supermarkets, and real-estate developers use spatial analytics to delineate market areas, estimate sales potentials, and select facility locations. Your law enforcement agency uses spatial analytics to discover crime hotspots and decide where to deploy more police officers. The infinite applications of spatial analytics range from calculating the probable evacuation area after a hazardous chemical spill and predicting the track and strength of a gathering hurricane to summarizing the land cover within a user-defined watershed and predicting a region's future development.

ArcGIS Online and ArcGIS for Server both differ from and complement each other in the way they provide online spatial analytics capabilities. ArcGIS Online provides ready-to-use spatial analytics functions and rich data to end users. ArcGIS for Server allows service providers to share their desktop analysis tools and data as web services for their clients to use.

Table 7.1 **Compare ArcGIS Online analysis with ArcGIS for Server geoprocessing services.**

	ArcGIS Online	ArcGIS for Server
Difficulty level	Low, easy to use. Non-professionals can use it.	Relatively high. Service providers need to know Model Builder, Python, or other languages to create services.
Target users	Non-professional users as well as professional users.	Professional users who provide GIS services and apps for end users.
Analysis services	Created and hosted. Ready for use.	You must create and host.
Data provided?	Yes (you can use your own data too).	No.
Cost	Costs credits to use.	Pay in advance and does not cost credits to use.
System requirements	ArcGIS Online organizational accounts with privileges to create content, publish hosted feature services, and perform spatial analysis.	Purchase, install, and host ArcGIS for Server for yourself or your organization.

Access to ArcGIS Online analysis

Depending on the scenario, you can use ArcGIS Online analysis in the following ways:

- Perform analysis in the map viewer of ArcGIS Online or Portal for ArcGIS. You can click the **Analysis** button on the map viewer toolbar or in the **Contents** pane under the layer you want to perform analysis on. This approach is mainly for app designers to perform analysis and share the results with end users via web maps and apps. End users only see the results and cannot run the analysis.

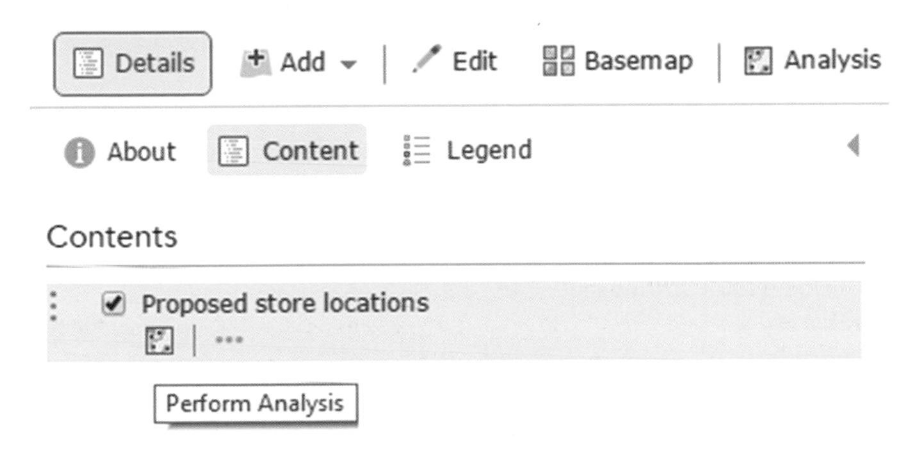

Perform analysis in ArcGIS Online or Portal for ArcGIS map viewer.

1
2
3
4
5
6
7
8
9
10

- Use the **Analysis** widget in Web AppBuilder for ArcGIS. This approach targets app configurators and designers. They can select and decide which types of analysis tools to make available to end users so that the end users can run the tools and experiment with different parameters. Section 7.2 demonstrates this approach.
- Use JavaScript, Python, and other languages to access ArcGIS Online Spatial Analysis Representational State Transfer (REST) API. This approach targets developers so they can write their own programs to chain multiple analysis tools together and automate the process. In contrast, users and app designers must run the analysis manually, one tool at a time.

ArcGIS Online analytics capabilities

ArcGIS Online provides analytical tools for administrator, publisher, and custom roles that have the privilege to create, update, and delete content, publish hosted features, and perform spatial analysis. These tools belong to one of the following six main categories:

- **Summarize Data:** Calculates total counts, lengths, areas, and basic descriptive statistics of features and their attributes within areas or near other features. This category includes the following tools:
    - Aggregate Points
    - Summarize Nearby
    - Summarize Within
- **Find Locations:** Finds areas that meet your criteria, which can be based on both attribute (over a designated population count) and spatial queries (within one mile of a river). This category includes the following tools:
    - Find Existing Locations
    - Derive New Locations
    - Find Similar Locations
    - Create Viewshed
    - Create Watersheds
    - Trace Downstream
- **Data Enrichment:** Retrieves detailed demographic data and statistics available for your chosen area(s) and generates reports that compare the input area with the county or state of the area.
- **Analyze Patterns:** Helps users identify, quantify, and visualize spatial patterns in your data. This category includes the following tools:
    - Calculate Density
    - Find Hot Spots
    - Interpolate Points

- **Use Proximity:** Helps users answer one of the most common questions posed in spatial analytics: "What is near what?" This category includes the following tools:
  - Create Buffers
  - Create Drive-Time Areas
  - Find Nearest
  - Plan Routes
  - Connect Origins to Destinations
- **Manage Data:** Helps users in the day-to-day management of geographic data and to combine data prior to analysis. This category includes the following tools:
  - Dissolve Boundaries
  - Extract Data
  - Merge Layers
  - Overlay Layers

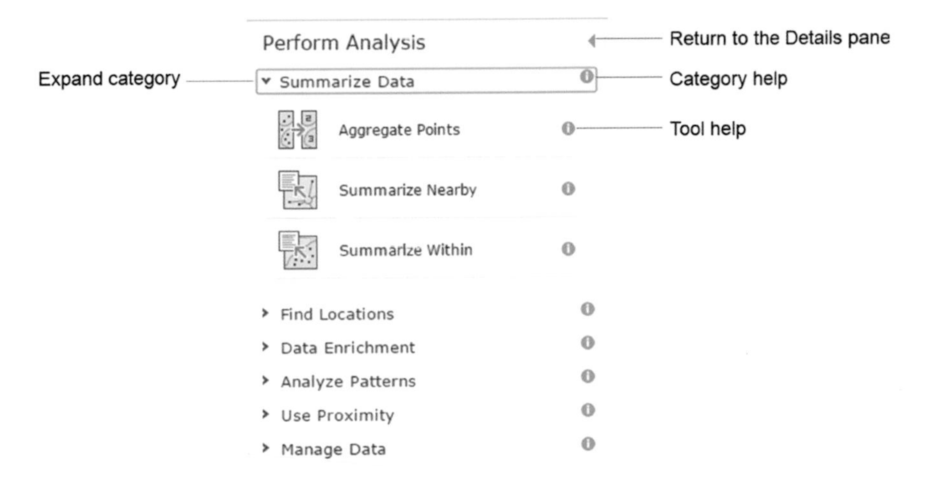

ArcGIS Online organizes its analytical tools in categories. You can find more information about each category by clicking the **Category** and **Tool** help buttons.

Workflow to use ArcGIS Online analysis

ArcGIS Online analysis includes four general steps (or two steps if you use the layers in the Living Atlas of the World). For your analysis, you can use data from the atlas, your own data, or data you discover from other people. If you use the atlas data only, you can skip the first two steps and start with the third step in the following list:

1. **Prepare data**: You may create your own layers or discover and use other users' layers. These layers are typically vector data from the following formats:
   - Feature service/layers (must be publically accessible)
   - Map service layers (the public must have access)
   - Comma-separated value (CSV) or gpx files

- Keyhole Markup Language (KML) layers and GeoRSS feeds
- Shapefiles (.zip)
- Map notes layers
- Route layers

2. **Add to map**: Add your data or the data you discovered into the map viewer. You can create data dynamically by using map notes layers or feature layers and then use the newly created data for analysis. Layer symbology is not required for analysis. If you are going to use layers from the Living Atlas, you do not need to add them to your map.
3. **Perform analysis:** You will determine the appropriate tool(s), specify appropriate parameters, and run the tool(s).
4. **Review and interpret results**: The results are often in the format of host feature layers or related tables. The results are automatically added to the web map with pop-ups configured. You can review and accept the results or adjust your parameters and run the tools again. As a part of your web map and app, the results can be shared with your web audience.

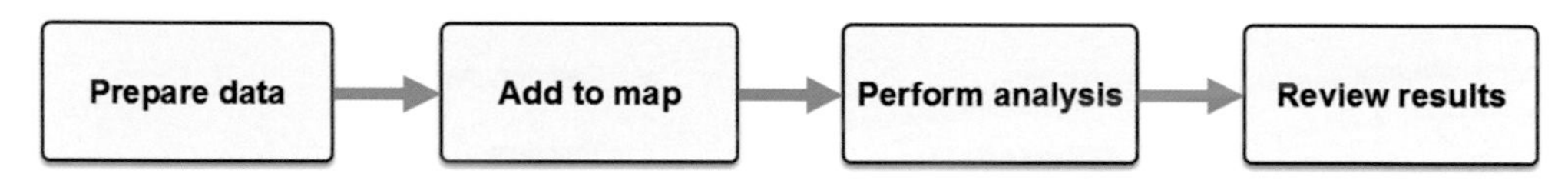

ArcGIS Online analysis includes four general steps.

Living Atlas of the World supports ArcGIS Online analysis

Web GIS differs from traditional desktop GIS-based analysis, which relies completely on the data you already have. Instead, your analysis can rely partly or even completely on the data layers and analysis capabilities available from ArcGIS Online. For example, your analysis can calculate driving directions based on the street network, historic traffic, or real-time traffic data in the atlas. You can enrich (or add to) your location of interest with the location's population, income, housing, consumer behavior, tapestry, and thousands of other variable values in the atlas. Collecting, converting, and hosting such live and historic traffic variables and commercial variables normally require a lot of time and money. By providing the data, ArcGIS Online makes the analysis much easier and more cost-efficient.

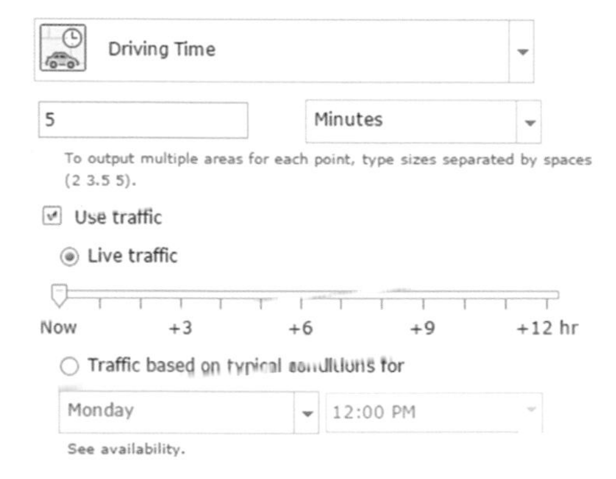

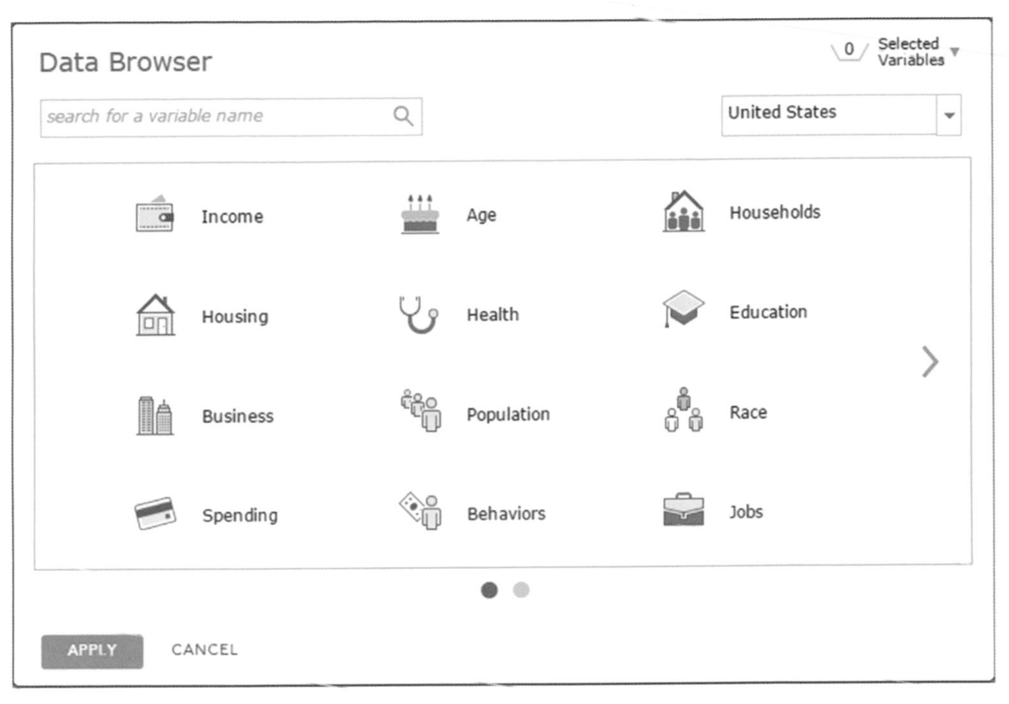

ArcGIS Online provides authoritative data and is a more cost-efficient way to perform spatial analysis than traditional desktop-based GIS, which requires you to collect your own data.

Raster analysis and image services

The Living Atlas includes a rich collection of imagery and landscape layers for raster-based analysis. ArcGIS Online provides powerful raster-based analysis capabilities based on ArcGIS for Server image services technology. Image services support server-side processing through default functions and raster function templates (RFT). Consider the following examples:

- An image service containing imagery, such as Landsat, can dynamically support a web app to create NDVI (normalized difference vegetation index) based on the red and near-infrared bands.

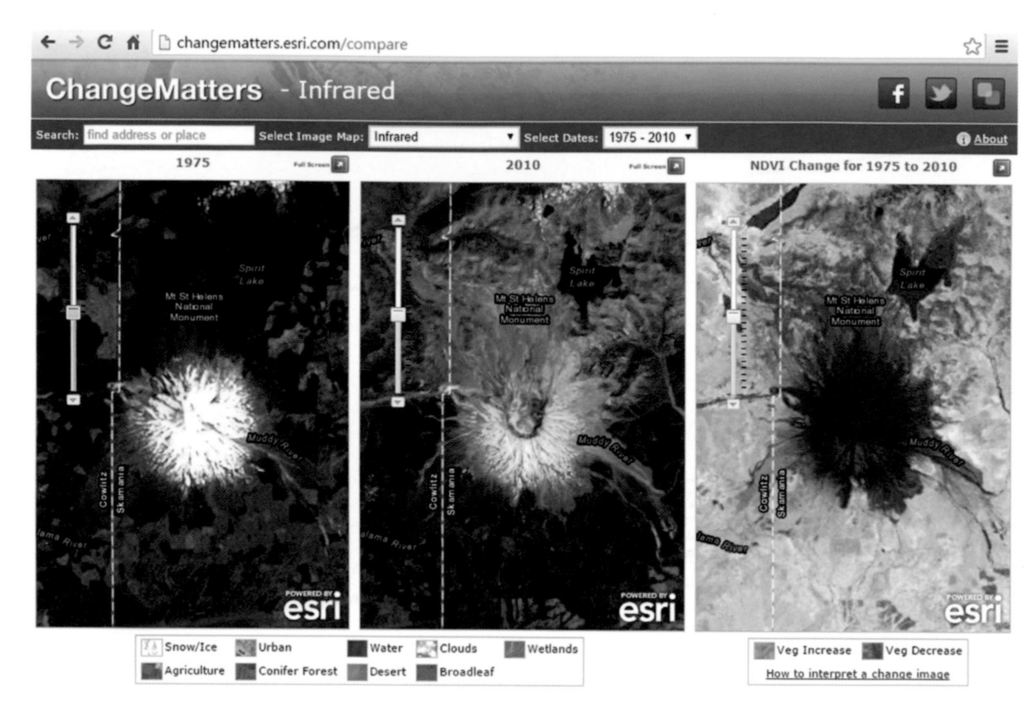

This image service supports raster-based analysis, for example, calculating a vegetation index based on multiple bands and calculating vegetation increase/decrease based on multiple years' vegetation indexes.

- An image service containing a digital elevation model (DEM) layer can support a web app to generate hillshade, slope, and shaded relief images dynamically.
- An image service containing multiple raster layers can support a web app to perform weighted overlay analysis dynamically.

You can use image services in many applications, including GeoPlanner for ArcGIS (**http://geoplanner.arcgis.com**). To try raster-based analysis in GeoPlanner, you can sign up for a free trial at **http://doc.arcgis.com/en/geoplanner/trial**. GeoPlanner for ArcGIS is a web-based geodesign app that helps users create, analyze, and report planning alternatives. GeoPlanner for ArcGIS integrates online data, analysis tools, weighted raster overlay, landscape content, sketching tools, and GeoEnrichment services to help you visualize and understand a plan or design. GeoPlanner for ArcGIS brings the power of ArcGIS Online and a geodesign workflow to land-based planning and design activities.

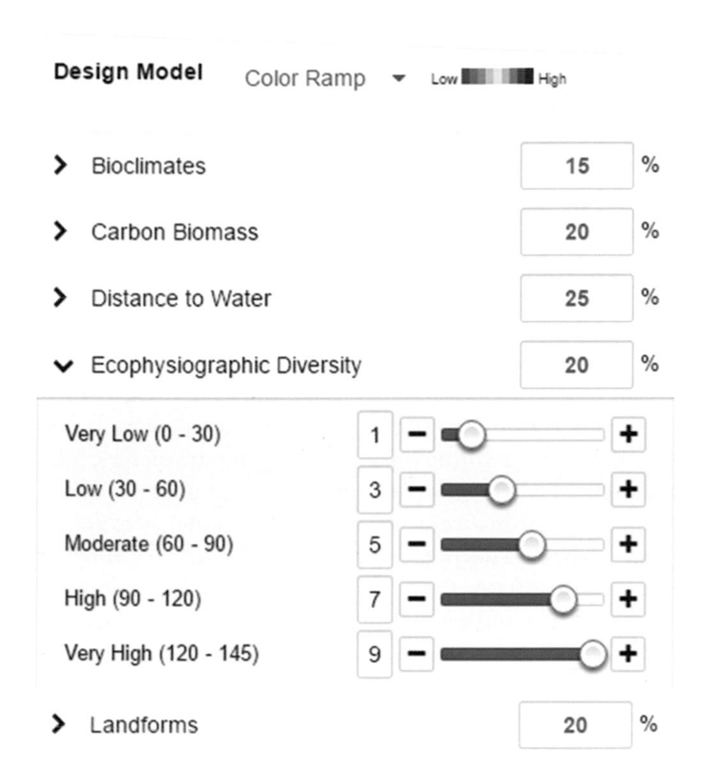

GeoPlanner supports suitability model analysis through weighted raster overlays. You can use the analysis to research sites for urban development, housing developments, habitat locations, and other projects.

Geoprocessing services

Geoprocessing services are the core technologies for serving spatial analytics functions with ArcGIS Online and ArcGIS for Server. Esri-published geoprocessing services basically power ArcGIS Online spatial analytics functions. If you want to make your own spatial analysis functions available online for use in web apps, you need to publish your analysis functions to ArcGIS for Server as geoprocessing services.

Typically, a geoprocessing service includes one or more geoprocessing tasks, each of which takes inputs, processes them, and returns meaningful and useful output(s) in the form of features, maps, reports, and files. If you have used geoprocessing in ArcGIS for Desktop, you can think of a geoprocessing service as a toolbox and its tasks as tools within that toolbox. The main difference between a geoprocessing service and a toolbox in your ArcGIS for Desktop is that when you execute a task in a geoprocessing service, the task runs on the server computer and uses the CPU of that computer.

Steps to create geoprocessing services

Creating a geoprocessing process requires you to author your tool or model, run your tool or model, and publish an execution plan, as shown in the figure.

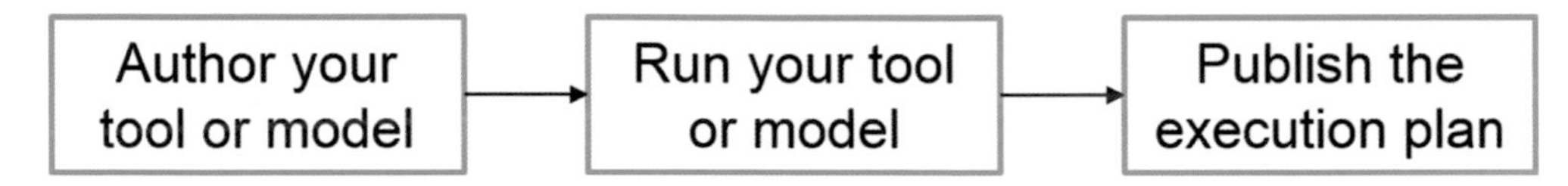

Three steps to using a geoprocessing service.

1. **Author your tool or model**: Creating a tool or model using ModelBuilder in ArcGIS for Desktop or using Python (or another programming language) involves using workflow logic, specifying the location of required data, and defining input and output parameters. The output parameters define what web clients can receive and end users can access.
2. **Run your tool or model:** ArcGIS for Desktop ensures that your tool or model works as expected. A successful run produces expected output in the ArcMap **Results** window.
3. **Publish the execution plan**: Publishing a geoprocessing service requires an administrator or publisher connection to an ArcGIS Server and a result in the **Results** window. You learned about administrator and publisher connections in a previous chapter. The previous step explains how to get a result. Right-clicking the result and clicking **Share As** > **Geoprocessing Service** opens a step-by-step wizard that defines the service and its task. This task is the same as the tool or model that created the result. For a geoprocessing service, a key parameter is the execution mode. Execution mode defines how the web client interacts with the server and gets results from the executed task. This mode can be asynchronous or synchronous. Both forms of this mode are supported by Web AppBuilder, ArcGIS API for JavaScript, and other clients.
    - **Synchronous:** If you set the service as synchronous, the client will wait for the task to finish to get the result. Typically, a synchronous task executes quickly—in five seconds or less.
    - **Asynchronous:** An asynchronous task typically takes longer to execute, and the client must periodically ask the server if the task has finished and to get the result. A web application using an asynchronous task must have logic implemented to check the status of a task and handle the result once execution is finished.

Python and ArcPy

You can use scripting languages such as Python to create a geoprocessing workflow as a task for your geoprocessing service. Python is a free, powerful, cross-platform, and open-source scripting language included in a typical ArcGIS for Desktop and ArcGIS for Server installation. Python often is used to automate workflows so that you do not have to perform them manually. As an interpreted language, Python does not need to be compiled. You can write Python script using any text editor (such as Notepad) or more sophisticated development environments. You can run a Python script inside ArcGIS or as a standalone.

ArcGIS extends Python by providing ArcPy, a module that facilitates data analysis, data conversion, data management, and map automation. This module offers users ease and convenience in using Python by including such features as code completion (in which, for example, users can type a keyword plus a dot to get a pop-up list of properties and methods supported by that keyword) and reference documentation for each function, statement, module, and class.

ModelBuilder

In addition to Python, you can use ModelBuilder to create a model that will become a task for your geoprocessing service. ModelBuilder comes with ArcGIS for Desktop, which you can use to create, edit, and manage models. Think of ModelBuilder as a visual programming language. Creating a model requires stringing together sequences of geoprocessing tools and connecting them with inputs and outputs.

For this tutorial, you will use ModelBuilder, which is the easiest way to create geoprocessing tools, compared to Python and other scripting languages.

This tutorial

Of the two site selection case studies in this tutorial, one study uses ArcGIS Online analysis, and the other study uses ArcGIS for Server analysis.

Case Study 1: Create a web app for selecting restaurant locations using ArcGIS Online analysis capabilities.

A company wants to open a new high-end, full-service restaurant. Company management must choose one of two possible recommended locations.

Functional requirements: The web app must have the following capabilities:

- Calculate the service area of each candidate.
- Obtain demographic information, including population, disposable income, sales potential, and leakage/surplus of restaurant service for each service area.
- Compare the demographics of the two recommended locations, and help the company choose the location.
- Obtain the block group IDs of the service area of the selected location for a direct mail marketing campaign.

Data:

- The two candidate locations
- All other data derived from the Living Atlas of the World

System requirements:

- ArcGIS Online publisher- or administrator-level account (with the privileges to run analysis and create hosted layers)

Case Study 2: Create a web app for selecting a factory location using the ArcGIS for Server geoprocessing service.

A company wants to build a factory in the US state of Alabama. You are contracted to build a web GIS app that helps company executives select possible factory sites.

Site criteria:

- The new factory should be close to desirable areas determined by company executives.
- This factory will use a lot of water and should be built close to a river.
- Factory products will require railroad transportation, so the factory should be close to a railroad.

Functional Requirements: Your web app should allow company executives to specify the following parameters:

- A point to indicate a location of interest
- A distance that delimits the factory's distance from the location of interest
- A distance that defines how close the factory should be to rivers
- A distance that defines how close the factory should be to railroads

Data:

- A file geodatabase containing the following three feature classes:
  - Main rivers in Alabama
  - Selected railroads in Alabama
  - An empty point feature class
- A map document named **Site_Selection.mxd to** display river and railroad layers
- A toolbox named **Planning.tbx**, which contains a **Select_Sites** tool

System requirements:

- ArcGIS for Desktop for designing your GP model
- ArcGIS for Server for publishing and hosting your geoprocessing service
- ArcGIS Online publisher or administrator account for the following purposes:
  - Creating apps using Web AppBuilder
  - Accessing ArcGIS Online analysis functions
- Portal for ArcGIS for managing geoprocessing services (optional)

Note: The tutorial section includes two independent case studies. The second study requires ArcGIS for Server. If do not have ArcGIS for Server, you will skip the second study.

7.1 Create a web app with selected analysis tools

You learned how to create web maps (in the map viewer) and web apps (using the Web AppBuilder) in earlier chapters. Therefore, this section will be brief.

1. **Sign in to ArcGIS Online (http://www.arcgis.com) or your Portal for ArcGIS.**

2. **Click Map to go to the map viewer.**

3. **On the toolbar, click the Basemap button, and choose the Streets basemap.**

The **Streets** basemap emphasizes transportation features and easily allows you to see the service areas that must be calculated.

4. **In the toolbar, click the Add button, and choose Browse Living Atlas Layers.**

5. **In the Browse Living Atlas Layers window, search for tapestry, add the 2015USA Tapestry Segmentation layer to your map, and click Close.**

The tapestry layer helps you determine the location of potential customers.

6. **In the Contents pane, turn off the tapestry layer.**

You will turn the layer on as needed in your web app.

7. **Save your map as Map for restaurant location selection, and set the keyword tag as tapestry.**

8. **On the toolbar, click the Share button, select Everyone, and click Create a Web App.**

9. **Click the Web AppBuilder tab.**

10. **In the Create a New Web App window, set the title as Restaurant Location Selection, and click Get Started.**

11. **Under the Theme tab, choose Box Theme and a layout.**

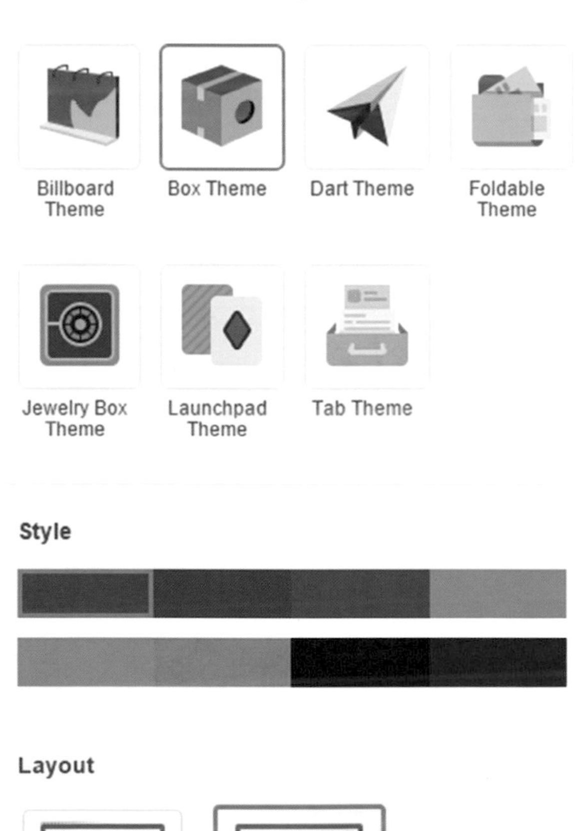

You can select a different theme, but the theme must have the **Attribute Table** widget. You will need this widget in the analysis steps later.

12. **Click the Widget tab.**

13. **Enable the Attribute Table widget by pointing to the widget and clicking the Eye button.**

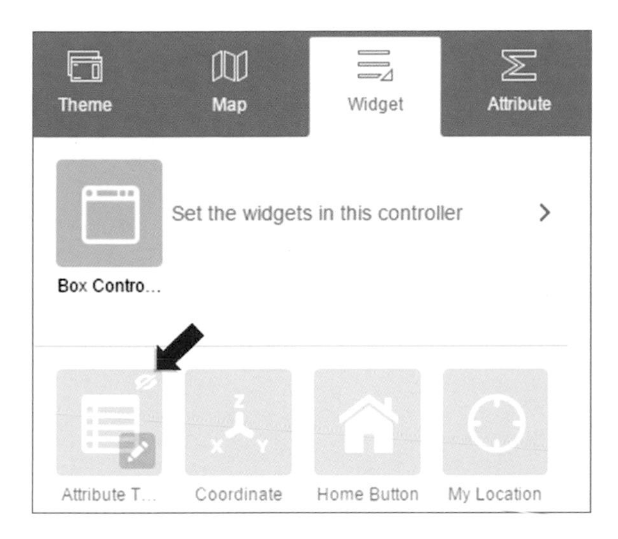

14. **Click Set the widgets in this controller** link.

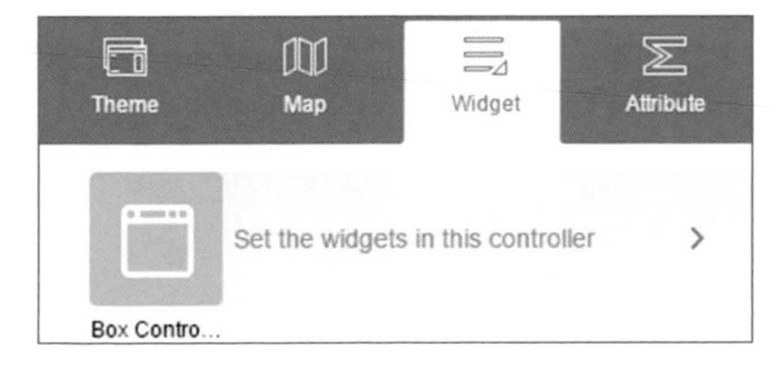

You will see the **Legend** and **Layer List** widgets in the box controller. You will add widgets so users can search for addresses, specify potential restaurant locations, and perform analyses.

15. **Click the Plus button to add a widget. In the Choose Widget window, choose the Search widget and click OK.**

16. **In the Configure Search window, click OK to accept the defaults and close the window.**

17. **Click the Plus button to add another widget. In the Choose Widget window, choose Draw, and click OK.**

18. **In the Configure Draw window, select Add the drawing as an operational layer of the map, and click OK.**

19. **Select the option Add the drawing as an operational layer of the map, as illustrated.**

✓ Add the drawing as an operational layer of the map.

Selecting this option will allow you to use the locations you draw on the map as input for your analysis. Otherwise, you must add the potential restaurant locations to your web map or to a CSV file, feature layer, or other supported format.

Next, you will add the **Analysis** widget and configure the necessary tools.

20. **Click the Plus button to add a widget. In the Choose Widget window, choose Analysis and click OK.**

21. **In the Configure Analysis widget, change the widget title to Restaurant Location Analysis.**

22. **Select the Create Drive-Time Areas tool.**

23. **To the right of that tool under Settings, click the Set tool details button, and perform the following actions:**

    - **Set The tool display name as Step 1: Calculate service area.**
    - **Clear Show option to use the current map.**

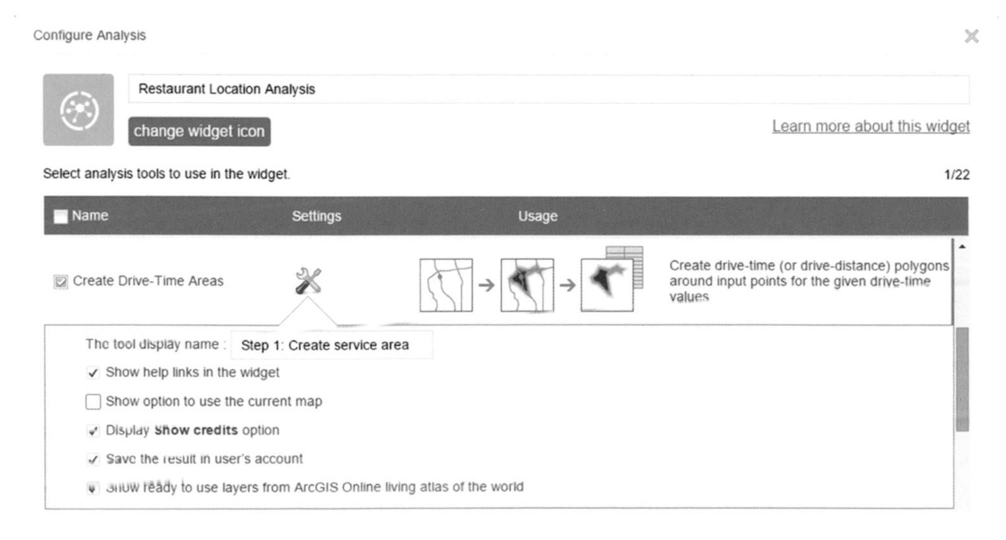

Using this tool will answer the following question: If most customers are willing to drive up to 15 minutes to reach my restaurant, what geographic area would those 15 minutes cover?

Your analysis will consider two possible restaurant locations, even when they are not included in the current map extent. If you have many candidate locations and want to analyze only points that fall in the map extent, you should leave **Show option to use the current map** selected.

24. **Select Enrich Layer, click the Set tool details button, and perform the following actions:**

   - **Set The tool display name as Step 2: Obtain service area demographics.**
   - **Clear Show option to use the current map.**

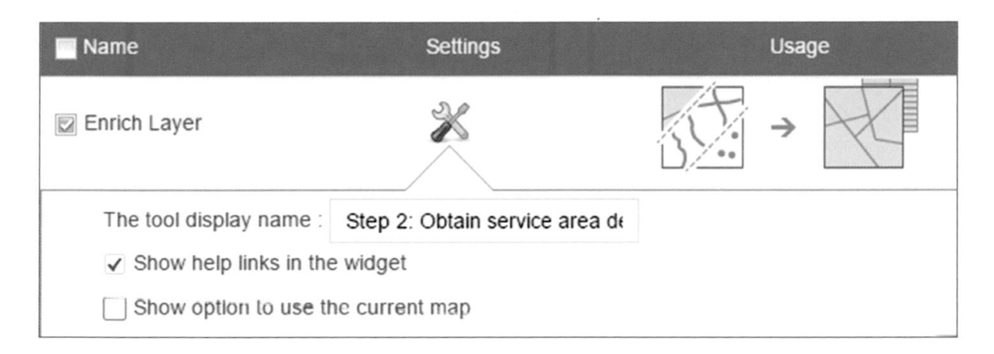

Selecting **Enrich Layer** allows users to run the ArcGIS Online geoenrichment analysis to find the people, places, and business facts for the input areas.

25. **Select Summarize within, click the Set tool details button, and perform the following actions:**

   - **Set The tool display name as Step 3: Find block groups for direct mail marketing.**
   - **Leave Show option to use the current map selected.**

This tool will allow users to determine what block groups overlap the service area of a restaurant.

26. **Click OK to close the Configure Analysis window.**

27. **Click Save to save your web app.**

You have created a web app offering analysis capabilities. Company executives can use this app to select their desired restaurant location.

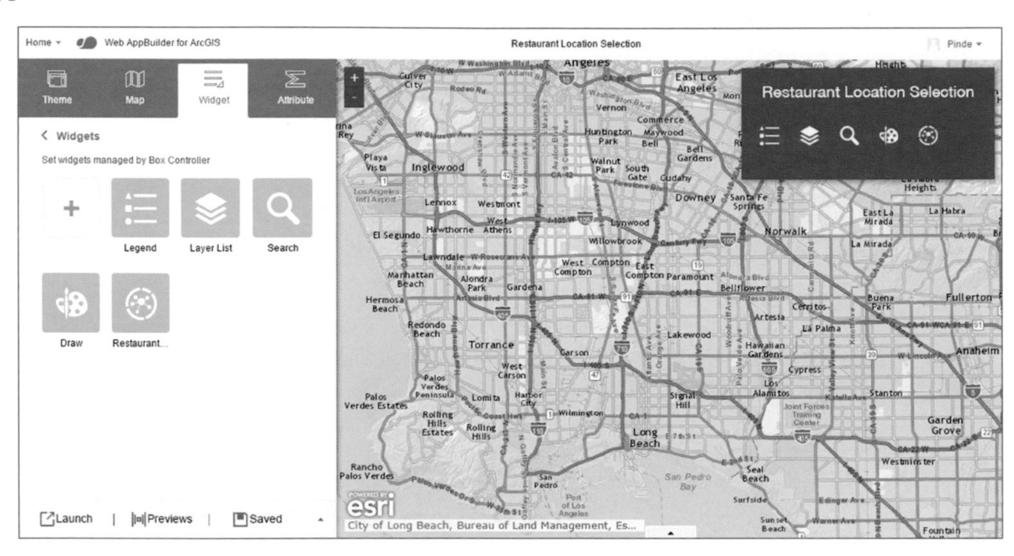

7.2 Perform analysis using the Web AppBuilder Analysis widget

In this section, you will specify two candidate restaurant locations and run the three analysis steps you just configured. If you are continuing from the previous section, you will click **Launch**. Otherwise, you can go to your **My Content** list, find the app, and view it there.

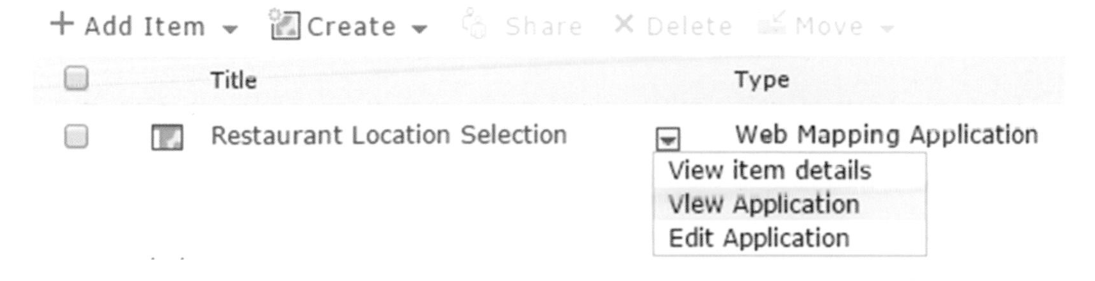

1. **Launch the app you just created.**

Next, you will find the first candidate restaurant location and draw a point there.

2. **Click the Search button, search for 826 W Valley Blvd, Alhambra, CA, 91803 in the box at the bottom of the screen.**

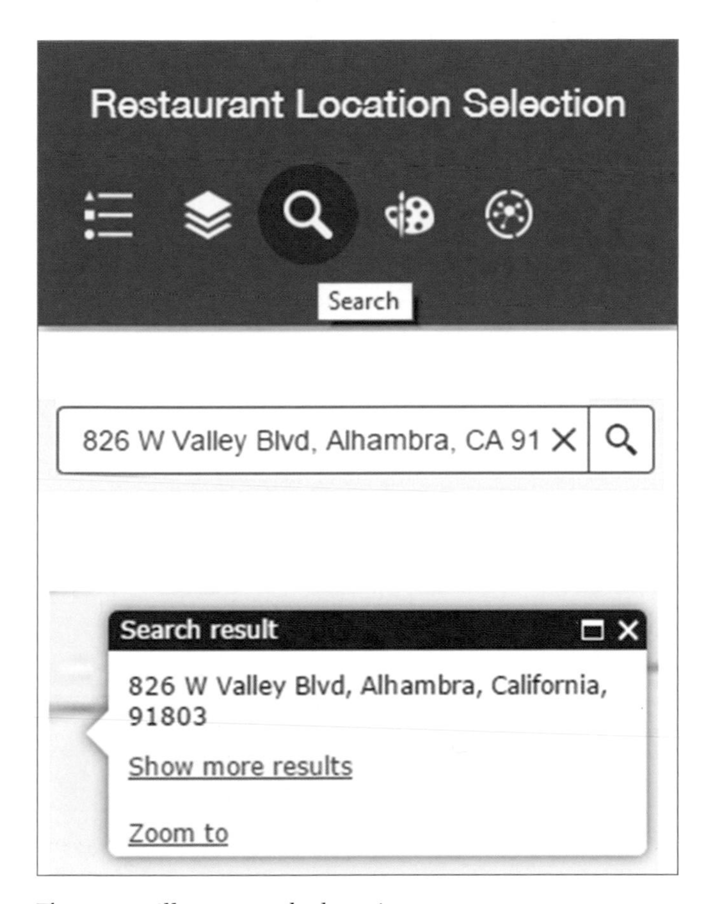

The map will zoom to the location.

3. **Click the Draw widget to draw a point at the location of the address.**

If you cannot find the widget buttons, they probably have minimized to a **More** button ···, and you will click the button to bring the tool buttons back.

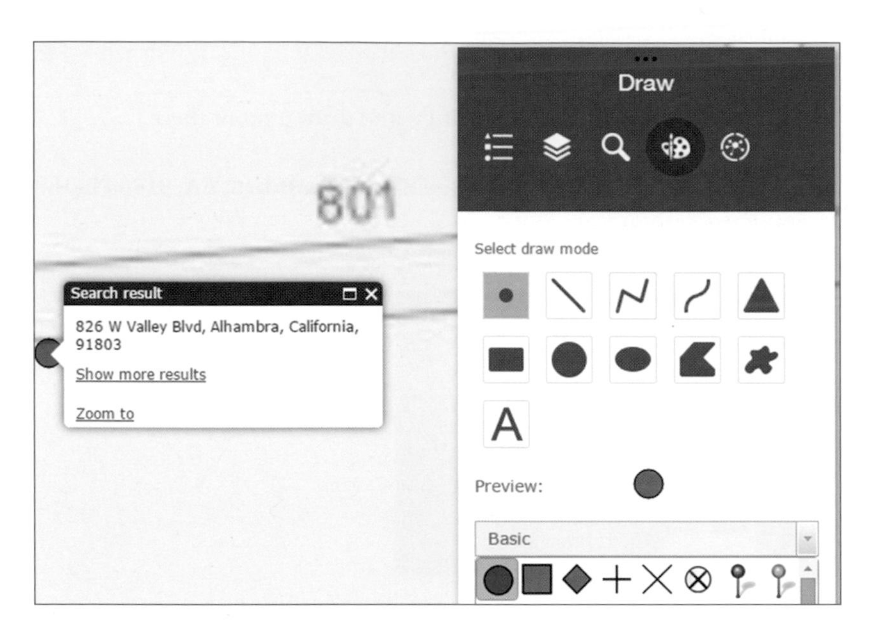

4. **Repeat the above two steps to add the next candidate location at 9773 Baseline Rd, Rancho Cucamonga, CA 91730.**

Now that two input locations have been added, you are ready to run the analysis.

5. **Zoom out the map to include both candidate locations.**

6. **Click the Restaurant Location Analysis widget.**

The **Restaurant Location Analysis** widget appears, showing the three steps you configured.

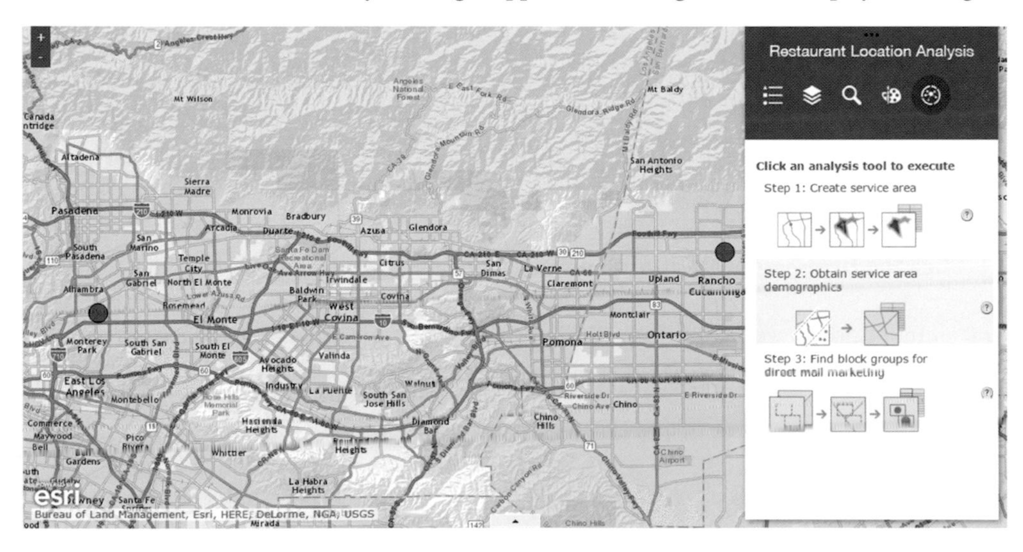

7. Click **Step 1**, and specify the following parameters:

   - For **Choose point layer to calculate drive-time areas around**, choose **Points** in the list.
   - For **Measure**, choose and set the **Driving Time** to **15 Minutes**.
   - Select **Use traffic**, choose **Traffic based on typical conditions for Friday 6:00 pm**.
   - For the **Result layer name**, specify **Service area of 15m**.
   - Click **Run Analysis**.

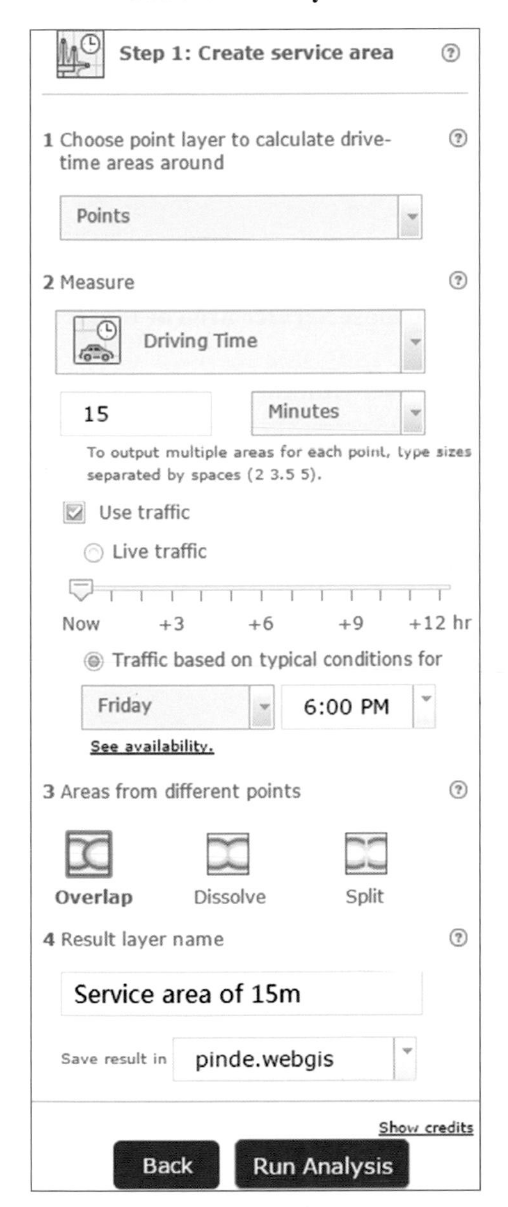

1
2
3
4
5
6
7
8
9
10

After you run the analysis, you will see that the service areas are added for the two locations. Next, you will get the demographics information for those areas.

8. **Click Home to go to the homepage of the Analysis widget.**

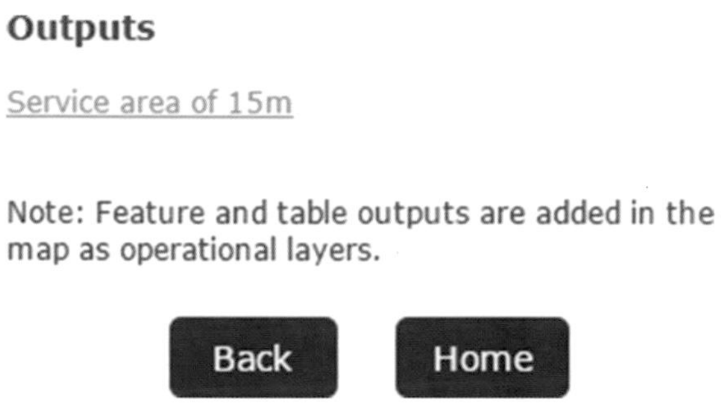

Outputs

Service area of 15m

Note: Feature and table outputs are added in the map as operational layers.

Back Home

9. Click **Step 2**, and specify the following parameters:
    - For **Choose layer to enrich with new data**, choose **Service area of 15m** from the list.
    - Click **Select Variables** to open the **Data Browser** window.

10. In the **Data Browser** window, perform the following actions:
    - Click **Income**, and select **2015 Average Household Income (Esri)**.
    - Click **Disposable Income**, expand **2015 Disposable Income (Esri)**, and select **2015 Average Disposable Income (Esri)**.
    - Click **Back** and **Back**.
    - Click **Population**, select **2015 Total Population (Esri)**, and **2015–2020 Population: Annual Growth Rate (Esri)**, and click **Back**.

11. Still in the **Data Browser** window, in the **Search for a variable name** text box, type **Restaurants**, and press enter. In the **Data Browser** window, expand **2015 Retail MarketPlace**, and select **2015 Retail Sales: Full-Service Restaurants**, **2015 Retail Sales Potential: Full-Service Restaurants**, and **2015 Leakage/Surplus Factor: Full-Service Restaurants**.
    - Click **Apply** to close the **Data Browser** window.

12. **Change the Result layer name to Service Area Demographics.**

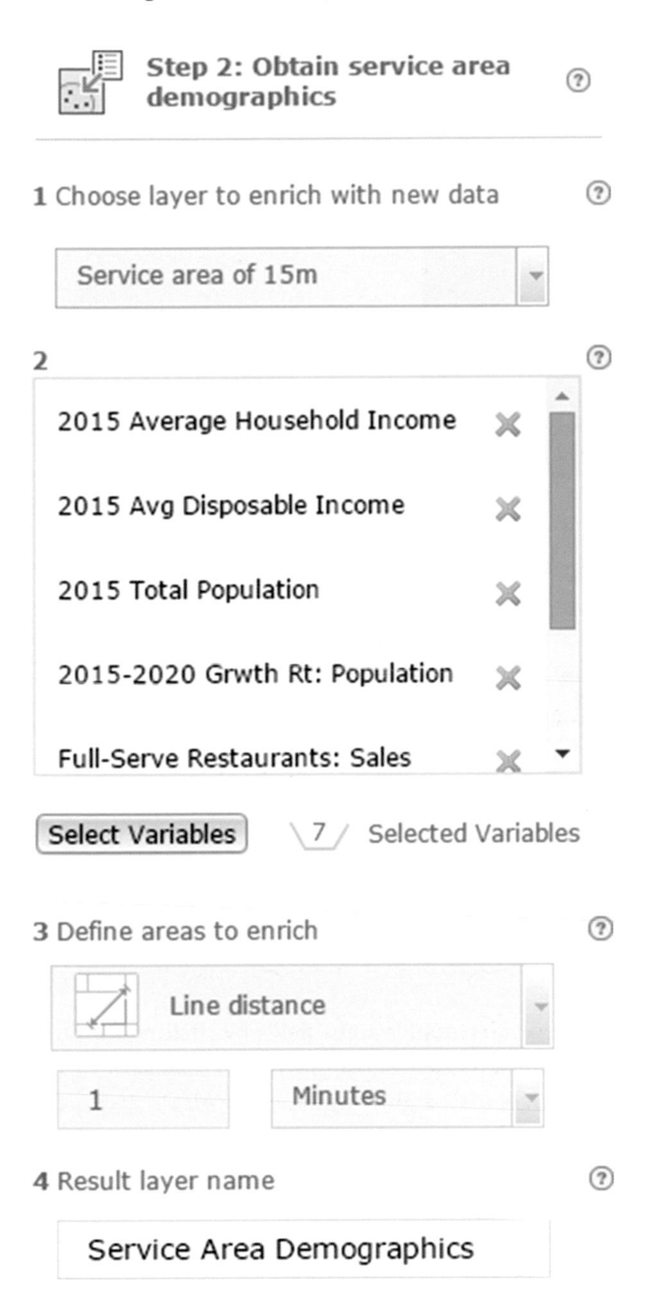

13. **Click Run Analysis.**

When you complete the analysis, the **Output** layer appears on the map as an operational layer.

14. **Click the Layer List widget, click the arrow next to the Service Area Demographics — Service Area Demographics layer, and choose Open Attribute Table.**

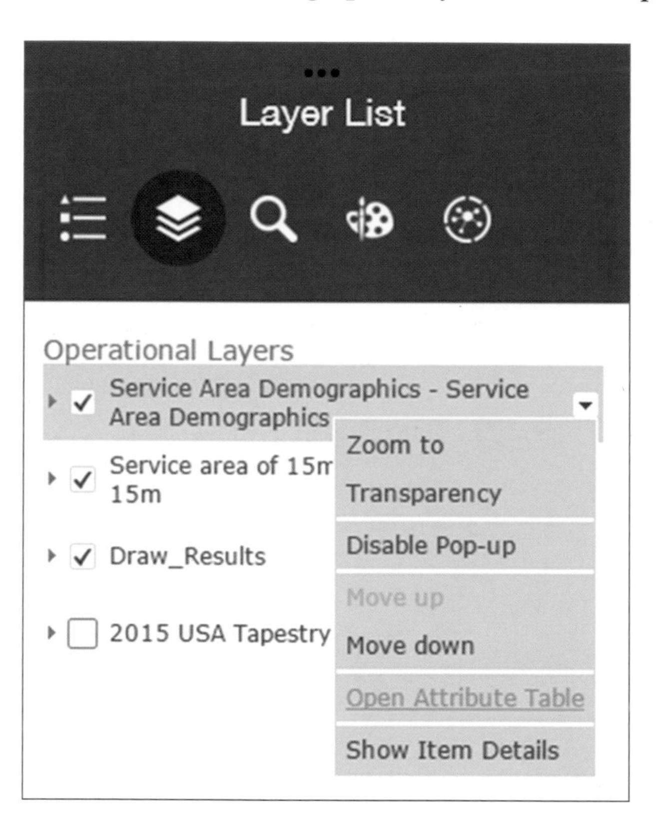

The **Attribute table** widget opens. Optionally, you can hide some fields by clicking the **Options** button in the upper left of the table. Your table values may differ slightly from the figure because the candidate locations you drew on the map might also slightly differ.

15. **Compare the demographics of the two service areas, and choose your candidate. Assuming other related factors are similar, you may want to choose the Rancho Cucamonga candidate (the second point you placed) because the population in that candidate's service area has a higher income and growth rate, and the area has less restaurant competition (as indicated by the Full Service Restaurants: L/S value, which is the leakage/surplus indicator. The indicator value ranges from –100 to 100, with a positive number indicating a leakage, a negative number indicating a surplus, and a large number indicating less competition).**

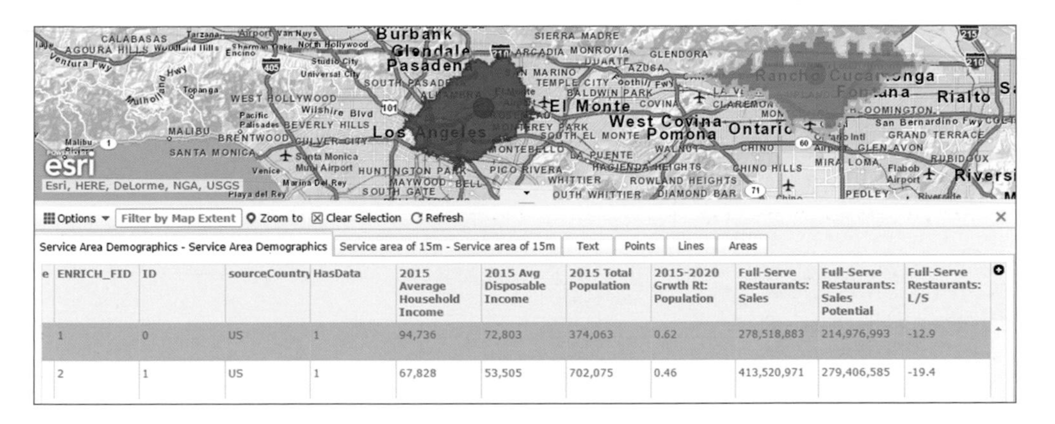

The tutorial is simplified for illustration purposes. In addition to the demographics obtained here, you can consider other factors such as restaurant niche, number of vehicles passing by, targeted ages and ethnicity groups, cost to establish the restaurant, and available funding.

Next, you will determine the list of block groups that the restaurant service area overlaps so that you can distribute flyers for promotional or survey purposes.

To limit the next analysis to the service area of the selected candidate, you can either zoom the map to include only the desired service area, or you can use a filter. This tutorial will use a filter.

16. In the **Attribute Table** widget, click the **Options** list, and choose **Filter**.

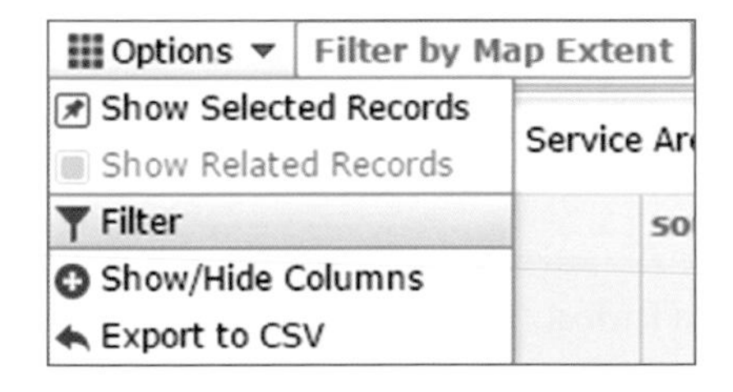

17. For **Add a filter expression**, select **Facility ID (Number)**, leave the operator as is, click **Unique**, and then select **2** from the dropdown list.

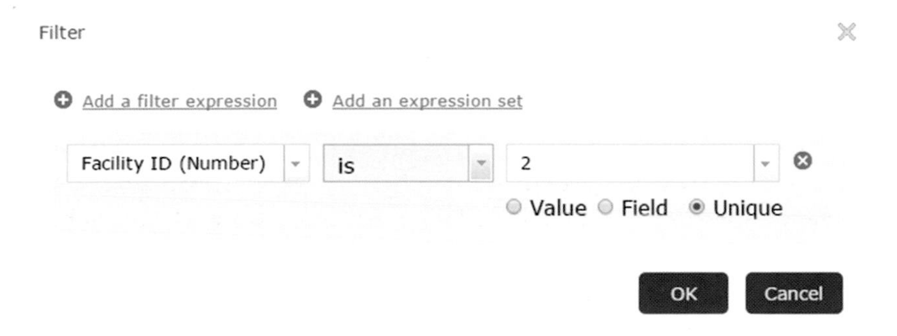

18. Hide the **Attribute table** widget by clicking the **Close** button icon at the upper-right corner of the widget.

19. Click the **Analysis** widget again. Click **Home** if you do not see the list of three steps.

20. Click **Step 3.**

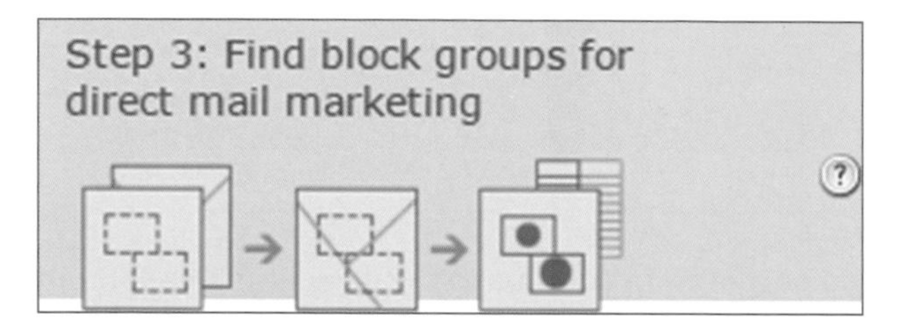

21. Specify the following parameters:

   - For **Choose area layer to summarize other features within its boundaries**, choose **Service Area Demographics.**
   - For **Choose layer to summarize**, select **Choose Living Atlas Analysis Layer.**

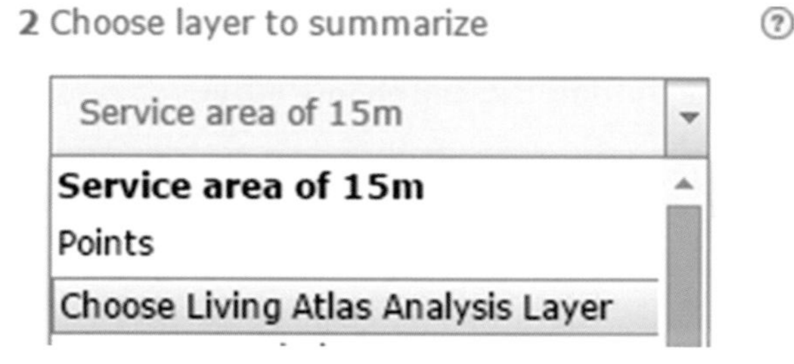

   - In the **Choose Living Atlas Analysis Layer** window, browse to find and then choose **USA Census BlockGroup Areas.**
   - For **Choose field to group by**, choose **FIPS**, which is the ID field of the block group layer.
   - For **Result layer name**, set the name as **Overlapping block groups.**

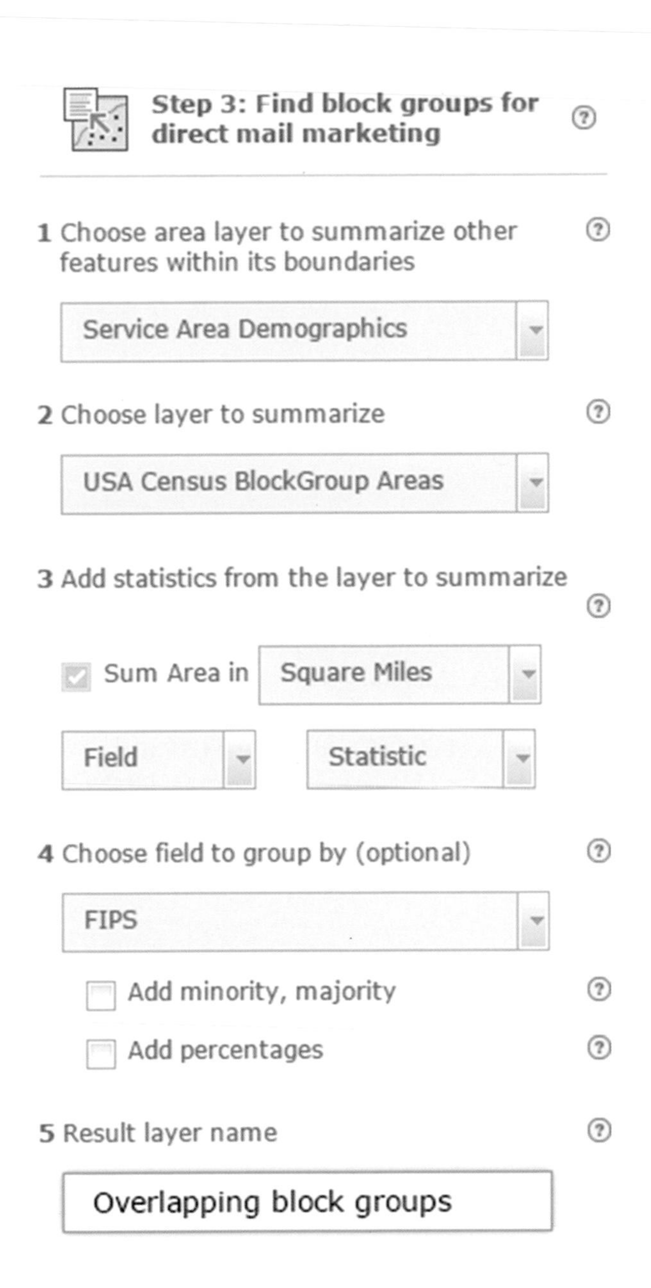

22. **Click Run Analysis.**

When you complete the analysis, the result automatically adds to the map.

23. **Click the Layer List widget, click the arrow next to the Overlapping block groups — GroupBySummary table, and choose Open Attribute Table.**

Operational Layers
Overlapping block groups - Overlapping block groups
Service Area Demographics - Service Area Demographics
Service area of 15m - Service area of 15m
Draw_Results
2015 USA Tapestry Segmentation
Overlapping block groups - GroupBySummary
Open Attribute Table
Description

You will see the FIPS of the overlapping block groups in the attribute table.

24. **In the Attribute Table widget, click the Options list, choose Export All to CSV. When the pop-up asks to you confirm your choice, click OK, and save the CSV file.**

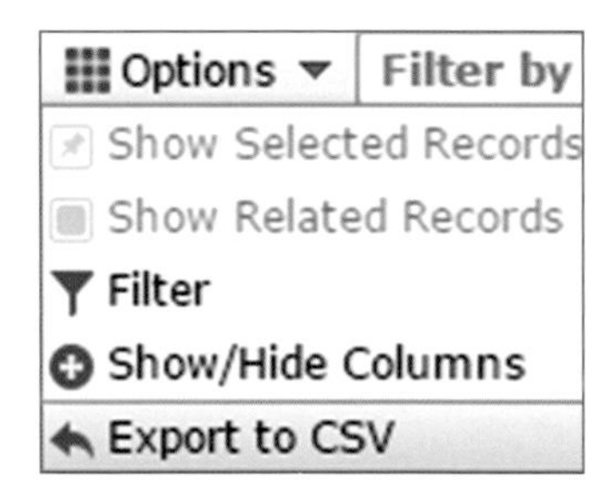

Having obtained the list of block group IDs in your selected restaurant service area, you can now arrange direct mail to these block groups. You can also turn on the tapestry layer to understand the socioeconomic status and spending behavior of each block group. This information will help you tailor the survey/marketing messages to each block group.

25. **Close Web AppBuilder for ArcGIS.**

In this simplified case study, you created a web app that enabled the company to perform restaurant location analysis. You could refine the analysis by obtaining demographic data for more details.

Note: The remaining sections teach you how to create and use geoprocessing services with ArcGIS for Server.

7.3 Design a geoprocessing model

This book does not focus on designing geoprocessing tools and for that reason does not discuss many details about ModelBuilder. Instead, the tutorial starts with a mostly completed model.

1. **Start ArcMap.**

2. **In the Catalog window, navigate to the folder connection C:\EsriPress\GTKWebGIS, and browse to the Chapter7 folder.**

If you do not have the folder connection, refer to the section on folder connections in the previous chapter.

In the **Chapter 7** folder, you will see a file geodatabase (.gdb), a map document (.mxd), and a toolbox (.tbx).

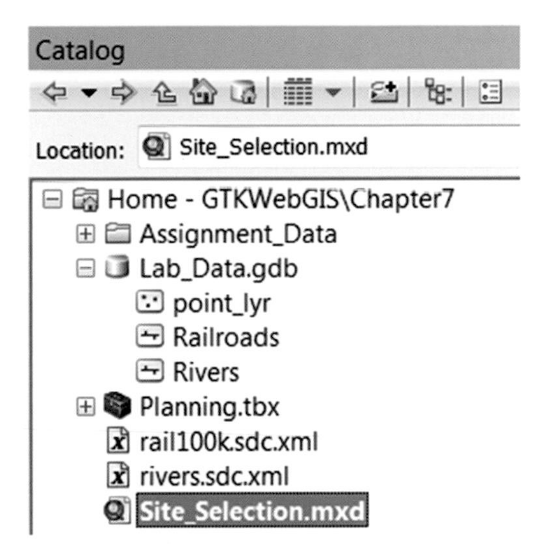

3. **Double-click to open the Site_Selection.mxd. The Railroads and Rivers layers appear on the map.**

4. **Expand the Planning.tbx file, right-click the Select_Sites model, and click Edit.**

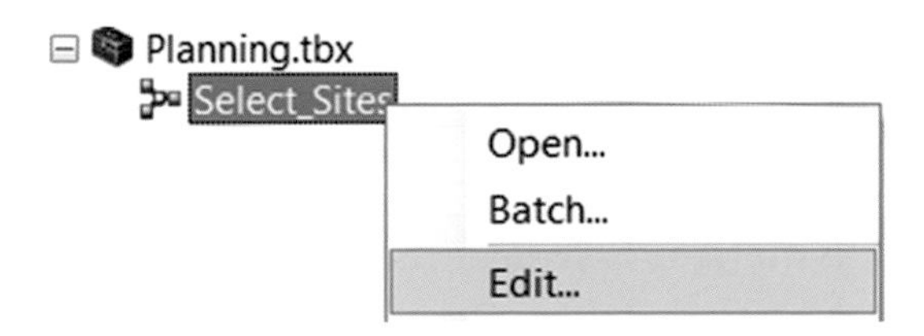

5. **Review the model elements.**

The model performs the following tasks:

- **Buffer1** generates a buffer around the location that users click (to be further configured).
- **Clip1** selects rivers that lie within a specified distance from the point of interest.
- **Clip2** selects railroads that lie within a specified distance from the point of interest.
- **Buffer2** generates a buffer around selected railroads.
- **Buffer3** generates a buffer around selected rivers.
- **Intersect3** intersects the two preceding buffers and finds common areas between them.
- **Dissolve** merges the adjacent little polygons into bigger polygons.

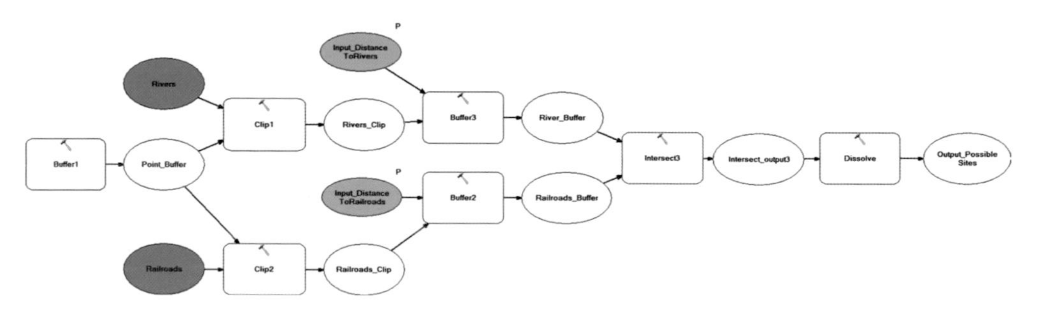

The model is not yet complete. To complete the model, you will define the input parameters for **Buffer1** first.

6. **Right-click Buffer1 and click Make Variable > From Parameter > Distance [value or field].**

This action allows users to specify the buffer distance surrounding the location of interest.

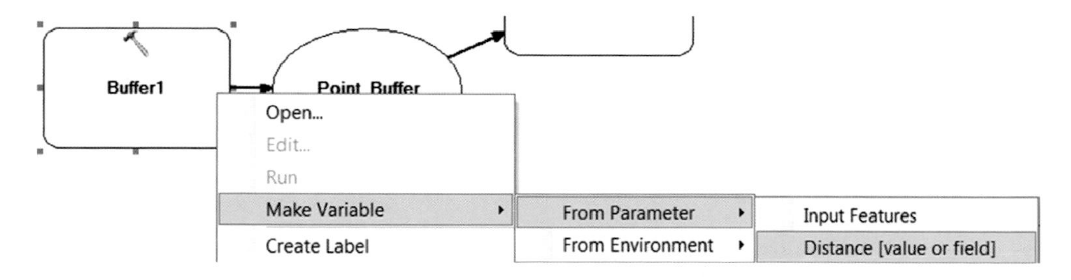

7. **Right-click Buffer1, and click Make Variable > From Parameter > Input Features.**

This action allows users to specify the location of interest.

8. **On the toolbar, click the Auto Layout button to rearrange the model elements layout.**

9. **Right-click the Input Features element, and click Properties (if you do not see this element, click the Full Extent button on the toolbar).**

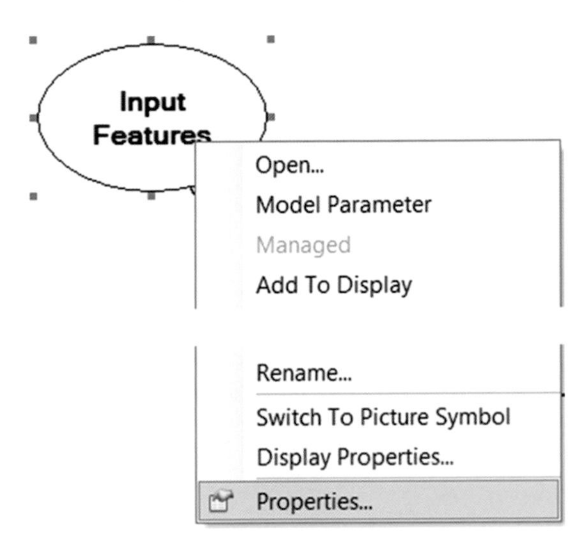

10. **In the Input Features Properties dialog box, click the Data Type tab, and set Select data type as Feature Set.**

11. **For Import schema and symbology from, navigate to the Lab_Data.gdb, select point_lyr, click Add, and then click OK.**

These actions ensure that users can specify only a point feature type (instead of a line or polygon type) for the location of interest.

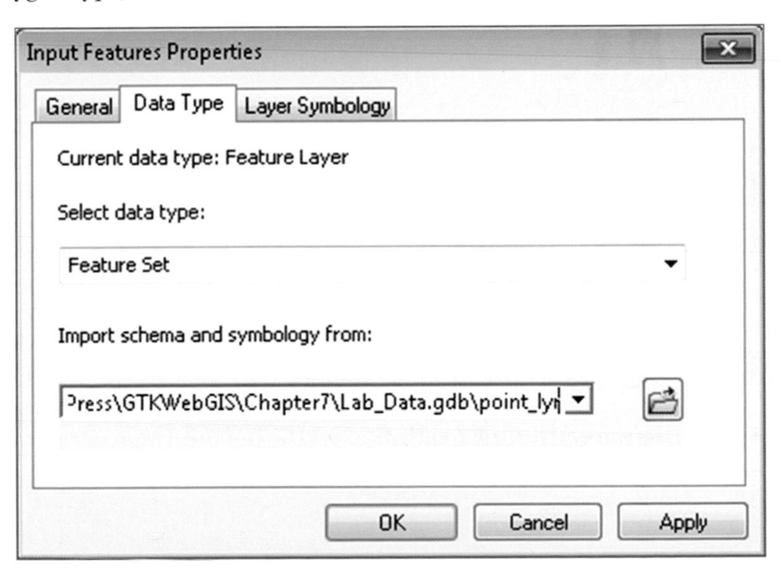

12. **Right-click the Input Features variable, click and specify Rename as Input_Location_Of_Interest, and click OK.**

Renaming the model parameters without special characters helps geoprocessing service users easily understand the meaning of the names and reference the parameters.

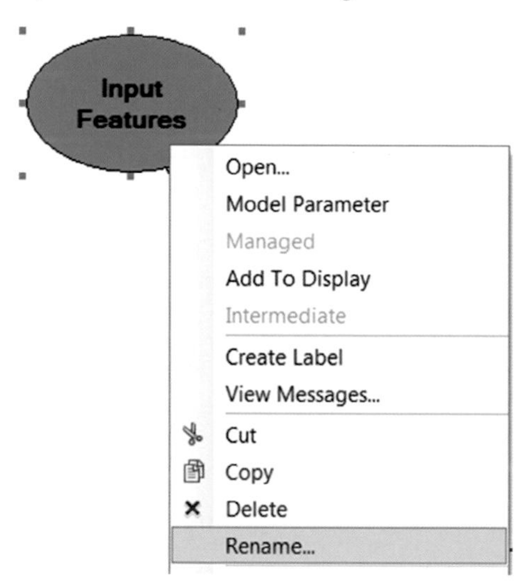

13. **Similar to the previous step, right-click the Distance [value or field] variable, click Rename, specify it as Input_DistanceOfInterest, and then click OK.**

14. **Right-click the Input_Location_Of_Interest variable and click Model Parameter.**

The letter "P" will appear to the upper right of this variable. This letter indicates that the variable is a model parameter. This parameter allows your users to specify new data or values for your model to process.

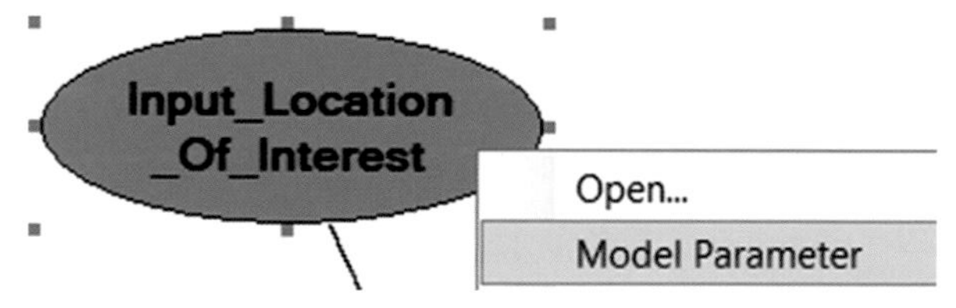

15. **Similar to the previous step, right-click the Input_DistanceOfInterest variable, and click Model Parameter.**

16. **Double-click the Input_DistanceOfInterest variable, set the default value of the variable to 60 Kilometers, and then click OK.**

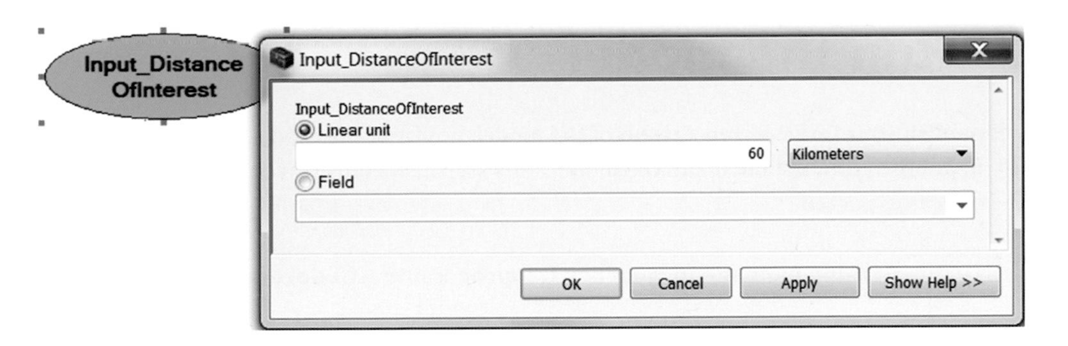

17. Right-click and set Output_PossibleSites as a model parameter.

Output_PossibleSites becomes the output parameter for your model.

18. Right-click Output_PossibleSites, and click Add To Display.

This action adds the model result to ArcMap so that the result displays on your map.

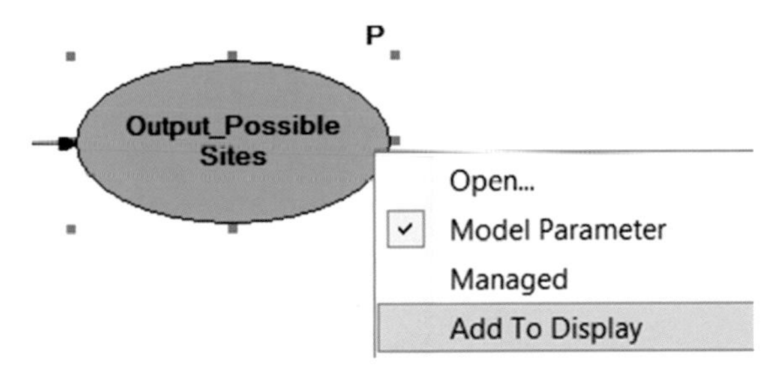

Your model is now complete and validated, with all the elements color-shaded. The round, blue-colored elements are inputs to the model. The rectangular, yellow-colored elements are tools. The round, green-colored elements are derived data, which is output from the model.

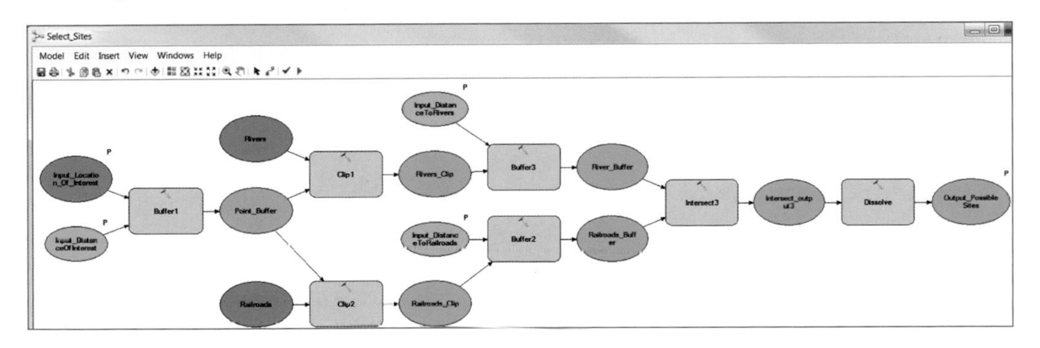

19. On the model toolbar, click the Save button to save your model, and close the ModelBuilder window.

7.4 Run the model

You will publish your service from a result of the model, not from the model itself. Therefore, before you publish, you first must run the model. This step also helps you confirm whether your model works as expected.

1. **On the ArcMap main menu bar, click Geoprocessing > Geoprocessing Options.**

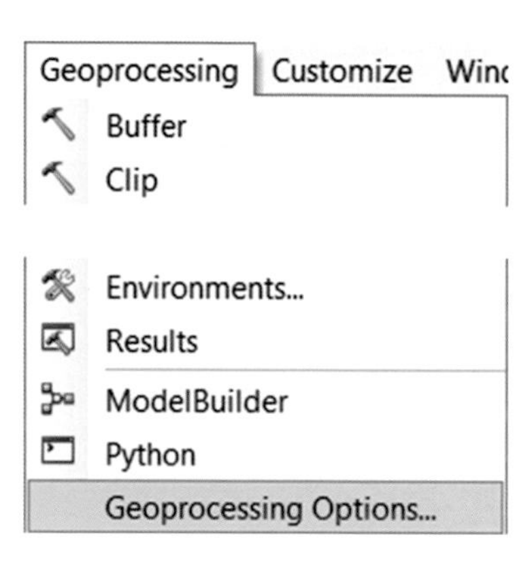

2. **In the Geoprocessing Options dialog box, select the check box for Overwrite the outputs of geoprocessing operations, and click OK.**

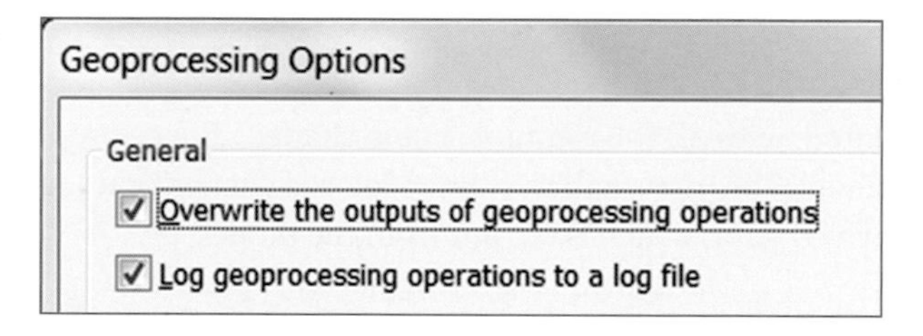

This option enables you to run your model many times without having to specify a different output feature class for each run.

3. **On the ArcMap main menu bar, click Geoprocessing > Environments.**

4. **Click Workspace, and for Current Workspace, navigate to C:\EsriPress\GTKWebGIS\Chapter7, click Lab_Data.gdb, and click Add. Set the same path and geodatabase for Scratch Workspace, and click OK.**

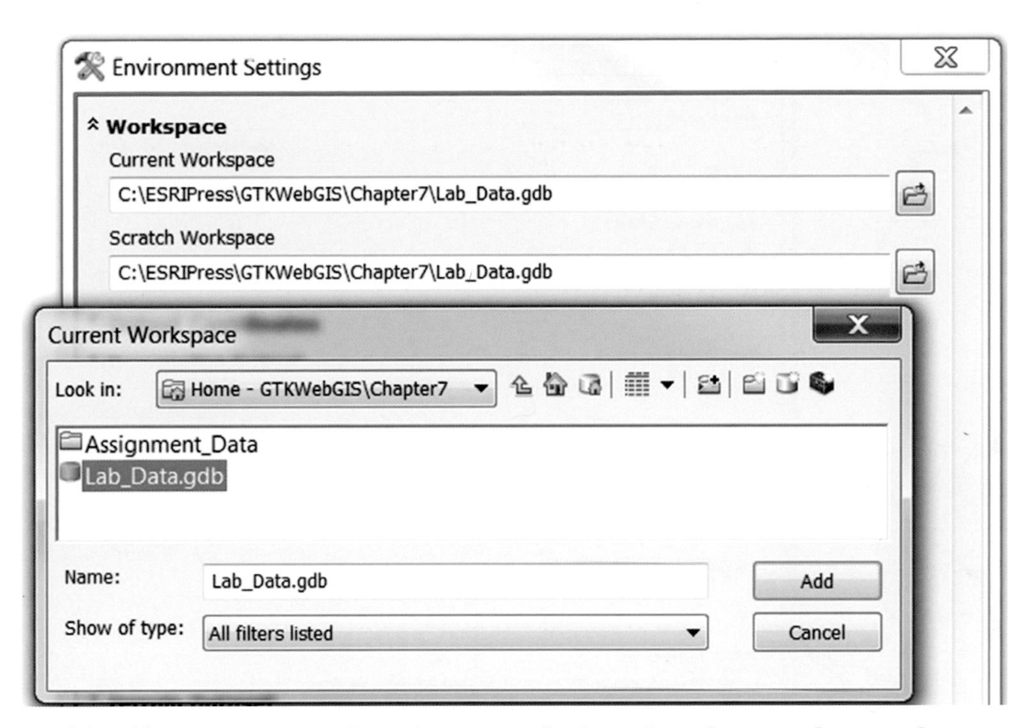

ModelBuilder writes intermediate datasets in the **Scratch** workspace—datasets that are useless once a model runs.

5. **In the Catalog window, double-click the Select_Sites model to run the model.**

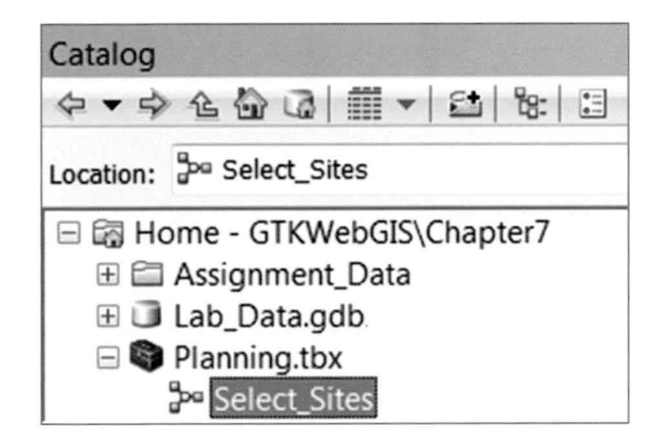

6. Specify the following inputs:

- **Next to the point symbol, click Input_Location_Of_Interest, and click a map location near rivers and railroads; otherwise, you might get an empty output.**
- **Leave the other parameters at their default values.**
- **Click OK to run the model.**

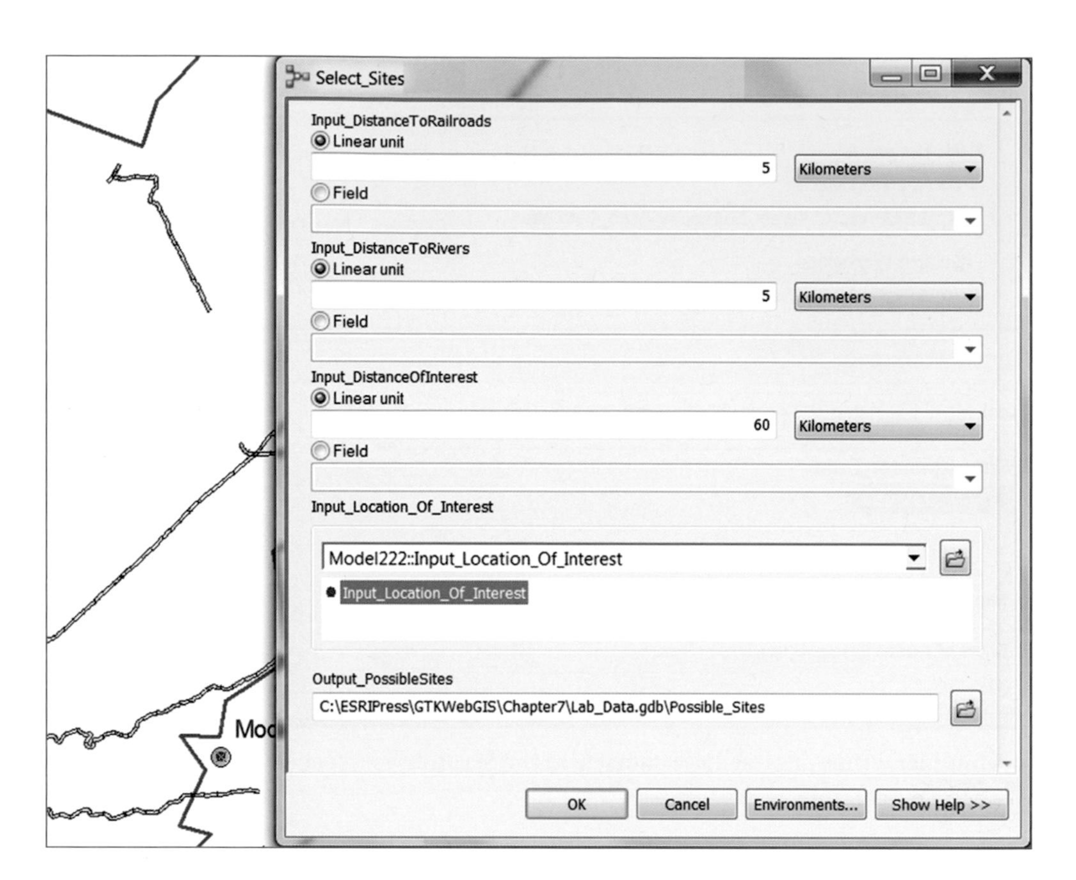

7. **Once you successfully run the model, close the dialog box manually if necessary.**

You will see the model results (**Possible_Sites**) added to ArcMap.

8. **Examine the results in ArcMap.**

These sites are situated close to the location you clicked on the map—and close to rivers and railroads.

7.5 Publish the execution path as a geoprocessing service

1. On the ArcMap main menu bar, click **Geoprocessing** > **Results**.

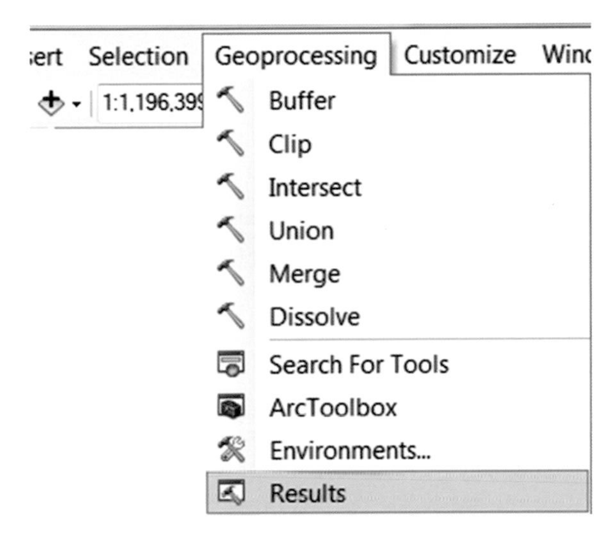

2. In the **Results** window, expand **Current Session**, and expand the **Select_Sites** tool result.

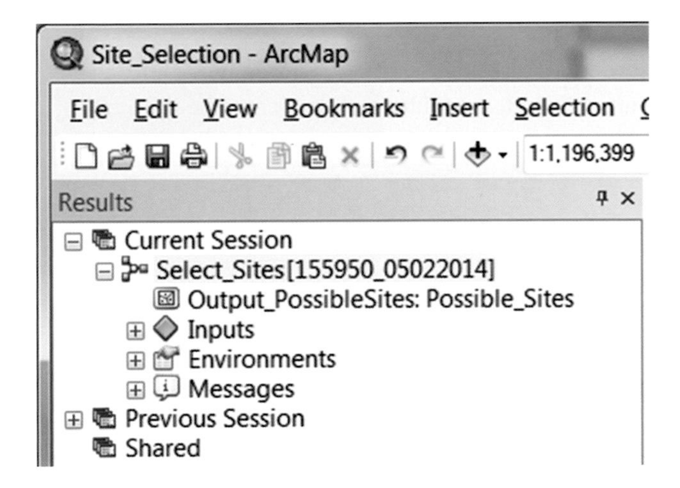

Here, you will find information about the inputs, environments, and messages generated when you ran the model.

1
2
3
4
5
6
7
8
9
10

3. **In the Results window, right-click the Select_Sites tool, and click Share As > Geoprocessing Service.**

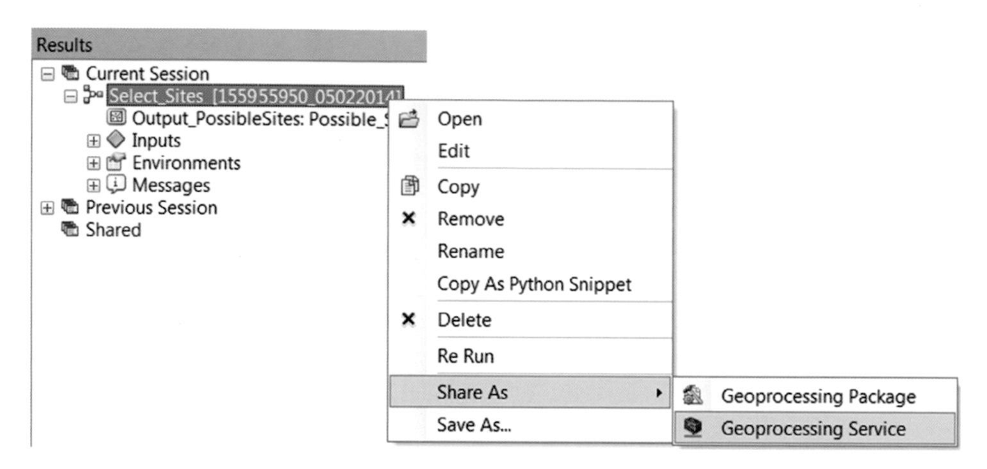

4. **In the Share as Service dialog box, perform the following tasks:**
   - **Select Publish a service, and click Next.**
5. **From the Choose a connection list, click the connection to your ArcGIS for Server (you can add a server connection if necessary by following chapter 6, section 6.1, "Connect to your GIS server"), specify a service name (for example, Planning), and click Next.**
6. **If you don't have a folder, choose Create new folder, and specify a folder name. If you have a folder, choose Use existing folder, and click your folder name. Click Continue.**

In a classroom setting where many students share one ArcGIS for Server, make sure the concatenation (or string combinations) of each student's folder and service name is unique.

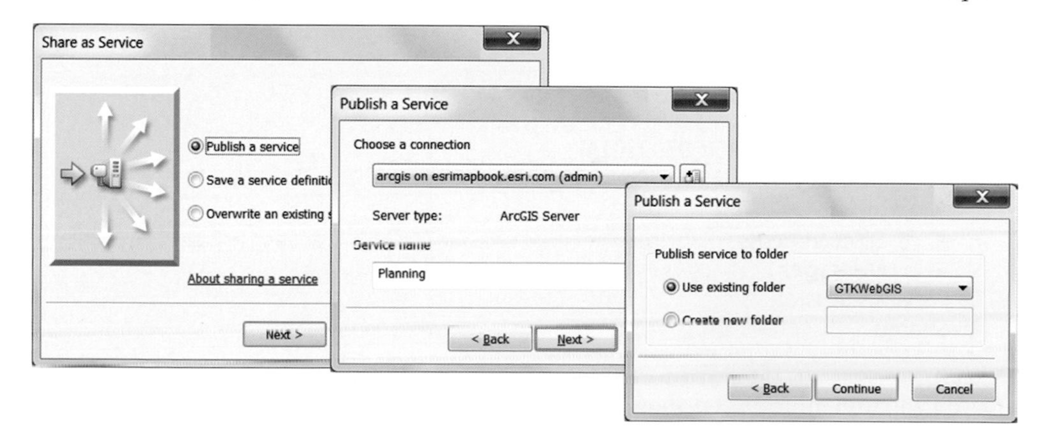

7. **In the Service Editor window, click the Parameters tab, and set Execution Mode as Asynchronous, because this task may take some time to run if your server is slow.**

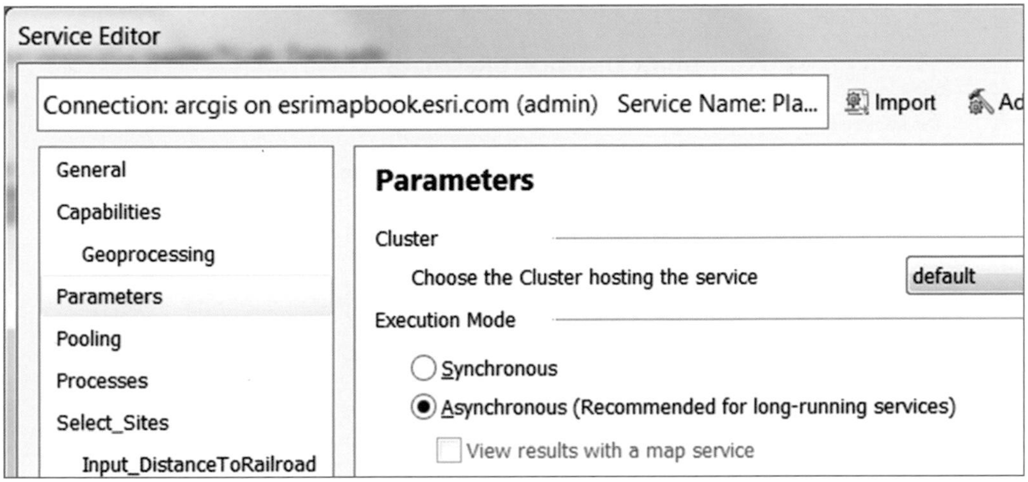

Under the **Asynchronous** option, you will see the option for **View results with a map service.** This option returns a geoprocessing task result as a map image (for example, a jpeg file). The returned map's symbology, labeling, transparency, and other properties are the same as those of the output layer in your current ArcMap session.

You can use a map service when your geoprocessing task generates too many features for your web clients to draw efficiently and when your task results in raster images.

8. **Click the Select_Sites tab to specify the following properties of the input and output parameters:**

- **Click Input_DistanceToRailroad, and set its description as Enter the distance to railroads.**
- **Click Input_DistanceToRivers, and set its description as Enter the distance to rivers.**
- **Click Input_DistanceOfInterest, and set its description as Enter the distance of rivers and railroads to the location of interest.**
- **Click Input_Location_Of_Interest, and set its description as Enter the location of interest.**
- **Click Output_PossibleSites, and set its description as Resulting sites meeting the criteria.**

Service Editor
Connection: arcgis on esrimapbook.esri.com (admin) Service Name: Pla... Import Ad
General
Capabilities
Geoprocessing
Parameters
Pooling
Processes
Select_Sites
Input_DistanceToRailroad
Input_DistanceToRivers
Input_DistanceOfInterest
Input_Location_Of_Intere:
Output_PossibleSites
Input_DistanceToRailroads
Name: (required) Input_DistanceToRailroads
Description: (required) Enter the distance to railroads
Type: Required Optional
Input mode: User defined value
Default value: 5 Kilometers

9. Click the **Item Description** tab to define the following service metadata:

- For **Summary**, type **Planning for the new factory.**
- For **Tags**, type **site selection, Getting to Know Web GIS.**
- For **Description**, type the following sentence: **This service provides a tool to select those sites that are close to the area users are interested in, close to rivers, and close to railroads.**

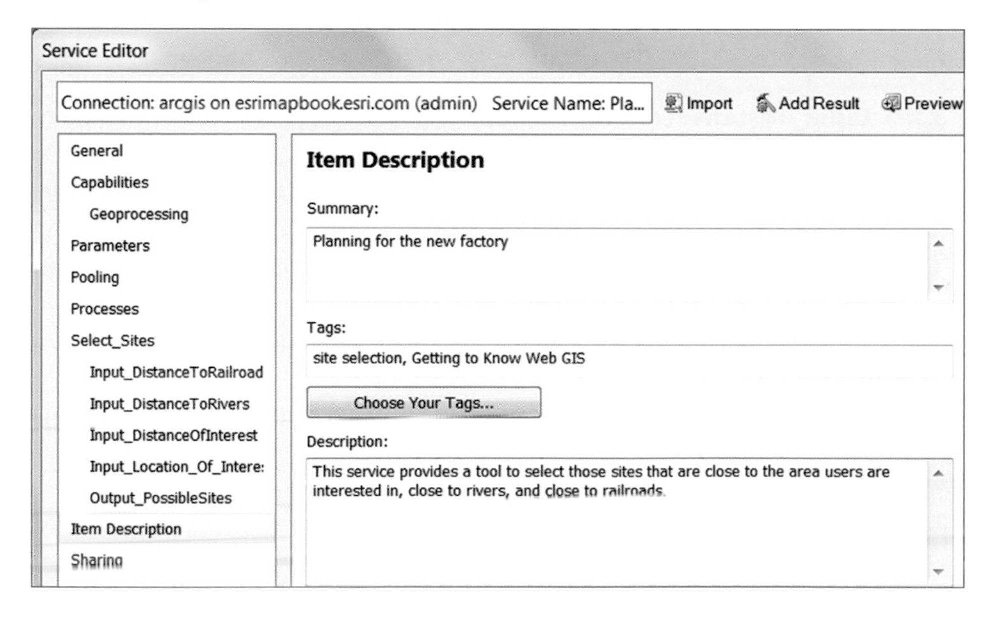

10. **On the toolbar of the Service Editor window, click the Analyze button ✓.**

Your **Prepare** window should look like the following figure (this window may be behind the **Service Editor** window):

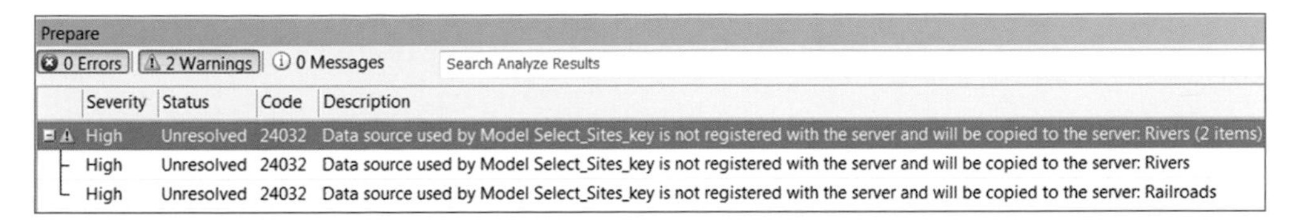

- The analysis should find no error messages.
- You may get two high-severity warnings: "Data source is not registered with the server," and "Data will be copied to the server." You can ignore these warnings and let the data be copied to the server.

11. **In the Service Editor, click the Publish button. If a prompt asks you to copy the data to the server, click OK (this may take a couple minutes). When a message window announces that your service is published, click OK.**

7.6 Explore your geoprocessing service in the Services Directory

ArcGIS REST Services Directory can help verify whether your geoprocessing service is published and working correctly. The directory can also help developers understand and test your geoprocessing service.

1. **If your ArcGIS for Server is federated with your Portal for ArcGIS, go to your Portal for ArcGIS in a web browser, sign in, and complete steps 1 through 4. Otherwise, go to step 5.**

You will need to sign in with the same account you used to connect to your ArcGIS for Server in ArcMap when you made the GIS server connection.

2. **Click My Content, and find the geoprocessing service you just published in the content list (Planning). Click the title of the service to see its item details page.**

1
2
3
4
5
6
7
8
9
10

3. **On the item details page, click Share to share your geoprocessing service with everyone.**

4. **Under the ArcGIS Web API REST Connection section, find the URL of the geoprocessing service. Click the URL link to go to its page in the ArcGIS REST Services Directory. Once you complete step 4, continue to step 8.**

5. **If your ArcGIS for Server and Portal for ArcGIS are not federated, start a web browser, navigate to the Services Directory of ArcGIS for Server, and complete steps 5 through 7.**

The URL is something like **http://your_server_name/arcgis/rest/services** or **http://your_server_name:6080/arcgis/rest/services** (*http* can be *https* depending your server configuration). If you are not clear about the URL, ask your instructor or system administrator. If you are on a computer installed with ArcGIS for Server, and if your computer is running Windows 7, you can click **Start** > **All Programs** > **ArcGIS** > **ArcGIS for Server** > **Services Directory** to go to your server **Services Directory**.

6. **On the home page in the Services Directory, click your folder if you specified a folder when you published the service.**

7. **Find the geoprocessing service you just published, and click the service.**

Note: The GPServer notation confirms that your service is a geoprocessing service.

Services:

- GTKWebGIS/Planning (GPServer)

8. **You can see the Select_Sites task in this geoprocessing service. Click the task.**

This action opens the REST page of this task.

9. **Observe the task description and the input and output parameters that you specified while publishing your service. Write the page URL, which is the REST URL of the geoprocessing task, on the following blank line:**

____.

You will need this URL when executing this geoprocessing task in section 7.7.

10. **Notice that Supported Operations include Submit Job, which indicates this is an asynchronous task.**

If this were a synchronous task, the supported operations would be **Execute Task** instead.

11. **Click Submit Job.**

This action opens the **Submit Task** page.

12. **For Input_Location_Of_Interest, scroll down in the text area, and find "features": []. Between the brackets, copy and paste or type the following text:**

```
{"geometry":{"x":-9642522,"y":3959878,
 "spatialReference":{"wkid":102100}}
}
```

Home > services > GTKWebGIS > Planning (GPServer) > Select_Sites > *submitJob*

Submit Job (Select_Sites)

Input_DistanceToRailroads: (*GPLinearUnit*)

```
{
 "distance": 5,
 "units": "esriKilometers"
}
```

Input_DistanceToRivers: (*GPLinearUnit*)

```
{
 "distance": 5,
 "units": "esriKilometers"
}
```

Input_DistanceOfInterest: (*GPLinearUnit*)

```
{
 "distance": 60,
 "units": "esriKilometers"
}
```

Input_Location_Of_Interest: (*GPFeatureRecordSetLayer*)

```
 "features": [
{"geometry":{"x":-9642522,"y":3959878,
  "spatialReference":{"wkid":102100}}
}
],
```

Options:

Output Spatial Reference:

Process Spatial Reference:

ReturnZ: True False

ReturnM: True False

Format: HTML

Submit Job (GET) Submit Job (POST)

This text specifies a point in JSON (JavaScript Object Notation) format. Note that JSON is case sensitive.

13. **For the other input parameters, leave the default values, and then click the Submit Job (POST) button.**

Because the task is asynchronous and you are using the REST API directly, you will need to check the job status repeatedly and retrieve the result when the job is completed.

14. **Click the Check Job Details Again link repeatedly to check the job status.**

Job Details: j6632de91214b4262b7e9592c4bfb214c

Job ID: j6632de91214b4262b7e9592c4bfb214c

Job Status: esriJobSubmitted

Check Job Details Again Cancel Job

However, ArcGIS web clients such as Web AppBuilder, ArcGIS API for JavaScript, and ArcGIS Runtime SDKs have built-in capabilities to check the status of a submitted job repeatedly without users having to do so manually.

15. **Once the task is completed, click Output_PossibleSites to see the analysis result.**

Job ID: j6632de91214b4262b7e9592c4bfb214c

Job Status: esriJobSucceeded

Results:

- Output_PossibleSites

The resulting feature set is in JSON format, which your web clients can use to display on maps. Do not be concerned about the long string of the result. This result is for your web clients to use in drawing their maps of possible sites. The JSON result represents a polygon, which symbolizes the possible site for the new factory. The result returned no error messages or warnings. This result confirms that your geoprocessing service has been created successfully and is running properly.

7.7 Create an app using your geoprocessing service

This section does not review the details of Web AppBuilder. Refer to chapter 5 for details.

1. **Sign in to your Portal for ArcGIS or ArcGIS Online (http://www.arcgis.com), and click My Content.**

2. **Click Create > App > Using the Web AppBuilder.**

3. **Specify a Title (for example, New Factory Site Selection), Tags (for example, site selection), optionally a Summary, and then click OK.**

4. **Choose a theme, for example, Launchpad.**

5. **Click the Map tab, and then click Choose Web Map.**

You will search for and use an existing web map that shows Alabama railways and rivers of Alabama.

6. **Click the Public tab, select ArcGIS Online, type Rivers and Railways Getting to Know Web GIS owner:pinde.webgis in the search box and click the search button. Select the only web map you see, and then click OK.**

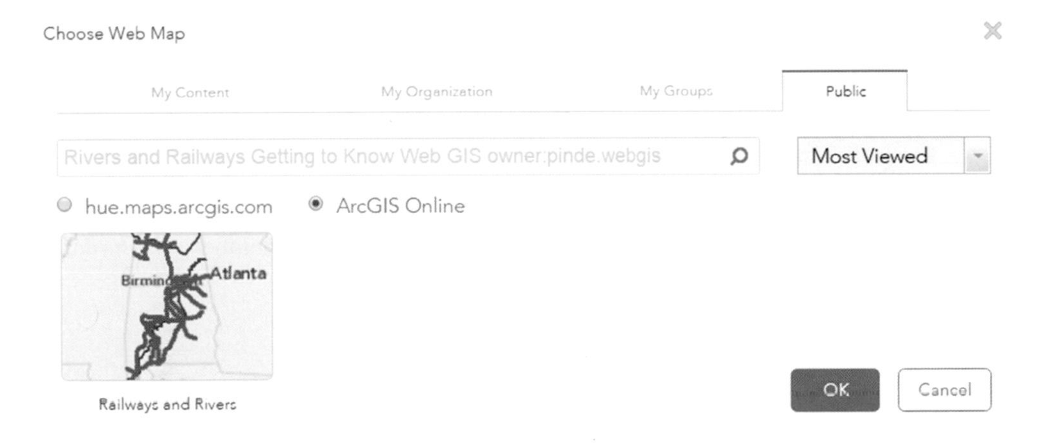

7. **Click the Widget tab, and then click the first placeholder. The map zooms to Alabama and shows the main rivers and railways that the geoprocessing service will use.**

8. **In the Choose Widget window, select the Geoprocessing widget, and click OK.**

9. **In the Configure Geoprocessing window, perform the following tasks:**

   - **Set the widget name as Site Selection.**
   - **For Task URL, click the Set button to open the Set GP Task window.**

10. **In the Set GP Task window, select Add Service URL, type or paste the task URL that you wrote down earlier in section 7.6. Click Validate, and then click OK.**

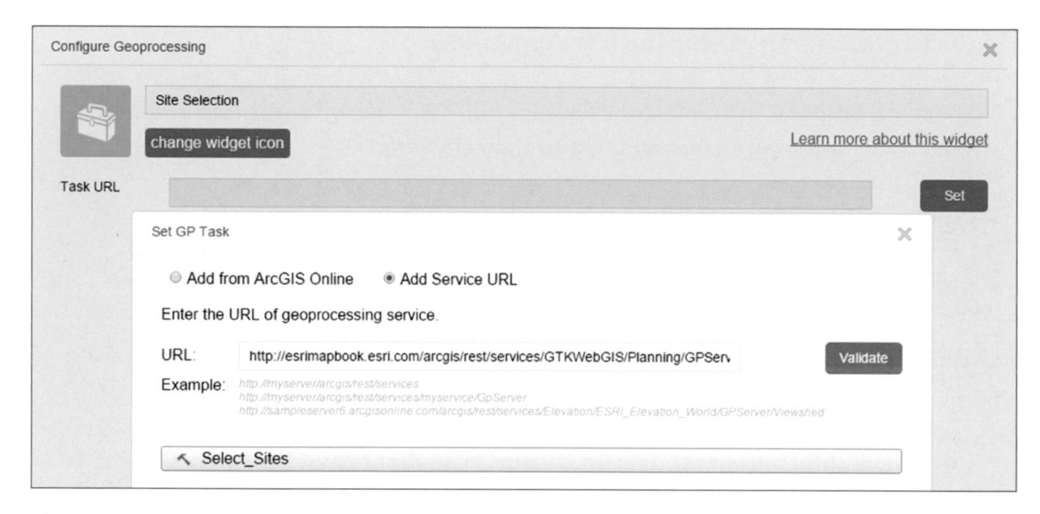

The **Geoprocessing widget** automatically detects the input and output parameters of your task.

- **Expand and click the Output parameter.**
- **Change Label and Tooltip of the output to Possible Sites.**
- **Change the output display renderer to use A Single Symbol, and then choose the polygon fill color and outline color.**

Optionally, you can click each of the input parameters and change the names of their labels and tooltips for readability. If you do not change the names, the parameter labels may look slightly different in the next several steps.

11. **Click OK to close the Configure Geoprocessing window.**

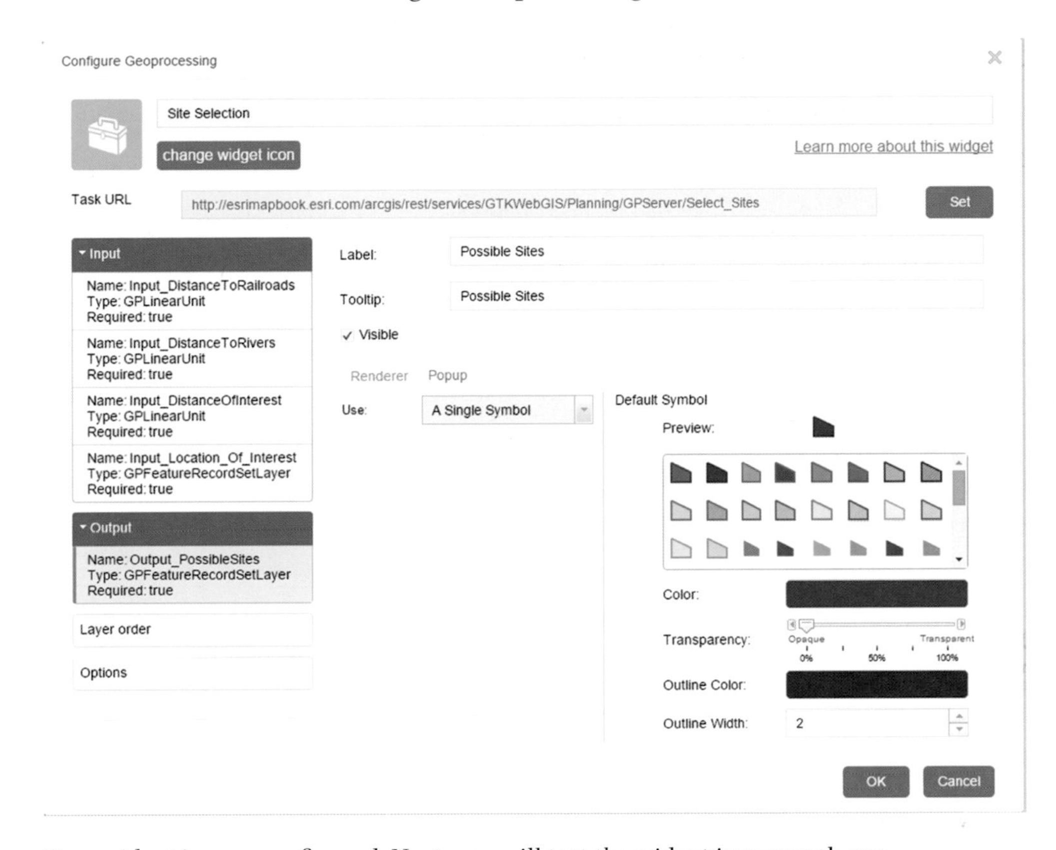

Your widget is now configured. Next, you will test the widget in your web app.

12. **In your app, click the Site Selection button. Use the default values for the input parameters, or specify new values, for example, 8 kilometers for Distance to Railroads, 9 kilometers for Distance to Rivers, and 50 kilometers for Distance of Interest.**

13. **Under Location of Interest, click the point icon. Click a location close to rivers and railroads on the map to specify your area of interest, and then click Execute.**

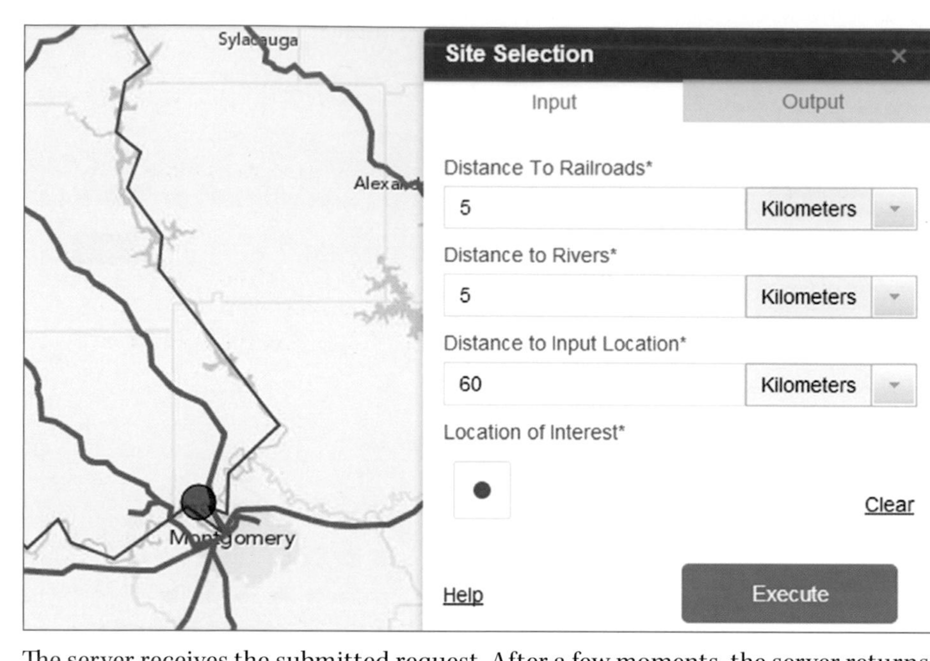

The server receives the submitted request. After a few moments, the server returns a resulting set of possible sites for the new factory and displays the sites on the map.

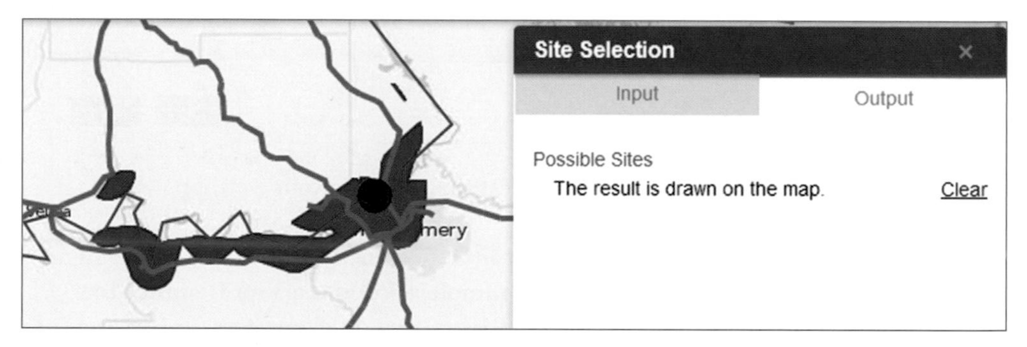

14. Click the Save button to save your web app configuration.

In sections 7.3 through 7.7, you finished a model in a toolbox, published the model as a geoprocessing service, explored the model in the ArcGIS REST Services Directory, and used the model in a web app. With this app, executives and other nonprofessional users can explore different scenarios, and (in the case of this tutorial) select the best site for a new factory.

In this tutorial, you actually finished two case studies, the first one using ArcGIS Online and the second one using ArcGIS for Server. You can consider the differences between the two approaches and compare what you find earlier in the chapter with table 7.1.

QUESTIONS AND ANSWERS

1. **ArcGIS Online analysis costs credits. Can I determine how many credits my analysis will cost before I run the analysis?**

   Answer: You will find a **Show credits** link next to the **Run Analysis** button in the **Perform Analysis** pane of ArcGIS Online map viewer as well as in the **Analysis** widget of Web AppBuilder. Click the **Show credits** link to see the credits that will be used.

2. **ArcGIS Online analysis costs credits, but ArcGIS for Server analysis does not. Can I say it costs less to use ArcGIS for Server to do analysis?**

   Answer: Not necessarily. ArcGIS Online and ArcGIS for Server fit different needs. See table 7.1.

   If ArcGIS Online analysis and data meet your needs, ArcGIS Online is often more cost efficient than purchasing/installing ArcGIS for Server, collecting the data, and creating and hosting your own geoprocessing services.

   If you need to share your specific models and specific data as web services, you will need to use ArcGIS for Server.

3. **Can I combine or chain multiple ArcGIS Online analysis tools and run the tools automatically without having to run each tool manually?**

   Answer: Yes

   To combine or chain multiple ArcGIS Online analysis tools and run the tools automatically without having to run each tool manually, you must write your own programs using JavaScript, Python, or other language. For example, ArcGIS API for JavaScript provides components to integrate ArcGIS Online analysis tools. You can use the output of one tool as the input for another tool and thus can chain multiple tools.

4. As I consider ModelBuilder or Python, which one should I use to author my model or tool?

Answer: The answer varies depending on what you want to accomplish and on your current skill sets.

Use ModelBuilder if you are new to both and need to get the work done quickly.

- ModelBuilder is much easier and quicker to learn than Python.
- ModelBuilder excels at visually and intuitively depicting the workflow for most tasks that you want to accomplish.
- Even if you prefer to learn Python, ModelBuilder can be a better place to start. ModelBuilder can export some models to Python scripts, so the skeleton of the script is already written for you.

If you already know or have more time to use Python, go ahead and use Python.

- Python is a scripting language. You can accomplish more complex workflows and exert more fine-grained control. For example, simple text manipulation is difficult using ModelBuilder, whereas Python makes the task easier.
- With additional libraries or modules, Python can better integrate with other software tools, such as Microsoft Excel, statistical Package R, or procedures in an RDBMS (relational database management system).
- In addition to using your tool to create a geoprocessing service, you may also want to run your Python script outside ArcMap or schedule the script to run at a certain time.

ASSIGNMENTS

Assignment 7A: Use ArcGIS Online analysis to purchase your new home.

You are in the market to look for a new home. You are considering many factors, such as the distance between your home and work or your children's schools, neighborhood income, median house value, median family income, unemployment rate, and crime rate. List your consideration factors and criteria, and create an app that uses ArcGIS Online analytics to help you narrow the area in which you would like to buy your new home.

Data:

- Provide the locations of your and your family members' offices or schools yourself.
- Use layers from the Living Atlas of the World.

Tips:

- You can add your and your family members' work or school locations using a CSV file, or you can add the locations interactively using a map notes layer in the map viewer.
- Browse the Living Atlas layers to see what layers are helpful for your decision. Add those layers to your web map.
- You can experiment with the analysis tools to see what tools are needed for your analysis. You will select these tools in the Web AppBuilder analysis widget.
- You will probably need the **Enrich layer** tool. Use the data browser to see what variables will help you choose your new home.
- Once you have your web map ready and know what tools you need, use Web AppBuilder to create a web app that anyone can use to decide where to buy new homes.

What to submit: Send an email to your instructor with the subject line **Web GIS Assignment 7A: Your name,** and include the following information:

- The layers in Living Atlas that you find useful for your home search
- The variables in the **Enrich Layer Data Browser** window that you selected
- The URL of the web app you created using Web AppBuilder
- Screenshots of your analysis processes and your final choice of neighborhood areas

Assignment 7B: Create a web app to clip, zip, and ship GIS data.

A survey and mapping bureau would like to improve its data-sharing workflow. Instead of extracting and copying data manually to serve its customers, the bureau wants to automate the workflow with a web app.

Data: The data is located at **C:\EsriPress\GTKWebGIS\Chapter7\Assignment_Data**. It contains the following data:

- **data.gdb,** which has the earthquake and hurricane layers that your users can download
- **natural_disasters.mxd,** which displays these data layers
- **ExtractData.tbx,** which contains **ExtractData,** a to-be-completed model for clipping, zipping, and shipping data

Requirements: This web app should allow its users to select the layers they need, draw the area of interest, select the desired data projection and format, and have the data clipped and zipped for them to download.

- Open **natural_disasters.mxd** in ArcMap.
- Edit the **ExtractData** model, and set **Layers** to **Clip, Feature Format, Area of Interest,** and **Output_zip File** as model parameters.
- Run the **ExtractData** model, and publish the result of the model as a geoprocessing service.
- Examine your geoprocessing service task in ArcGIS REST Services Directory.
- Create a web app using Web AppBuilder:
  - Use the web app you created in the chapter 6 tutorial. This web app displays the earthquake and hurricane layers so that users can see where the earthquakes and hurricanes are.
  - Add the **Geoprocessing** widget, and configure it to point to the **ExtractData** task in your geoprocessing service.

Note: This assignment is similar to the ArcGIS for Server Clip and Ship Geoprocessing Service example in ArcGIS Help, where you can find some **useful** instructions **(http://server.arcgis.com/en/server/latest/publish-services/windows/gp-service-example-clip-and-ship.htm).**

What to submit: Send an email to your instructor with the subject line **Web GIS Assignment 7B: Your name,** and include the following information:

- The URL to your geoprocessing task
- The URL to your web app

Resources

ArcGIS Online Help document site

"Living Atlas of the World," http://doc.arcgis.com/en/living-atlas.
"Perform Analysis," http://doc.arcgis.com/en/arcgis-online/use-maps/perform-analysis.htm.
"Use the Analysis Tools," http://doc.arcgis.com/en/arcgis-online/use-maps/use-analysis-tools.htm.

ArcGIS for Server Help document site

"A Quick Tour of Creating Tools with ModelBuilder," https://desktop.arcgis.com/en/desktop/latest/analyze/modelbuilder/a-quick-tour-of-creating-tools-with-modelbuilder.htm.
"A Quick Tour of the Geoprocessing Service Examples," http://server.arcgis.com/en/server/latest/publish-services/windows/a-quick-tour-of-the-geoprocessing-service-examples.htm.
"Authoring Geoprocessing Tasks with ModelBuilder," http://server.arcgis.com/en/server/latest/publish-services/windows/authoring-geoprocessing-tasks-with-modelbuilder.htm.
"Authoring Geoprocessing Tasks with Python Scripts," http://server.arcgis.com/en/server/latest/publish-services/windows/authoring-geoprocessing-tasks-with-python-scripts.htm.
"Geoprocessing Service Example: Clip and Ship," http://server.arcgis.com/en/server/latest/publish-services/windows/gp-service-example-clip-and-ship.htm.

Esri Tutorials and Trainings

"A Place to Play," https://learn.arcgis.com/en/projects/a-place-to-play.
"Gain Geographic Insight with ArcGIS Online Analysis Tools," http://training.esri.com/gateway/index.cfm?fa=catalog.webCourseDetail&courseid=2717.
"I Can See for Miles and Miles," http://learn.arcgis.com/en/projects/i-can-see-for-miles-and-miles.
"Introduction to Geoprocessing Scripts Using Python," http://training.esri.com/gateway/index.cfm?fa=catalog.courseDetail&CourseID=50131064_10.x.
"Manage Driver Routes," http://learn.arcgis.com/en/projects/lets-get-the-chow-mein-on-the-road/lessons/manage-driver-routes.
"Python for Everyone," http://training.esri.com/gateway/index.cfm?fa=catalog.webCourseDetail&courseid=2520.

"Restaurant Report Cards," http://learn.arcgis.com/en/projects/restaurant-report-cards.

"Sharing Analysis Workflows on the Web Using Geoprocessing Services," http://training.esri.com/gateway/index.cfm?fa=catalog.webCourseDetail&courseid=2507.

"Spatial Analysis with ArcGIS Online," http://training.esri.com/gateway/index.cfm?fa=catalog.webCourseDetail&courseid=2793.

Chapter 8
Mobile GIS and real-time GIS

This chapter covers two different but related topics—mobile GIS and real-time GIS. Mobile devices are becoming the pervasive client platform for web GIS, largely because of the fundamental human need for convenience and the proliferation of smartphones and tablets. Real-time GIS is another important topic in web GIS because of the large volume of real-time data collected from mobile phones, individual sensors, sensor networks, and the Internet of Things. For mobile GIS, this chapter introduces three options for building apps—browser-based, native-based, and hybrid-based; illustrates the use of Collector for ArcGIS mobile app for collecting data; and teaches the use of AppStudio for ArcGIS to build native apps. For real-time GIS, this chapter teaches the use of Operations Dashboard for ArcGIS to monitor near real-time data.

Learning objectives

- *Understand the three approaches to building mobile apps.*
- *Use Collector for ArcGIS to collect GIS data.*
- *Use AppStudio for ArcGIS to build native apps.*
- *Understand real-time GIS and its related technologies.*
- *Use Operations Dashboard for ArcGIS to monitor near real-time data.*

This chapter in the big picture

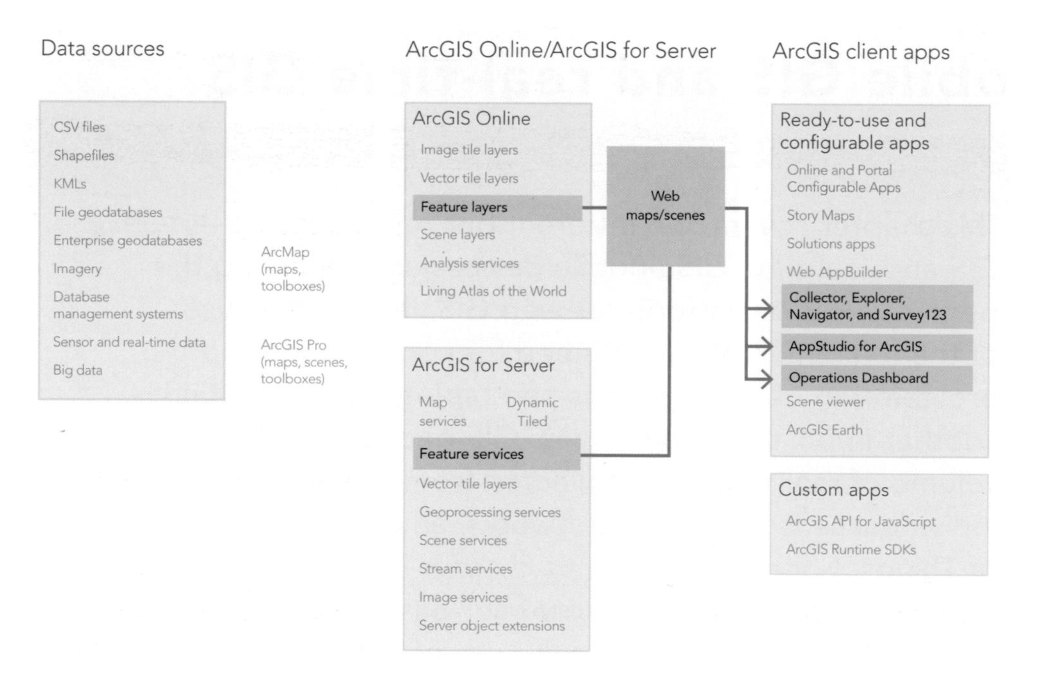

The ArcGIS platform offers many web GIS apps and many ways to build apps. The green lines in the figure highlight the technologies that this chapter teaches.

Mobile GIS advantages and applications

Mobile GIS refers to GIS for use on mobile devices and has the following advantages over traditional desktop GIS:

- **Mobility:** Mobile devices are not hindered by wire or cable. These devices extend GIS to areas where wiring is infeasible or costly and users most need GIS.

- **Location awareness:** You can use GPS, cellular networks, Wi-Fi networks, Bluetooth technology, and other technologies to pinpoint the current location of a mobile device. You also can use the compass, gyroscope, and motion sensors of a mobile device to determine the device's direction, tilt angle, and moving speed.
- **Ease of data collection:** Mobile GIS can replace existing paper-based workflows, which are prone to errors that arise when surveyors draw on the map in the field and enter data manually back in the office. Mobile GIS can replace such paper datasheets, reduce costs, and improve the accuracy of your data.
- **Near real-time information:** The live connection of mobile networks greatly enhances the temporal dimension of GIS. This dimension gives mobile GIS the potential ability to monitor the spatial and temporal aspects of the world around us.
- **Large volume of users:** Mobile devices create a pervasive platform for GIS and deliver GIS to the hands of billions of people.
- **Versatile means of communication:** Integrating with voice, short message, photo, video, email, and many social networking apps, mobile devices facilitate collaboration and communication among professionals and consumers.

Mobile GIS has broad application for consumers and organizations. Individual consumers use mobile GIS to learn what is nearby, where to eat, and how to get there. Organizations need mobile GIS to complete such tasks as field mapping, data query, field inspection and inventory of assets, tracking assets and field crews, field surveys, incident reporting, and parcel delivery.

Location-based services and augmented reality

Mobile GIS is often related to two popular types of apps and frontiers: location-based services (LBS) and augmented reality (AR).

- LBS refers to information services that integrate the locations of mobile devices to provide added value to mobile users. A desktop GIS app allows you to click a point of interest (POI) to get its information. With LBS, you essentially become the mouse cursor on a map of 1:1 scale—the real world. As you enter an area or get close to a POI, the mobile apps on your phone know where you are and push to you information about this area or POI.
- AR combines information from the database, especially from the web, with information from human senses. AR is often related with mobile GIS because a mobile device can retrieve information based on your location, the direction you are facing, the tilt angle of your view, and what you may see (captured by your camera), and it can display or overlay the retrieved information on your phone. For example, an AR-based tour guide app can provide you information about the building in front of you if you point your phone to the building. The app can even superimpose the building's historic pictures on its current photo. And although we cannot see through solid ground with our eyes, if you hold your phone downward, your phone's AR app can retrieve underground pipeline information and display an underground pipeline map to help you "see" underground.

1
2
3
4
5
6
7
8
9
10

Mobile app development strategies

Choosing a mobile app development strategy depends on the development team's skill set, the application's required functionality, the targeted platform(s), and the amount of funding available. Mobile application development includes the following approaches:

- **Browser-based approach:** This approach builds apps using HTML, JavaScript, and CSS (Cascading Style Sheets). Users access these apps via mobile web browsers. This strategy can potentially reach all mobile platforms. Browser-based apps typically are less costly and quicker to develop than native apps. However, browser-based apps cannot access the full capabilities of mobile devices, such as notifications. As such, the user experience of browser-based apps typically cannot compete with the experience of using native apps, which do not have the limitation of web browsers.
- **Native-based approach:** You must download and install native apps on your mobile device. For example, apps downloaded from App Store and Google Play are native apps. This approach requires native development skills, such as Objective-C or Swift for iOS, Java for Android, and .NET for Windows Phone. These apps typically have deep-level access to device hardware and other resources and typically have better performance than browser-based apps. However, native apps are often more expensive to develop than JavaScript apps, and one app cannot run on multiple platforms.
- **Hybrid-based approach:** This approach integrates native components and HTML/JavaScript/CSS to build native applications. You can achieve this integration in many ways. The simplest way embeds a web control into a native app to load HTML and JavaScript contents. More advanced methods include the use of frameworks such as Adobe PhoneGap to allow deeper integration with the native platform.

ArcGIS offers APIs that work for each of these approaches. ArcGIS API for JavaScript can support both browser-based and hybrid-based approaches, while the ArcGIS Runtime Software Development Kit (SDK) for Mobile (iOS, Android, Windows Phone, and a cross-platform app development framework called Qt) supports the native-based approach. Each of these APIs ultimately offers similar core functionalities that include editing, layers, graphics, and geometry, as well as access to ArcGIS web maps and web services and a variety of tasks such as querying, identifying, searching, and geoprocessing.

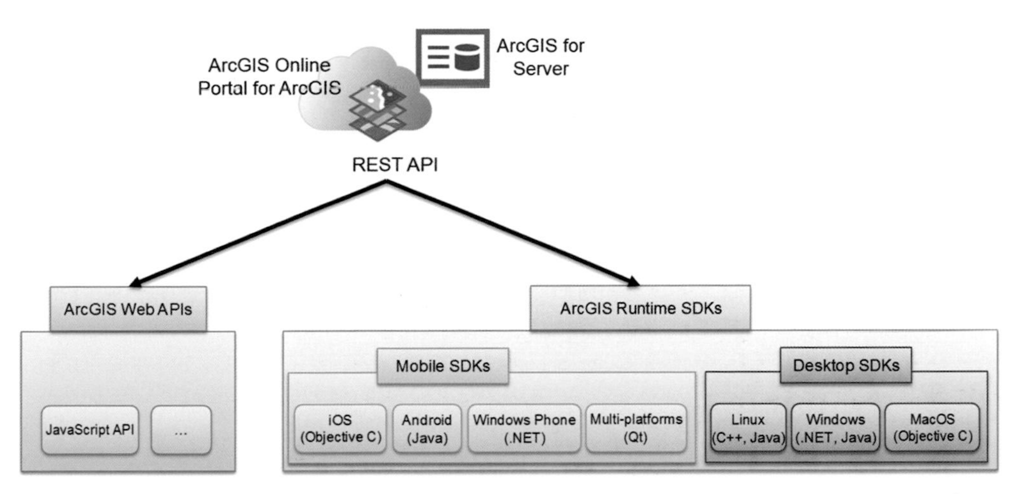

ArcGIS provides a suite of web APIs and runtime SDKs, which interact with ArcGIS Online, Portal for ArcGIS, and ArcGIS for Server via ArcGIS REST API.

Because ArcGIS Web APIs, Runtime SDKs for Mobile, and Runtime SDKs for Desktop share similar concepts, understanding one helps you to learn another. For example, you may refer to a sample code of ArcGIS API for JavaScript when you develop Android apps using Java.

ArcGIS native apps

ArcGIS provides the following native apps:

- **Collector for ArcGIS** helps you perform the following functions:
    - Collect and update information in the field, including attaching photos and videos.
    - Fill in data entry forms to collect attribute information.
    - Easily access and use your and your organization's web maps.
    - Perform online and offline data collection.
- **Survey123 for ArcGIS** helps you perform the following functions:
    - Design surveys with predefined questions that support domains and feature templates, default values, embedded audio and images, and dependency among questions (if the answer to one question is true, then show a related question; otherwise, do not show the related question). These predefined surveys will make it easier for users to quickly fill out the form and ensure data quality.
    - Capture field data using an intuitive form-centric data-gathering solution.
    - Store survey results in hosted feature layers that you can easily share with organization users.
    - Perform online and offline data collection.

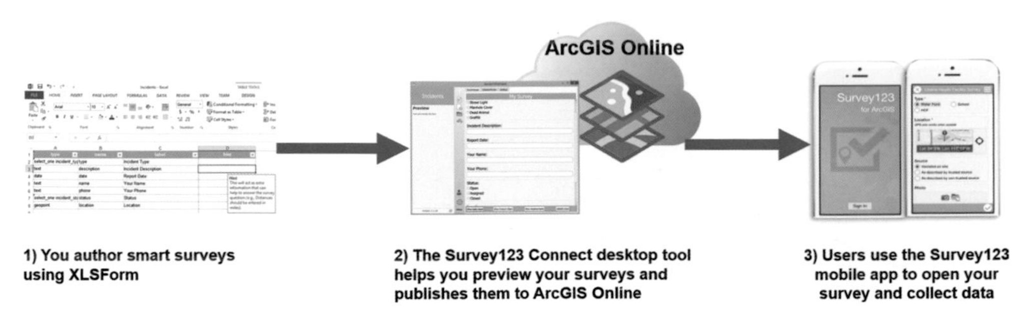
ArcGIS Online
1) You author smart surveys using XLSForm
2) The Survey123 Connect desktop tool helps you preview your surveys and publishes them to ArcGIS Online
3) Users use the Survey123 mobile app to open your survey and collect data

With Survey123 Connect and Quick Designer, you can design survey questions easily and upload your surveys to ArcGIS Online. With Survey123 mobile app, your surveyors or users can collect data quickly using an intuitive form-centric user interface.

- **Explorer for ArcGIS**, a native app, enables you to search and display web maps, search for places and features in web maps, share maps with other users as images and links, and give presentations with interactive maps.
- **Navigator for ArcGIS**, also a native app, is more than a consumer-level navigation app. In addition to common functions such as searching for locations, getting directions, and voice-guided, turn-by-turn navigation, Navigator for ArcGIS allows you to use your organization's road network data, access downloadable maps for offline navigation, and integrate with ArcGIS apps, such as Collector for ArcGIS.

AppStudio for ArcGIS

AppStudio for ArcGIS provides a template-based approach for building cross-platform apps without coding. You select an out-of-the-box template, configure it by following an easy-to-use wizard, link it to an existing web map/app or feature layer, specify your own branding, and lastly build the app for one or multiple platforms including iOS, Android, Windows, OS X, and Linux. AppStudio for ArcGIS lets developers and organizations quickly turn their existing web maps, apps, and layers into beautiful, consumer-friendly cross-platform native apps.

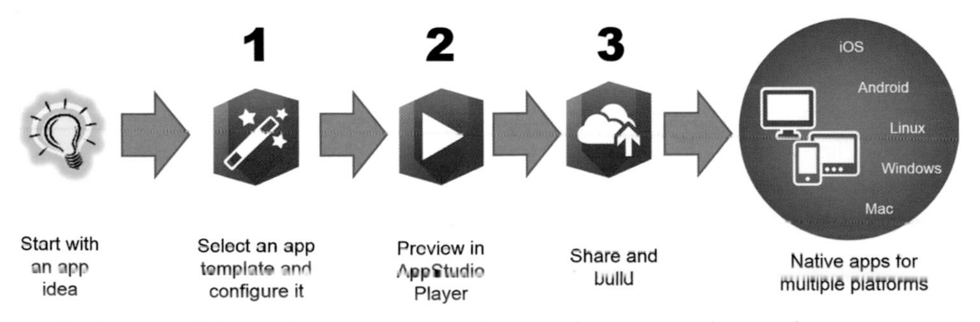
1
2
3
iOS
Android
Linux
Windows
Mac
Start with an app idea
Select an app template and configure it
Preview in AppStudio Player
Share and build
Native apps for multiple platforms

AppStudio for ArcGIS provides app templates. You can select one template, configure it, preview it, and build native apps for your target platform(s).

Real-time GIS

Real-time GIS refers to GIS that handles current and continuous data, which can be the latest position, altitude, speed, direction, temperature, pressure, concentration, or water level of various sensors and other objects. Real-time GIS supports better decision making by providing data at the moment events happen. Real-time GIS can provide better situational awareness, enhance emergency response, allow live tracking of resources or other targets, and facilitate the collection of big data.

Live data is often collected from mobile phones, individual sensors, and networks of sensors. **Sensor network** refers to spatially distributed autonomous sensors that can cooperatively pass their data through the network, often wirelessly, for live data access and analysis. **Internet of Things (IoT)** is the network of physical objects or "things" embedded with sensors and network connectivity, which enable these objects to collect and exchange data. The fast development of IoT will generate a huge amount of real-time data and demand real-time GIS to locate and track objects and manage, search for, display, and analyze the data.

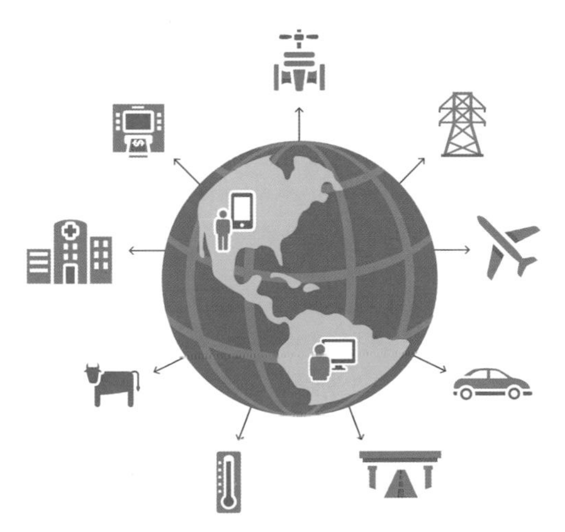

Sensor network (top) and IoT (bottom) provide live data about the world and need real-time GIS to visualize, monitor, and analyze the data.

The ArcGIS platform includes many products for real-time GIS. Mobile client apps such as Collector for ArcGIS and Survey123 can collect data in real time. ArcGIS Online and Portal map viewer can create web maps referencing live data. Operations Dashboard and Web AppBuilder can display live data. ArcGIS GeoEvent Extension for Server can receive and analyze real-time data. This extension can connect many types of streaming data, perform continuous data processing and analysis, and send updates and alerts when specified conditions occur, all in real time.

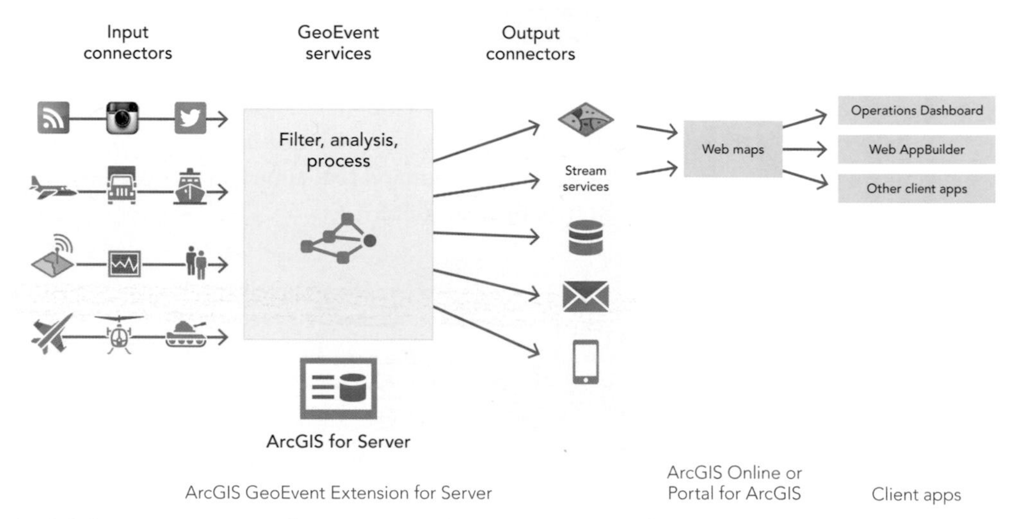

ArcGIS GeoEvent Extension for Server provides input connectors for taking in real-time data and processors for performing real-time filtering and analysis. The extension also provides output connectors for generating results, which can be used in web maps and apps.

Deliver real-time data from servers to clients

ArcGIS supports both the poll and push ways to deliver real-time data.

- **Poll** is the traditional approach in which a server stores new event data in a web file or a feature service. Clients periodically (such as every 30 seconds) poll or refresh the web file or the feature service to retrieve the latest data. For example, you can configure the layers in your web map to refresh at an interval between 6 seconds and 1 day.
- **Push** is a new way to serve data in near real time using HTML5 WebSocket protocol. For example, GeoEvent Extension stream services can push data to the client side. Stream services are especially useful for visualizing real-time data feeds that have high data volumes or that have data that changes at unknown intervals. To view stream service layers, you can add these layers to your web map.

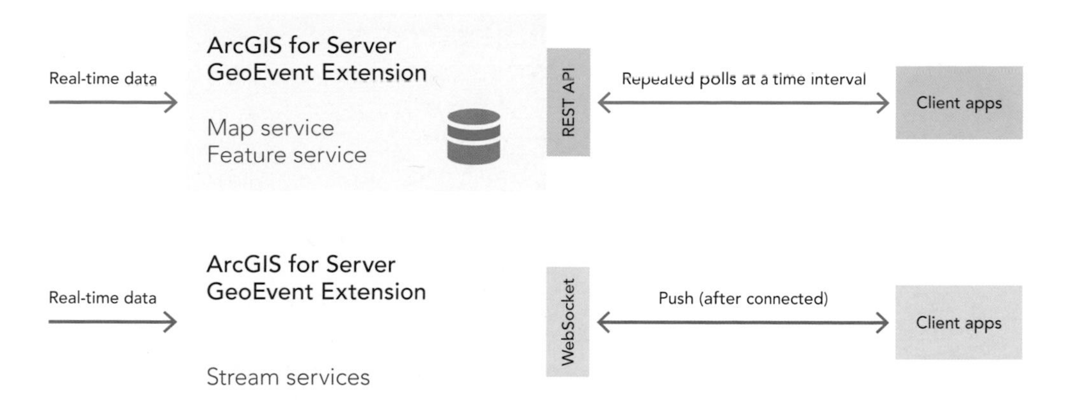

Using the poll technology method (top), real-time GIS data is first saved in a feature class and then exposed as a feature or map service for the client to poll periodically. Using push technology (bottom), the real-time GIS data is pushed out to the client immediately as stream services via a WebSocket.

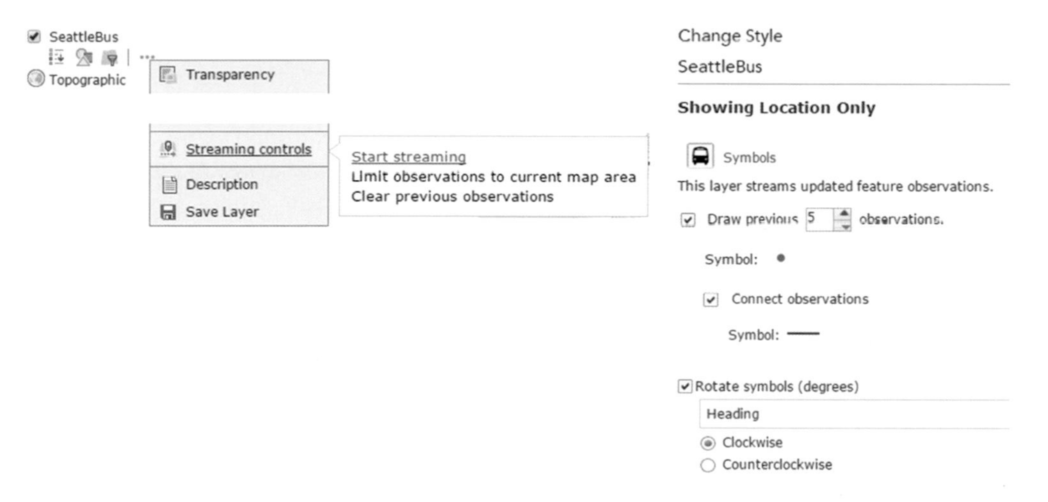

In ArcGIS map viewer, you can add a stream service layer and filter the layer (left). You can also configure the layer's style (right)—including symbols, rotation angles, and the number of previous observations you would like to show on the map.

Operations Dashboard

When you drive a car, you rely on the car's dashboard to monitor such conditions as current speed, fuel level, gearshift position, and engine status. Similarly, you need a dashboard to gain situational awareness when you are responding to emergencies, managing workforces, and performing many other operations. Esri created Operations Dashboard for ArcGIS for such needs.

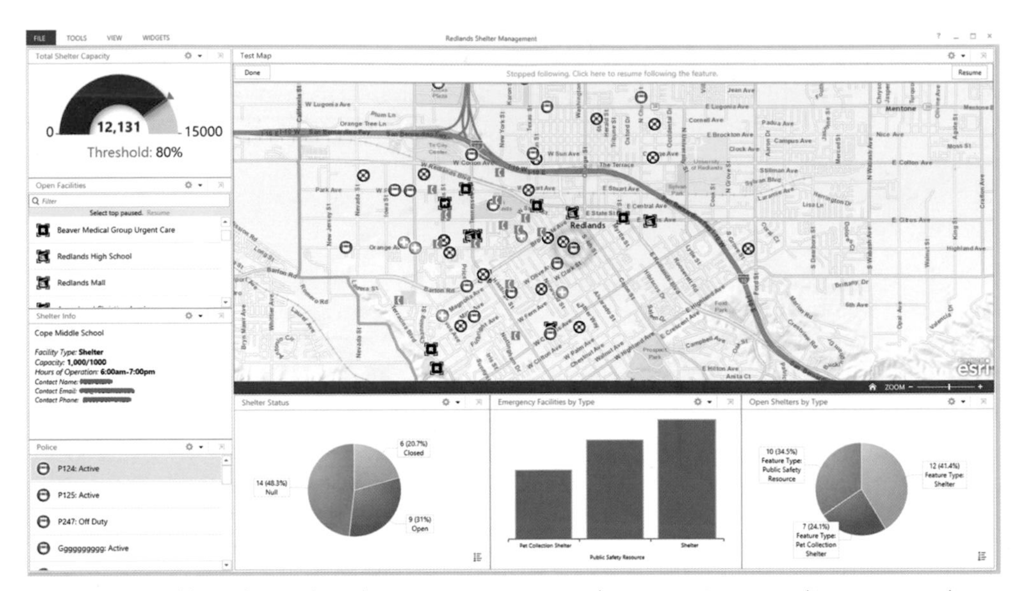

Operations Dashboard provides a live common operational picture using maps, lists, gauges, charts, indicators, and more widgets.

Operations Dashboard is an ArcGIS Online app that provides a common view of the systems and resources you manage. The dashboard integrates web maps and a variety of data sources to create comprehensive operational views. The views include charts, lists, gauges, and indicators, which update automatically as underlying data changes. These live views allow you to monitor and track live events in real or near real time. Operations Dashboard has Windows and web browser editions. The Windows edition creates and displays operations views and displays multiple views. The browser edition runs on both desktop and tablet devices but is primarily for displaying views.

This tutorial

This tutorial teaches both mobile GIS and real-time GIS technologies.

- Section 8.1 creates a web map for use with Collector for ArcGIS and for displaying near real-time data.
- Section 8.2 teaches Collector for ArcGIS.
- Section 8.3 uses Operations Dashboard to monitor live data updates.
- Sections 8.4 and 8.5 introduce you to AppStudio for ArcGIS so that you can create native apps.

System requirements:

- An ArcGIS Online publisher or administrator account
- An iOS or Android smartphone or tablet (for section 8.2)

- An Android device or a desktop computer (Section 8.5 explains how to deploy native apps to devices. Deploying apps to an iOS device is quite involved and requires that you pay an annual fee as a registered iOS developer and that you have certificates to identify you and your device. Deploying apps to an Android device is relatively easy. Therefore, section 8.5 focuses on deploying apps to Android devices and desktop computers.)

8.1 Prepare your web map

Collector for ArcGIS requires a web map and a feature layer. You or your organization must own the web map. The feature layer can belong to anyone, as long as you have access to edit the layer. You previously learned how to publish feature layers and create web maps. Thus, this section will be brief. In this section, you will simply copy an existing web map and then save the map to use in the remaining sections of this chapter.

1. **Open your browser, navigate to ArcGIS Online (arcgis.com), and sign in with a publisher or administrator account.**

2. **In the search box, search for 311 incidents samples map owner:GTKWebGIS.**

You will not see any matches because ArcGIS Online searches only in your organizational contents by default. Next, you will remove this restriction.

3. **Unselect the Only search in your organization's check box on the left, and you will see the result.**

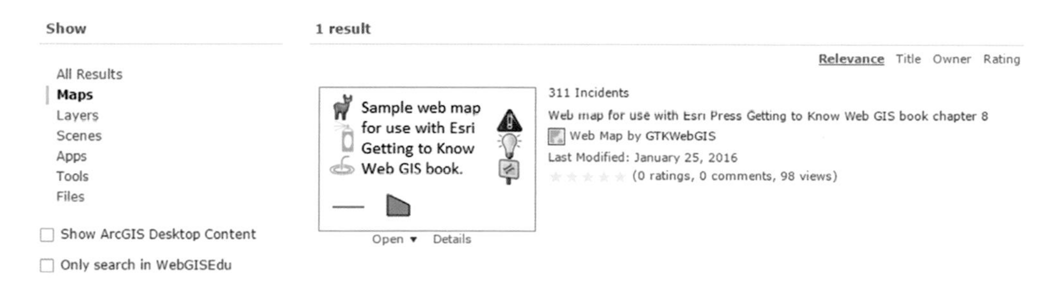

4. **Under the thumbnail, click Open to open the found map in the map viewer.**

This web map has three layers: point, line, and polygon feature layers, which allow you to practice how to collect point, line, and polygon data in the next section.

5. **Click the Content button, and in the Contents pane, point to the Incidents (Points) layer, click the more options button ···, and choose Refresh Interval. Note that the layer is set to refresh every 0.1 minutes, or 6 seconds.**

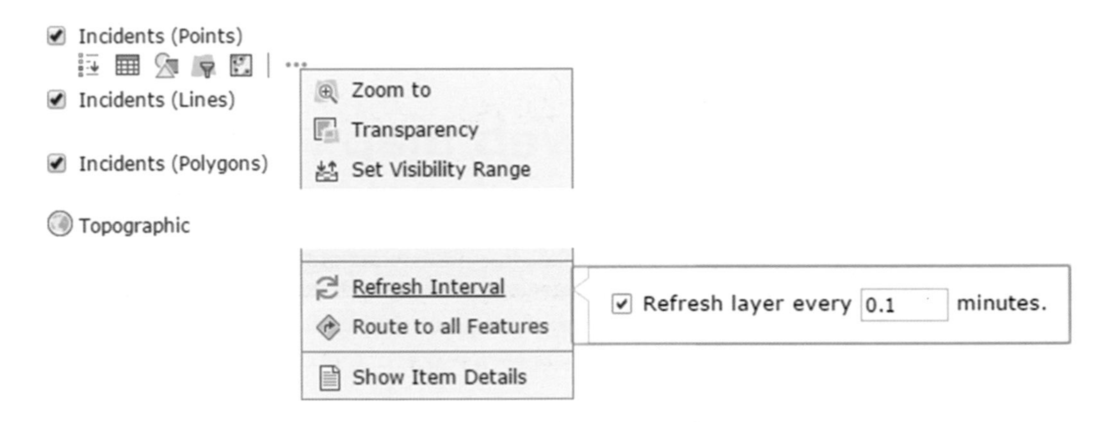

With the refresh option turned on, you can see the layer's latest data, added by you and other users, in near real time. The same refresh interval is also set on the other two incidents layers.

6. **On the map viewer toolbar, click Save > Save As.**

7. **In the Save Map window, for Title, remove -Copy at the end. If you have a web map with the same title, specify a different one.**

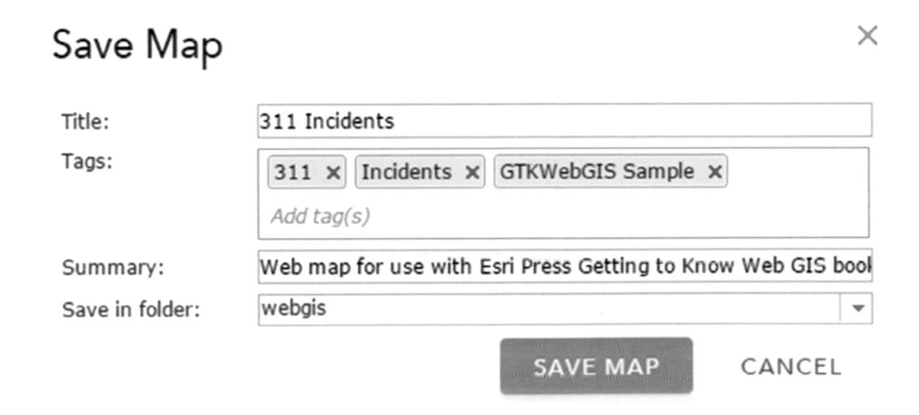

8. **Click Save Map.**

9. **Click Share, share your map with everyone, or your organization only, and click Done.**

8.2 Collect data using Collector for ArcGIS

This section requires the Collector for ArcGIS mobile app. To install this app on your smartphone or tablet (iOS or Android), go to the App Store or Google Play store, search for **Collector for ArcGIS**, and install this app. This section follows the instructions for the iPhone version. You should find similar instructions for other phones but may find different tool layouts for tablet versions.

1. **On your smartphone or tablet, tap the Collector app to start the app.**

Collector

Tap **ArcGIS Online**, and sign in with your account. (If a prompt asks you to allow Collector for ArcGIS to access your location, click **Allow**.)

2. **You can also enable location services for Collector in Settings > Privacy > Location Services > Collector > while using the app.**

3. **Find and tap the 311 Incidents web map you created in the last section.**

4. **Familiarize yourself with the functions of the buttons. Tap More to see the additional buttons.**

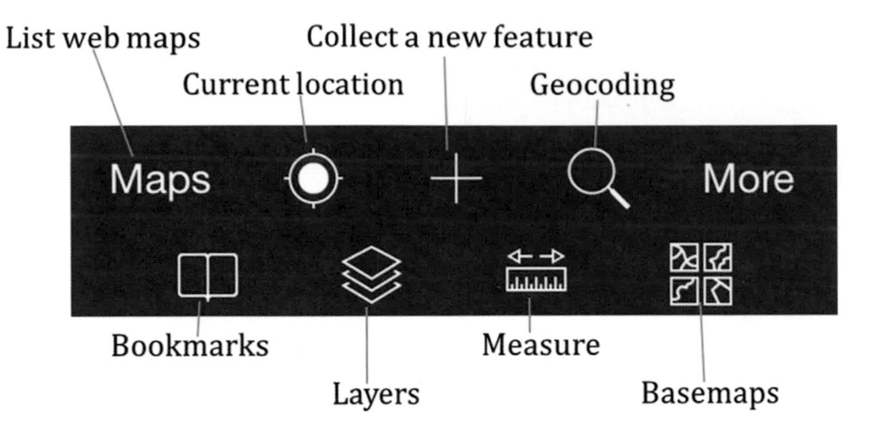

5. Tap the **Collect a new feature** button + to create a new incident. In the **Collect a new feature** panel, tap **Pothole** to choose this type of incident.

6. In the **Attribute** pane, notice that the incident **Type** is already set as **Pothole**, with a red exclamation warning icon indicating the pothole symbol. Specify the remaining attributes. For example, for **Description**, specify **Big pothole!**; for **Date**, choose today; for **Your Name**, specify your name; and for **Your Phone**, specify a phone number.

You can set the values as illustrated in the screen capture or make up your own.

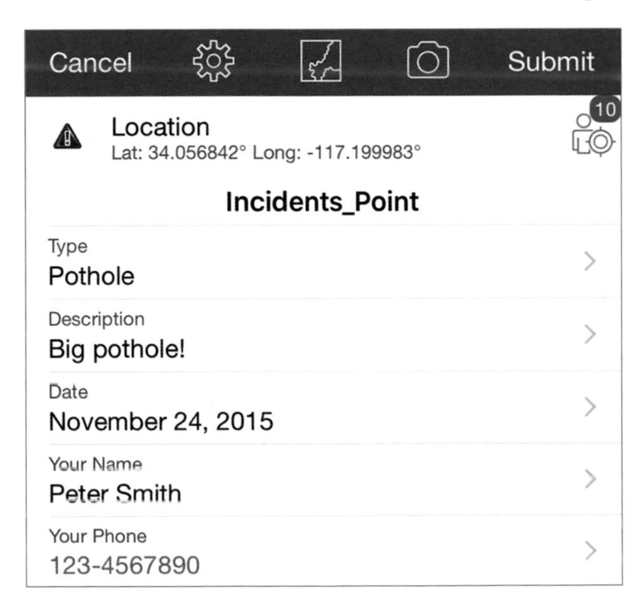

7. **Tap the Map button to see the location of this incident.**

If you enabled location services for Collector for ArcGIS, your map should have opened with the extent reflecting your current location. You may see some data (points, lines, polygons) collected by other users. The ability to locate your current location is convenient if you are at the place of the incident. Also, notice the icon with a number. This number indicates the margin of error at this moment. This margin of error is also indicated by the outer circle of the icon. If you are in an open space with GPS enabled on your mobile device, the margin of error may be only a few meters. If you are indoors, the margin of error can be much higher and the accuracy considerably lower, which may require you to change the **Required Accuracy** setting. Next, you will manually specify a location.

8. **Tap and hold your finger on the map to see a magnifier. Move the magnifier so that the cross is at a new location of your choice, and release your finger.**

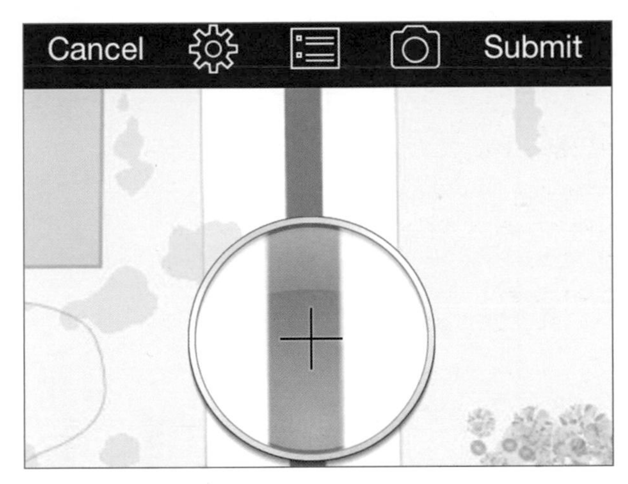

On the map, you will see a red dot, which marks the location of the incident.

9. **On the toolbar, tap the add attachment icon to attach a photo or video to the incident.**

10. **Tap Add > Take Photo or Video, and take a photo or video. Once you are happy with the picture, tap Use Photo; otherwise, retake the picture.**

You can tap **Add** if you want to attach more photos or videos to this incident.

11. **Tap Done to finish adding attachments.**

12. **Tap Submit on the toolbar to post any attachments, and save the incident.**

You have collected a point, and the point is displayed in the correct symbol. Next, you will collect lines using both manual and stream modes.

13. **While you are in the map mode, tap the Collect a new feature button to create a new incident. Choose a line feature type, such as the Street to Resurface incident type.**

14. **Specify attributes for this incident. For example, for Description, specify Lots of cracks!; for Date, choose today's date; for Your Name, specify a name; and for Your phone, specify a phone number.**

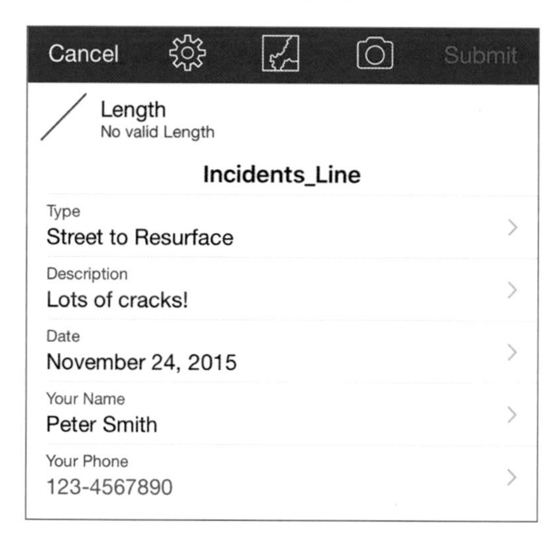

15. **Tap the Map icon to switch to the map mode.**

16. **Add a line manually by tapping on the map to specify the nodes and vertices.**

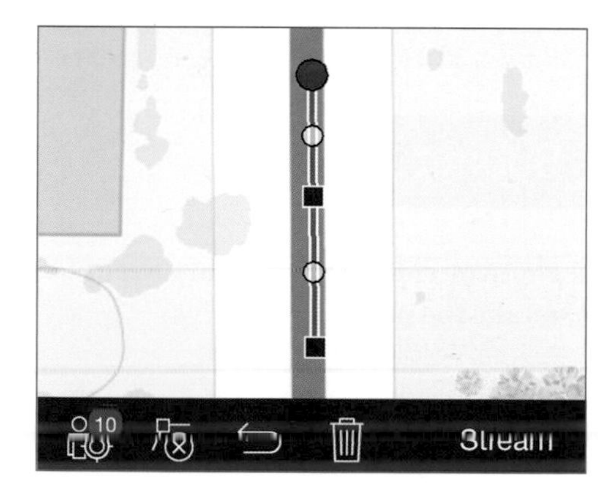

You can also collect lines or polygons by streaming your GPS location. First, you will review your streaming-related settings.

17. **Tap the Settings button ⚙ to review the Collect Settings panel, including Required Accuracy and Streaming Interval. Change the values as needed.**

If you are indoors, for example, the **Required Accuracy** often needs to be 20 meters or more, but if you are outdoors in open spaces, the **Required Accuracy** can often be set at 5 meters or less. You can often set the **Streaming Interval** at 5 seconds for walking.

- **Required Accuracy** sets the accuracy of data that you must meet for the GPS to add a point.
- **Streaming Interval** sets how often a vertex is added to the feature that you are creating. The smaller the time interval, the more detailed the shape.

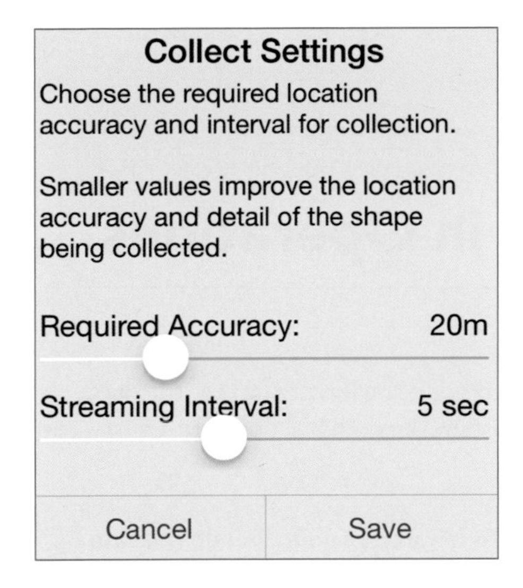

If you change the values, tap **Save**. Otherwise, tap **Cancel**.

Optionally, you can add a line using GPS streaming. You can tap **Stream**, and take your phone while you walk or drive along the line formation (such as a street or trail) where you wish to collect data. The GPS locations of your phone are collected and drawn on the map as you move.

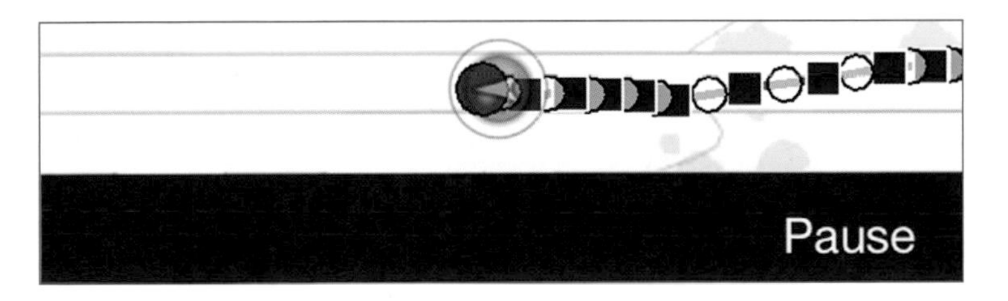

18. **In the same way that you did for points, tap the Add Attachment icon to attach photos or videos relevant to the line incident.**

19. **Tap Submit on the toolbar to save the line incident.**

You just collected a line feature. Similarly, you can collect a polygon feature using either the manual or stream mode.

You are now familiar with Collector for ArcGIS and can collect point, line, and polygon features. The collected data is saved to the feature layers that hold all the incident data. You can see the data you collected appearing in your web map if you have the map opened in the map viewer. In the next section, you will monitor the data you collected in Operations Dashboard.

8.3 Monitor live data in Operations Dashboard

In this section, you will monitor the live incidents collected to the feature layers. The data includes incidents that you and others collect. You will view the data live on maps, charts, gauges, and lists.

1. **In a web browser, go to ArcGIS Online (http://www.arcgis.com), but do not sign in.**

2. **In all content, search for 311 incidents operation view owner:GTKWebGIS.**

Sign In   view owner:pinde.webgis ×

3. **On the search results page, under the thumbnail of the only result, click Details.**

The item details page appears.

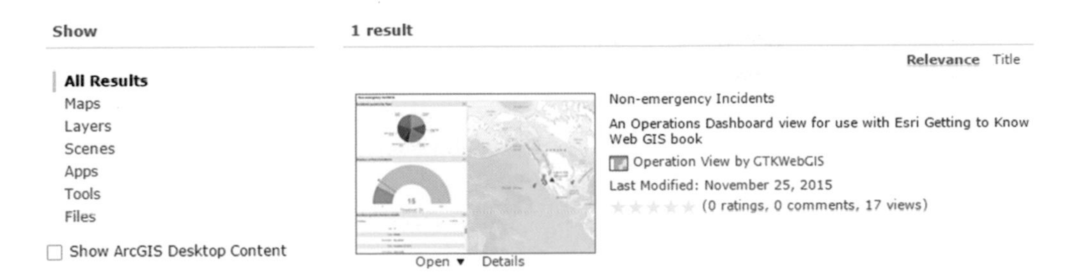

If you signed into ArcGIS Online, you will need to unselect the **Only search in your organization** check box if you want to see the result.

4. **On the details page, click the arrow next to Open, and choose Open in browser app.**

If you are using a Windows computer, you will see two options. The other option, **Open in Windows app,** requires the Windows version of Operations Dashboard, which is a free download. The following tutorial uses the browser version.

When the operation view opens, you will see a map displaying the incidents collected by you and other readers. You will also see **Pie Chart, Gauge**, and **Feature Details** widgets. You will explore these widgets in the following steps.

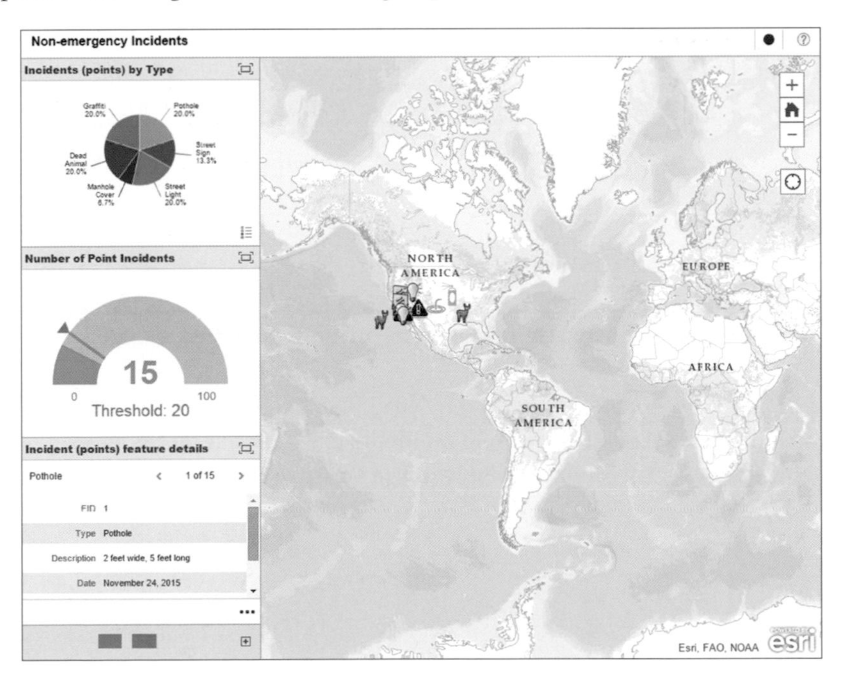

5. **Examine the Gauge widget.**

The **Gauge** widget compares the total number of point incidents with a threshold. When the count goes above the threshold, the green color will become red to alert viewers that there are more incidents to be processed.

6. **In the Pie Chart widget, click the pothole slice.**

A popup appears, displaying the count and the incident type of the slice you clicked and a list of actions you can perform on incidents of this type.

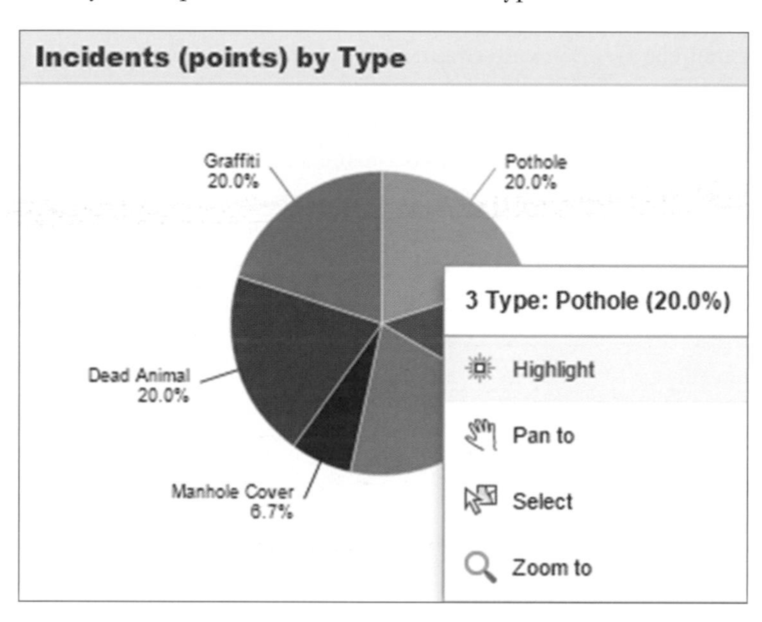

7. **In the Pie Chart pop-up, click Highlight to see the pothole incidents flash on the map.**

8. **Still in the pop-up, click Select. The pothole incidents are selected on the map, and their attributes and attachments display in the Feature Details widget.**

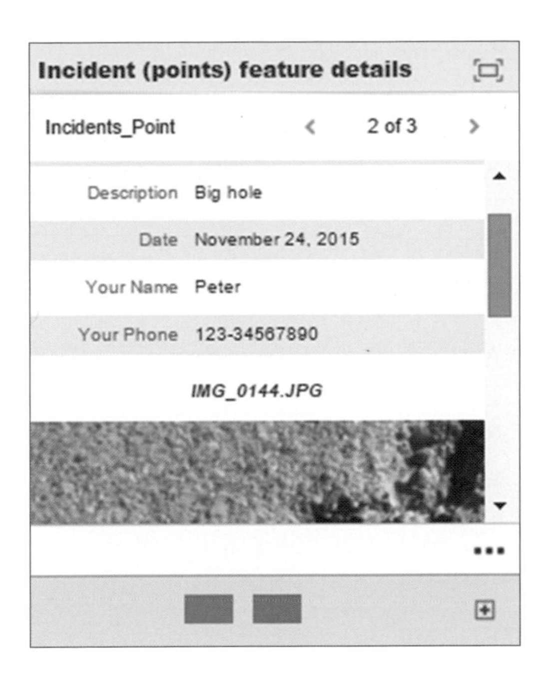

9. **On the map toolbar, click the Unselect button ☒ to unselect the potholes.**

An operation view can have multiple panels that you can toggle.

10. **At the bottom of the window, click the gray rectangle in the panel position indicator.**

The second panel appears. The second panel has three **List** widgets, showing the incidents in the point, line, and polygon layers, respectively.

11. **Click an incident in the Incidents (points) list, and then click the Feature Actions (or More) button ⋯.**

A context menu with a list of actions appears.

You will see a **Follow** action in the list. The incidents in this tutorial do not move, so **Follow** does not really apply here because the incidents in this section do not move. But for features such as police cars and ambulances, you can **Follow** the incidents moving around the map. Vehicles that the map follows flash on the map, which enhances their visibility.

12. **Click the Plus button next to the panel position indicator to show both panels.**

You have learned the basic functions of the basic widgets in Operations Dashboard. Next, you will use the dashboard view to monitor live data updates.

13. **Use Collector for ArcGIS on your mobile device to add one or more incidents as points.**

You learned how to collect points in the previous section. In the operation view, you will see any new incidents appear on the map, and you will see the updates in the **Gauge** and **Chart** widgets.

You previously configured the incident layers to refresh every 6 seconds. Therefore, the data you collected using Collector for ArcGIS should appear in the operation view within 6 seconds.

Notes:

- You will see the data that you and other readers collected during this tutorial. This data is visible because the different web maps all reference the same incident feature layers.
- The operation view you just explored was preconfigured for you. Operations Dashboard has more widgets and supports more flexible configurations. To create operation views, you must download and use the Windows version of Operations Dashboard. You can go to **http://doc.arcgis.com/en/operations-dashboard**, and click **Author Views** for details.

8.4 Create native apps using AppStudio for ArcGIS

AppStudio for ArcGIS was in beta version when this new section was written. For that reason, this section tried to avoid screenshots and details that may change in future releases.

1. **Start a web browser, go to http://appstudio.arcgis.com, and sign in.**

2. **Click My Apps.**

3. **Click Create New App.**

You will see several available templates. The number of templates will grow with future releases.

4. **Find the Quick Report template, and click Start with this template.**

5. **In the App Info page, set the Title as Incidents Reporter.**

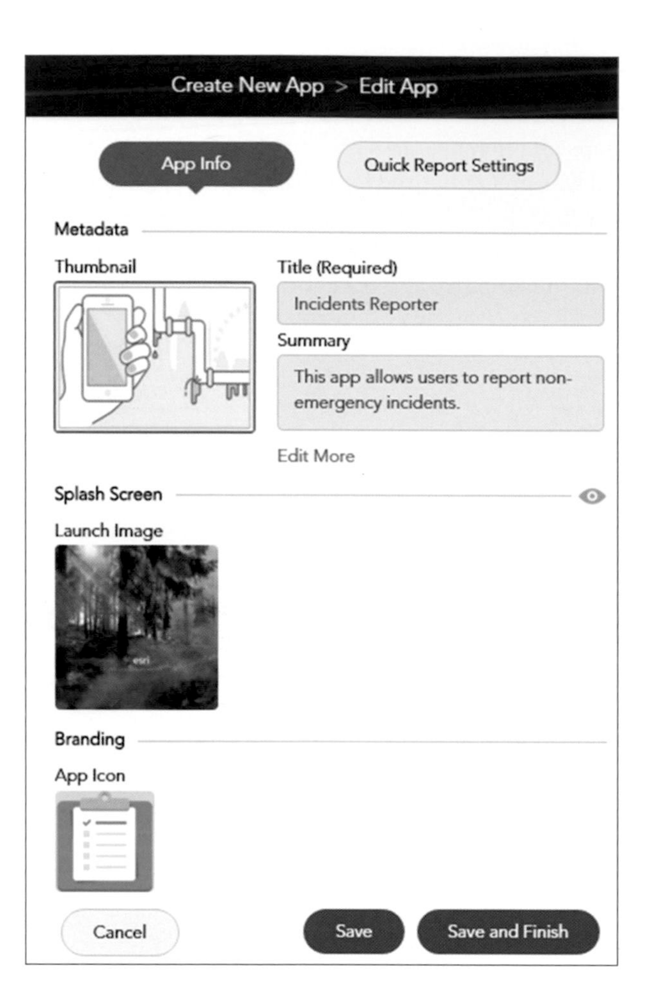

Optionally, you can change the **Thumbnail, Launch Image, App Icon,** and other settings.

6. Click **Save** to save your settings.
7. Click **Quick Report Settings** at the top.
8. Click **Choose Feature Service.**
9. In the **Choose Feature Service** window, click the **Public** tab.
10. Search for **311 incidents gtkwebgis owner:GTKWebGIS.** Click the only result, and then click **Next.**

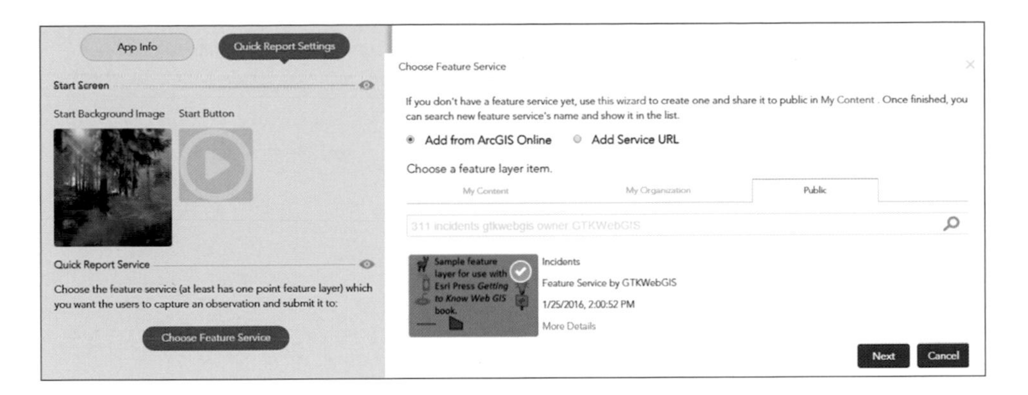

11. Click **Incidents**, and from the expanded list, click **Incidents_Point**, and then click **OK.**

12. Click **Save.**

Next, you will preview your configured app.

13. In the upper-right corner of your screen, click **View the App Live**, and read the instructions in the pop-up.

    - **On your mobile device, go to Google Play or App Store, search for AppStudio Player for ArcGIS, and download and install the app.**
    - **On your mobile device, run the AppStudio Player for ArcGIS, sign in with the same ArcGIS Online account you used at the beginning of this section.**

You will see a list of apps you created.

- **From the list, tap the app you just created, and choose Download Now.**
- **Tap the app again, choose Open** App, **tap the Start button, click New Report and preview the app.**

  You can add or select a photo or skip those options. You also can specify a location, choose an incident type, specify the incident attributes, and submit the incident report.

14. In the **AppStudio** web browser, click **OK** to close the **View the App Live** window.

15. Click **Save and Finish**, which brings up the **App Console** page.

16. Click **Build App.**

17. Choose the platform(s) you want for building your app. If you do not have a mobile device, click a desktop platform.

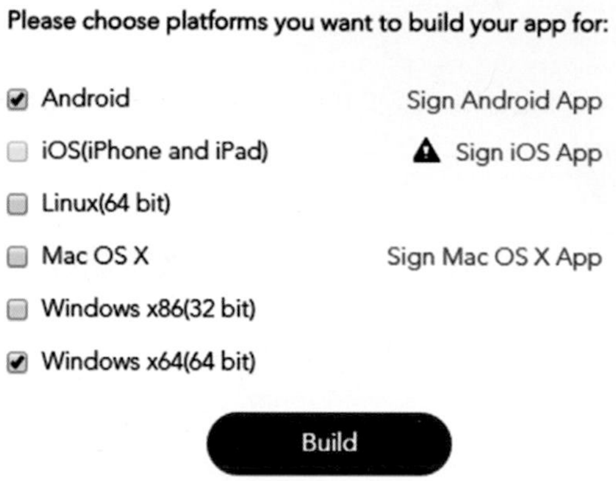

Please choose platforms you want to build your app for:

- [x] Android — Sign Android App
- [ ] iOS(iPhone and iPad) — Sign iOS App
- [ ] Linux(64 bit)
- [ ] Mac OS X — Sign Mac OS X App
- [ ] Windows x86(32 bit)
- [x] Windows x64(64 bit)

Build

You will see that the iOS option is disabled. The option needs additional files, including a certificate file and a provisioning profile. The certificate identifies you as the author of the app, and the provisioning profile indicates that the code you built is allowed to run on your mobile devices. To generate these files, you must register as an iOS developer at Apple's website, which involves an annual fee to Apple and is not covered in this book.

18. Click Build.

Your request to build the app enters a queue for processing. Building your app may take a few minutes.

8.5 Install and test your native app

Building your app will create an installation package for the platform you selected. After your build is complete, you must install the installation package on your device before you can run the app. It is relatively easy to deploy your app to an Android device. If you do not have an Android device, you can test the **Quick Report** app using a desktop version, which works similarly to the mobile native versions.

1. **Using the device on which you want to install the new app, start a web browser, go to http://appstudio.arcgis.com, and sign in.**

2. **Click My Apps, and click the app you just created in the previous section.**

The console page of the app will open. At the bottom of the console page, you will see a list of installation files that have been built for your app.

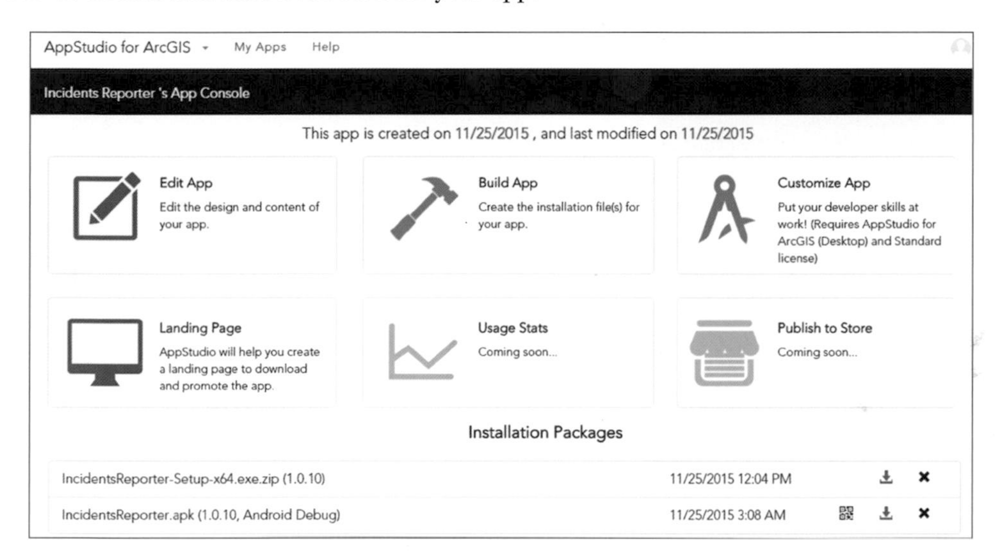

3. **On the app console page, click the Download button for the correct operating system of your device.**

4. **For desktop computers, find the downloaded file, launch the file, and then go to step 7 (ignore any warning saying that this file is an unrecognized app or is from an unknown publisher).**

5. **For Android devices, tap Apps, tap Downloads, find the installation file you just downloaded, and tap the file to run the file.**

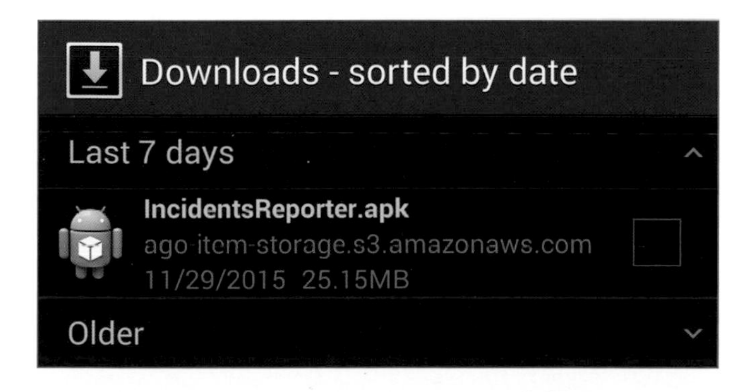

If you did not enable **Unknown sources** for your Android to **Install Apps Outside the Play Store**, you will see the message **Install Blocked**.

6. **Click Settings, and in the Unknown sources** window, select **Allow initial installation only, and click OK.**

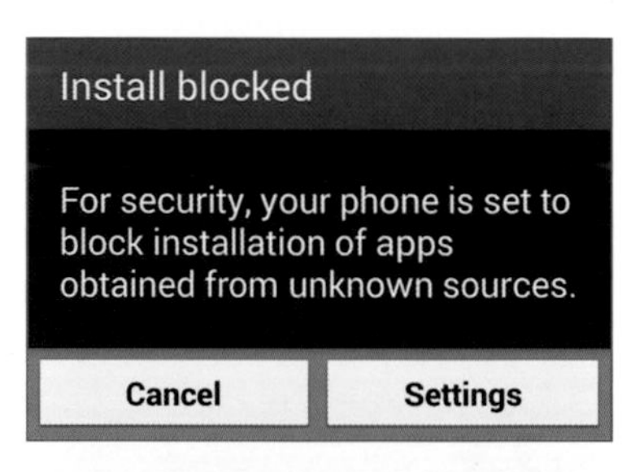

7. **In your Android or other device, follow the instructions to install the app.**

8. **In your Android or other device, find the Incidents Reporter** app you just **installed, and run the app.**

You will first see the initial splash image, and then the background image for the start stage, and the **Start** button. You configured these images and the button in the previous section.

9. **Tap or click the Start button.**

10. **Tap or click New Report.**

11. **Take or select a photo.**

12. **Add an incident location by using your current location or manually selecting a location on the map.**

The **My Current Location** button works on mobile devices but may not work on desktop computers.

13. **Tap or click Next: Add Details.**

14. **Pick an incident type first, and then fill out the incident attribute form, including incident description, a general location, the date, your name and contact information, and any other details you want to add.**

15. **Scroll to the bottom of the form, and click Submit.**

16. **Follow section 8.3, steps 1 through 4, to open Operations Dashboard to view the incident data you just added.**

Optionally, you can collect a few more incidents using your **Incidents Reporter** app and monitor the new incidents that appear in **Operations Dashboard**.

Because this app and the operation view both reference the same feature layer, the data collected in the mobile app also appears in the operation view.

Once you are satisfied with the app, you will typically authorize and submit your app to Google Play and App Store for your users to download. You also can release the app for enterprise deployment without going through Google Play and App Store. Refer to the AppStudio for ArcGIS website for details about how to deploy your apps.

In this tutorial, you used Collector for ArcGIS to collect point, line, and polygon data. You also created a native app using AppStudio for ArcGIS and collected data using your app. You used Operations Dashboard to monitor the data in near real time as you collected the data from Collector for ArcGIS and the Quick Report app.

QUESTIONS AND ANSWERS

1. **What are the differences between Collector for ArcGIS and Survey123?**

   Answer: Survey123 was in beta while this chapter was being written. Its features will evolve, but as of this writing, Collector for ArcGIS and Survey123 had the following main differences:

   - Collector for ArcGIS is map-centric, whereas Survey123 is form-centric.
   - Survey123 provides ways to design surveys and will create a feature layer automatically. Collector for ArcGIS does not provide ways to design surveys and simply uses pre-created feature layers.
   - Survey123 respects dependency among questions (if the answer to one question is true, then show a related question; otherwise, do not show the related question). At the time this chapter was written, Collector for ArcGIS did not yet respect dependency among questions.

2. **The field where I will collect data does not have a reliable Internet connection. Do Collector for ArcGIS and Survey123 support the offline mode?**

   Answer: Yes.

   - Collector for ArcGIS allows you to download maps and use cached basemaps on your device and can collect data in offline mode. You can sync your work with ArcGIS Online or Portal for ArcGIS once you regain connectivity.
   - Survey123 has a solution similar to most email systems, with an outbox folder to temporarily store the data you collected. You can send the data when you have an Internet connection again.

3. **How do I publish stream services?**

   Answer: You need to have ArcGIS Server and the GeoEvent extension to publish stream services. In ArcGIS GeoEvent Manager, choose the **Send Features to a Stream Service Output Connector.** Refer to GeoEvent extension website.

ASSIGNMENTS

Assignment 8A: Collect point, line, and polygon data using Collector for ArcGIS, and monitor your data collection using Operations Dashboard for ArcGIS.

Data:

- A web map: Search for the web map in ArcGIS Online using "recreation map gtkwebgis owner:GTKWebGIS." Note that the search type is **Maps**.
- An operation view: Search for the view in ArcGIS Online using "recreation view gtkwebgis owner:GTKWebGIS." Note that the search type is **Apps**.

Requirements:

- Save a copy of the recreation web map.
- Use Collector for ArcGIS to collect points, lines, and polygons into the web map. Put your name, nickname, or initials in the attributes.
- Use the recreation operation view to monitor the data updates.

What to submit: Send an email to your instructor with the subject line **Web GIS Assignment 8A**: Your name, and include the following information:

- One or two screen captures of your phone or tablet to show you were collecting data using GPS streaming
- A screen capture of the operation view to show the data you collected (with your name, nickname, or initials in the pop-up or feature details)

Assignment 8B: Create a native app using AppStudio for ArcGIS.

Requirements:

- Your app should use a template other than the **Quick Report**.
- Try to configure your app to use the web maps or apps you created previously.

What to submit: Email your instructor with the subject line **Web GIS Assignment 8B: Your name**, and include the following information:

- One or more screen captures of your app running on your mobile device, or in the preview mode of **AppStudio Player**, or on your desktop computer

Resources

ArcGIS Product Document Website

AppStudio for ArcGIS product information site, https://appstudio.arcgis.com.

ArcGIS GeoEvent Extension for Server help documents, http://server.arcgis.com/en/geoevent-extension.

ArcGIS GeoEvent Extension for Server product introduction, http://www.esri.com/software/arcgis/arcgisserver/extensions/geoevent-extension.

"Choosing the Right Platform" (ArcGIS for Developers Documentation) https://developers.arcgis.com/documentation/guides/choosing-the-right-platform.

Collector for ArcGIS product information site, http://doc.arcgis.com/en/collector.

Operations Dashboard for ArcGIS help documents, http://doc.arcgis.com/en/operations-dashboard.

Survey123 for ArcGIS product information site, http://survey123.esri.com.

Esri blogs

"Development Strategies for Targeting Android, iOS, and Windows Phone," by Lloyd Heberlie, http://blogs.esri.com/esri/arcgis/2014/01/23/development-strategies-for-targeting-android-ios-and-windows-phone.

"Who Says Building Native Apps Has to Be Challenging?" by Chris Le Sueur, http://blogs.esri.com/esri/arcgis/2015/09/10/who-says-building-native-apps-has-to-be-challenging.

Online training and tutorials

"Manage a Mobile Workforce," http://learn.arcgis.com/en/projects/manage-a-mobile-workforce.

"Monitor Real-Time Emergencies," http://learn.arcgis.com/en/projects/monitor-real-time-emergencies.

"Real-Time Dashboards," in *The ArcGIS Book*, Christian Harder (Ed.), http://learn.arcgis.com/en/arcgis-book/chapter9.

"Real-time GIS with ArcGIS GeoEvent Processor for Server," by Esri, http://training.esri.com/gateway/index.cfm?fa–catalog.webCourseDetail&courseid=2642.

"Simplify Field Data Workflows with Collector for ArcGIS," by Esri, http://training.esri.com/gateway/index.cfm?fa=catalog.webCourseDetail&courseid=2891.

"Survey123 for ArcGIS: Ask Questions, Get the Facts, Make Decisions," by Esri, http://training.esri.com/gateway/index.cfm?fa=catalog.webCourseDetail&courseid=2965.

"Utility Asset Inspection Using ArcGIS," by Esri, http://training.esri.com/gateway/index.cfm?fa=catalog.webCourseDetail&courseid=2913.

"Your GIS is Mobile," in *The ArcGIS Book*, Christian Harder (Ed.), http://learn.arcgis.com/en/arcgis-book/chapter8.

Esri videos

"2015 Esri Petroleum GIS Conference: Mobile," http://video.esri.com/watch/4509/2015-esri-petroleum-gis-conference-mobile.

"AppStudio for ArcGIS," http://video.esri.com/watch/4635/appstudio-for-arcgis.

"AppStudio for ArcGIS: The Basics," http://video.esri.com/watch/4705/appstudio-for-arcgis-the-basics.

"Collector for ArcGIS: An Overview," http://video.esri.com/watch/4697/collector-for-arcgis-an-overview.

"Development Strategies for Building Mobile Apps—The Great Debate," http://video.esri.com/watch/4290/development-strategies-for-building-mobile-apps-_dash_-the-great-debate.

"GeoEvent Introduction," by Morakot Pilouk, http://video.esri.com/watch/4827/morakot-pilouk_comma_-phd.

"Using the Operations Dashboard for ArcGIS," http://video.esri.com/watch/4476/using-the-operations-dashboard-for-arcgis.

Chapter 9
3D web scenes

Because we live in a 3D world, we typically find it easier to understand and analyze the world using 3D maps, which are referred to as scenes in ArcGIS. This chapter introduces the basic terminology of web scenes and explains how web scenes are supported across the ArcGIS platform. The tutorial section teaches you how to view and create web scenes using ArcGIS scene viewer, how to work with event data and create a scene in ArcGIS Pro, and how to share the scene to ArcGIS Online or Portal for ArcGIS. A single chapter cannot possibly describe the full range of capabilities contained in a sophisticated desktop product such as ArcGIS Pro. However, this chapter covers the basic, essential skills you will need to start creating 3D web apps.

Learning objectives

- *Understand web scenes in the context of web GIS.*
- *Understand basic scene terminology.*
- *View and create web scenes using ArcGIS scene viewer.*

- *Create scenes in ArcGIS Pro.*
- *Use comma-separated value (CSV) data in ArcGIS Pro.*
- *Configure 3D symbols in ArcGIS Pro.*
- *Configure labels and pop-ups in ArcGIS Pro.*
- *Share web scenes.*

This chapter in the big picture

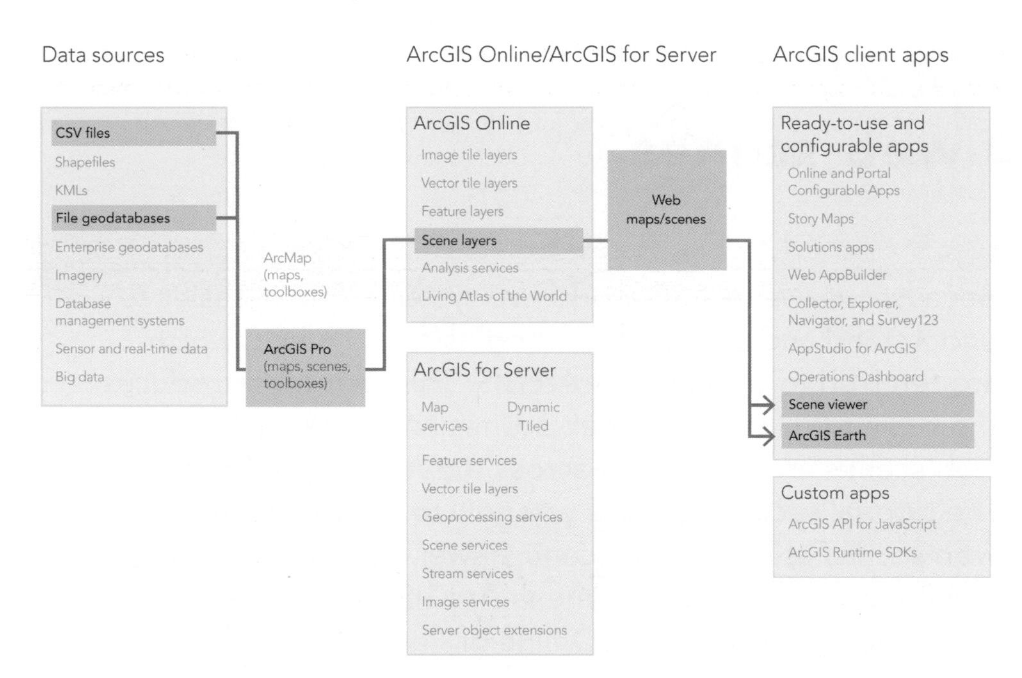

ArcGIS offers many ways to build web applications. The green line in the figure highlights the technology that this chapter teaches.

Why 3D GIS?

3D brings an extra dimension into 2D maps, including advantages in data visualization, analysis, and communication. Users often find 3D maps more interesting and intuitive to interpret than 2D maps. These advantages give 3D GIS wider applicability in storytelling, urban planning, architectural design, defense simulation, and even filmmaking. 3D GIS enables audiences to quickly understand the size and relative positions of objects. And 3D GIS enables designers to build flexible scenarios quickly and cost effectively to avoid costly mistakes in the building phase. Some 3D apps provide functions such as fly around and x-ray vision or radar vision, which allows users to see through buildings and ground surfaces. In addition to visualization, 3D GIS can offer powerful analytical functions, such as visibility or viewshed analysis, sunlight and shadow analysis, and vertical zoning violation detection.

Web scenes and scene services

As the counterpart of web maps, which are 2D maps, web scenes refer to 3D maps in an ArcGIS platform. Like a web map, a web scene can include many layers, including feature layers, map image layers, and importantly, scene layers. ArcGIS scene layers, or scene services, follow the open indexed 3D scene (i3d) format and provide a representational state transfer application program interface (REST) API to support client apps across all platforms.

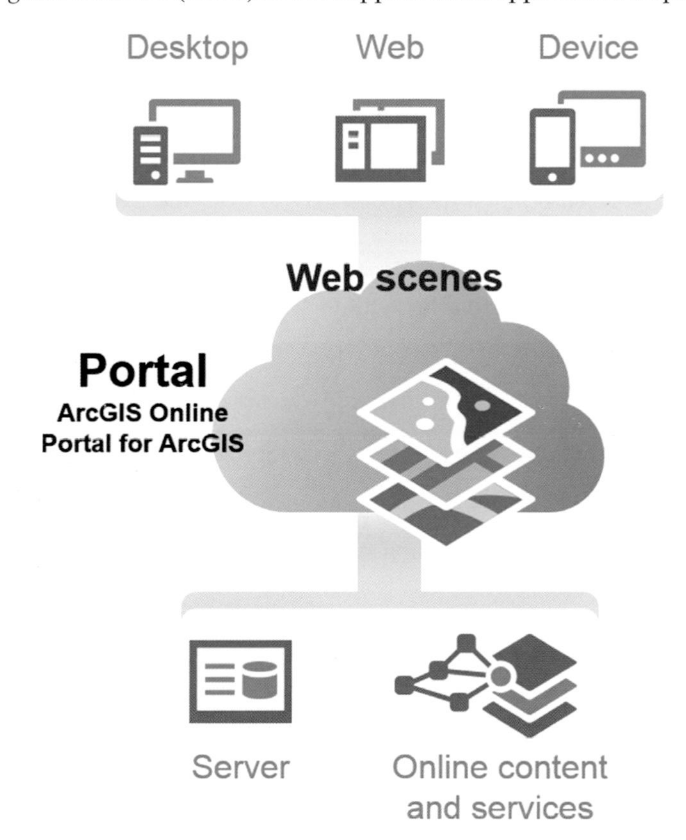

Web scenes are central to ArcGIS 3D capabilities. ArcGIS Online or Portal for ArcGIS can serve 3D web scenes, which users can view on desktop apps, mobile apps, and browser-based apps.

Scenes have four main types of elements:

- **Surfaces**: Provide ground elevation. ArcGIS provides an elevation service on which other map features can drape or stand. You can also use your own elevation surface.
- **Features**: Situated on, above, or below the surfaces. These features are the operational layers of your 3D app.
- **Textures**: Provide exterior or interior covers of your 3D features. Textures often use aerial imageries or cartographic symbols.
- **Atmospheric effects**: Such as lighting and fog.

When you work with web scenes, you can choose to display your data within one of the following two scene environments:

- **Global scene**: Displays features on a sphere. A global scene is best used for displaying phenomena that covers a large geographical area or wraps around the spherical surface of the earth.
- **Local scene**: Displays features on a planar surface. A local scene is best used for displaying or analyzing data at the local or city scale.

Based on the features and textures, scenes can be grouped into three main types:

- **Photorealistic:** Aims to re-create reality by using photos to texture features. These types of scenes often use imagery as the texture and are extremely well suited for showing visible objects such as a city.
- **Cartographic**: Takes 2D thematic mapping techniques and moves them into 3D. These types of scenes often use attribute-driven symbols (extrusion height, size, color, and transparency) to display abstract or invisible features such as population density, earthquake magnitudes, flight paths, zoning laws, solar impact, and air corridor risks.
- **Virtual reality**: Combines photorealistic and thematic techniques.

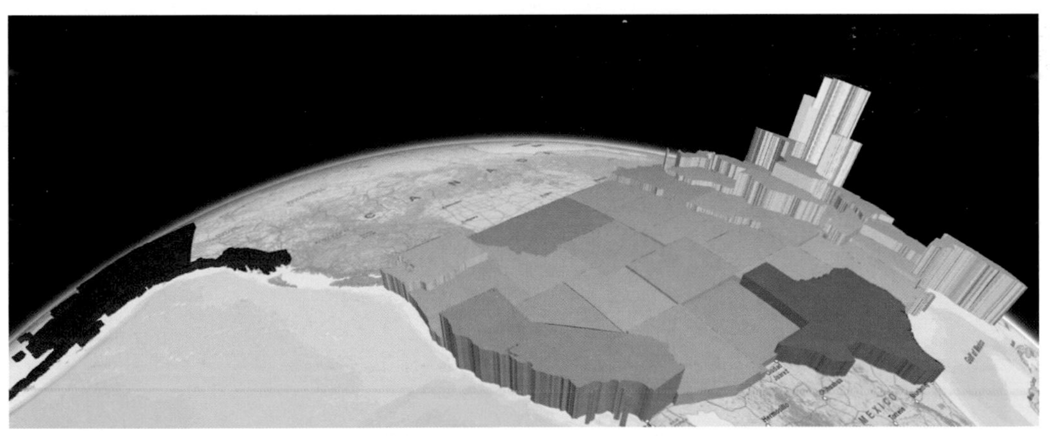

A global and cartographic scene displays US states, with colors symbolizing the total area and extrusion height symbolizing the population density of each state.

A local and photorealistic scene showing the city of Montreal, Canada.

3D across ArcGIS platform

ArcGIS platform provides a suite of 3D products to support the creation, visualization, analysis, and sharing of 3D scenes.

- **ArcGIS Pro**: This desktop app provides comprehensive tools for managing 2D and 3D data, authoring 2D and 3D maps, and sharing 2D maps and 3D scenes.
- **3D Analyst**: This ArcGIS Pro extension provides processing and analysis tools to work with GIS data in a variety of 3D formats.
- **CityEngine**: This desktop app provides advanced 3D creation capabilities. CityEngine can create photorealistic scenes and thus has been used extensively in urban design and even in filmmaking. CityEngine can also create rule packages for generating large numbers of 3D objects in a batch.
- **ArcGIS Online** and **Portal for ArcGIS**: These platforms can host scene layers for on-premises and public cloud 3D web GIS apps, manage access to web scenes and related layers, offer a scene viewer for creating and viewing web scenes, and provide Web AppBuilder for ArcGIS and other ready-to-use apps for displaying web scenes.
- **ArcGIS Earth**: ArcGIS Earth is a lightweight, easy-to-use 64-bit Windows desktop app that enables anyone in your organization to easily view and explore 2D and 3D data, including Keyhole Markup Language (KML) and ArcGIS services, whether behind a firewall or over the Internet. ArcGIS has the following additional capabilities:
  - Display KML layers and shapefiles in 3D.
  - Connect to Esri REST services, including feature, map, image, and scene services.

    - Use Esri basemaps and Living Atlas of the World content.
    - Measure, add place marks, create simple geometry, and more.
    - Save the view as an image and share the image with others via email.
- **ArcGIS API for JavaScript**: Develop browser-based custom 3D apps.
- **ArcGIS Runtime SDK** for iOS, Android, and Windows phones: Develop native-based, custom 3D apps.

ArcGIS Earth provides an interactive globe for exploring the world and working with a variety of 2D and 3D map data formats, including KML, ArcGIS scene layers, map layers, image layers, and feature layers.

Notes:

- The 3D creation capabilities of ArcGIS Pro are quickly evolving and are including more capabilities of CityEngine. In addition, ArcGIS Pro can publish new and scalable scene services to ArcGIS Online and Portal for ArcGIS. For these reasons, this chapter focuses on ArcGIS Pro for authoring scene services and on ArcGIS scene viewer for viewing web scenes.
- Browser-based ArcGIS scene products, including ArcGIS scene viewer, Web AppBuilder for ArcGIS, and ArcGIS API for JavaScript, are based on WebGL (Web Graphics Library), a web technology standard for rendering interactive 3D graphics in web browsers. Because WebGL is not yet supported on most of today's mobile browsers, browser-based ArcGIS scene products do not currently work on most of today's mobile devices. However, you can use ArcGIS Runtime SDKs for iOS, Android, and Windows Phone to create native apps that support 3D web scenes.

Create web scenes

ArcGIS Pro is the primary desktop tool for creating web scenes and scene services. You can also create web scenes using ArcGIS scene viewer. The former provides more advanced capabilities in configuring 3D symbols, but the latter is much easier to use. You can create web scenes by following these general steps:

1. Choose global or local scene.
2. Choose a basemap.
3. Add layers:
   - ArcGIS scene viewer can use web layers, including scene layers, elevation layers, feature layers, tiled/dynamic map service layers, and tiled image service layers.
   - ArcGIS Pro can use web layers and local data layers, including shapefiles, KMLs, layer packages, multipatch layers, and other layers in file or enterprise geodatabases.
4. Configure layers:
   - ArcGIS scene viewer provides limited options for 3D symbols.
   - ArcGIS Pro provides more and advanced options for 3D symbols. You can create 3D models manually or through procedural rules. For example, a rule can make use of feature geometry—such as a building footprint or attribute information, including the number of floors, roof type, and wall material type—to quickly generate a large number of 3D models.
5. Capture slides or bookmarks.
6. Save and share your scene:
   - For ArcGIS scene viewer, the scene is already saved in ArcGIS Online or Portal for ArcGIS. This share step enables you to select who can view your web scene.
   - For ArcGIS Pro, this share step enables you to publish the scene and its layers to ArcGIS Online or Portal for ArcGIS and select who can view your web scene.

1. Choose global or local

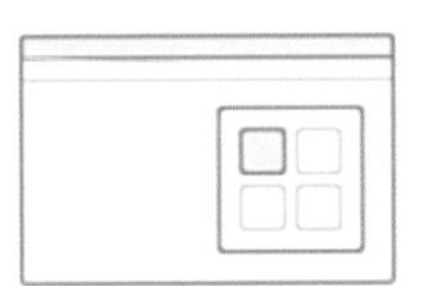
2. Choose basemap

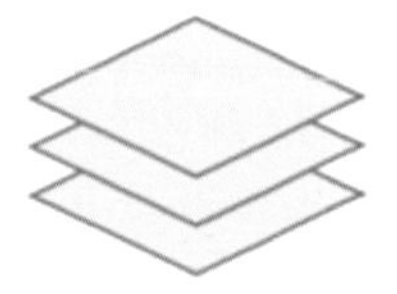
3. Add layers

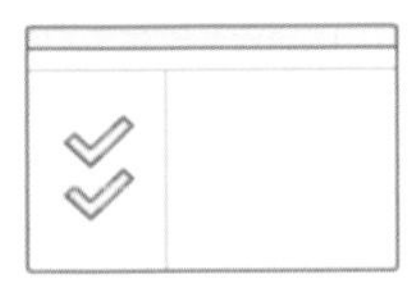
4. Configure layers

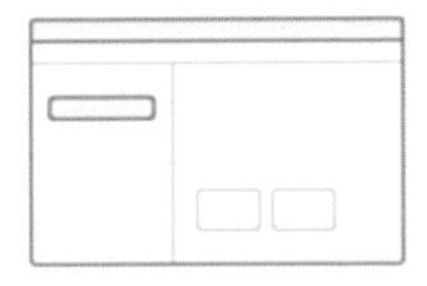
5. Capture slides/bookmarks

6. Save/share scene

General steps to create web scenes using ArcGIS Pro and scene viewer.

This tutorial

This tutorial teaches you how to create and explore web scenes.

- Section 9.1 uses ArcGIS scene viewer to create and explore web scenes. This section creates a web scene for displaying natural disasters, including earthquakes, hurricanes, tornadoes, and typhoons. The data you need is provided online as web layers.
- Sections 9.2 and 9.3 use ArcGIS Pro to create a web scene and publish it to ArcGIS Online or Portal for ArcGIS. The scene displays earthquake magnitudes and tectonic plate boundaries in 3D. For data, you will use the provided CSV file of earthquakes. Additional tectonic plate boundaries data is provided online as a web layer.

System requirements:

- ArcGIS Online for Organizations:
  - A publisher or administrator account is required.
- ArcGIS Pro, version 1.2 or higher:
  - A free trial is available to download at **http://pro.arcgis.com**.
  - Make sure your ArcGIS pro is licensed. If you created an ArcGIS Online trial account, you are an administrator, and your trial account comes with five ArcGIS Pro licenses. You can manage ArcGIS Pro licenses in ArcGIS Online (in My Organization > Manage Licenses > ArcGIS Pro. You can find more information at **http://arcg.is/1LcF5Aq**). Otherwise, contact your instructor or system administrator for your ArcGIS Pro license.
- A desktop web browser that supports WebGL:
  - Sections 9.1 and 9.3 will use ArcGIS scene viewer, which requires WebGL support by your web browser and graphics card.
  - Refer to **http://arcg.is/1PQ9bKb** for more information if your web browser cannot load ArcGIS scene viewer.

9.1 Explore and create scenes in ArcGIS scene viewer

In this section, you will learn the basics of how to use ArcGIS scene viewer, including navigation. You also will learn different ways to add various types of layers, configure layers, and create slides.

1. **Start a web browser, go to ArcGIS Online (http://www.arcgis.com), and sign in.**

2. **Click Scene in the top menu bar.**

Home Gallery Map Scene Groups My Content My Organization

ArcGIS Online scene viewer will open in your web browser.

Note: If you see an error message saying "the scene viewer cannot be opened in your web browser," that means your web browser does not support WebGL. Refer to **https://doc.arcgis.com/en/arcgis-online/reference/scene-viewer-requirements.htm** for more details on the hardware and software requirements.

3. **In the upper-right corner of your screen, click New Scene, and choose New Global Scene.**

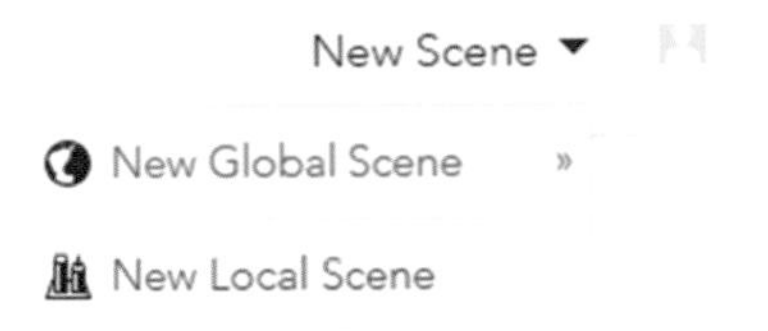

4. **In the upper-right corner of the scene, click the Basemap button to open the Basemap gallery, choose Imagery as your basemap, and click the X button to close the gallery.**

Next, you will search for a location and navigate the scene.

5. **To search for Grand Canyon National Park, click the Search button, and press Enter.**

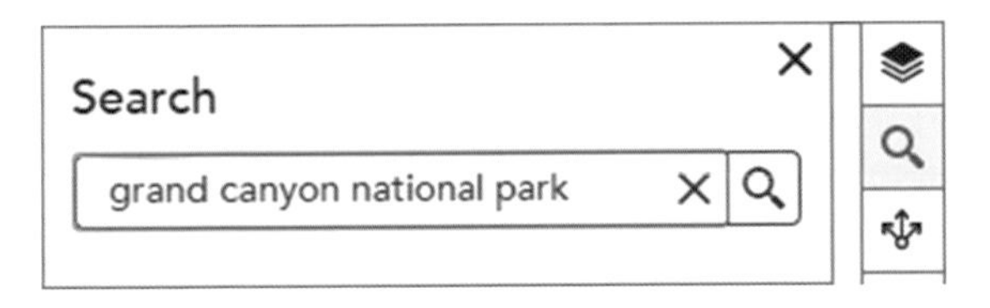

The scene will zoom to the Grand Canyon.

6. **Click the X button to close the search box.**

7. **If the Pan button is not selected by default, click this button.**

8. **Take the following actions to navigate and explore your scene:**

   - **Click and hold the left mouse button to pan to any section of the Grand Canyon.**
   - **Use your mouse scroll wheel to zoom in. If you do not have a mouse wheel, you can click the Zoom In button to zoom in.**
   - **Click and hold the right mouse button to rotate and tilt the scene.**

If you zoom in too close, the surface may disappear, and the view point may go below the surface. The scene should reveal 3D effects, similar to this illustration. For more information on navigating 3D scenes, refer to **http://arcg.is/1S1TLUe.**

9. **Click the Compass button to reorient your scene to the north.**

10. **Click the Home button to return to your initial view.**

Next, you will add several layers that have been provided to you as web layers.

11. **In the Contents pane, click Add Layers.**

12. **In the Layers search box, type in historic earthquakes and disasters owner:GTKWebGIS, and press Enter, or click the Search button.**

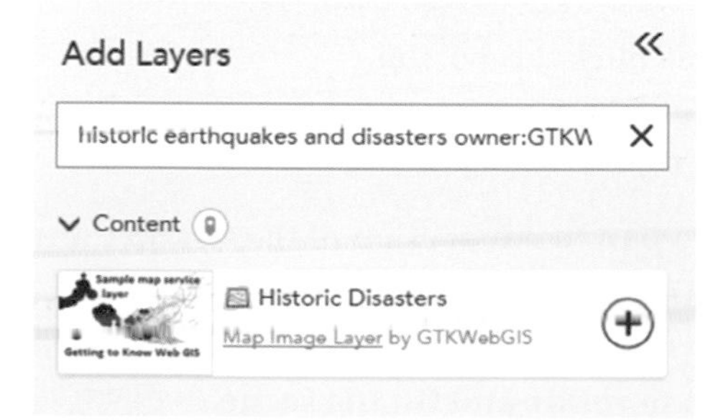

You will see the layer found in the search result. This layer is the same as the map service you published in the ArcGIS for Server chapter of this book.

13. **Click the Add button ⊕ to add this layer to the scene viewer.**

You will see the map image layer added to the scene. Map image layers appear draped on the ground surface. Next, you will find and add a feature layer.

14. **In the Layers search box, type Tornado owner:GTKWebGIS, press Enter, and click the Add button ⊕ to add this layer to the scene.**

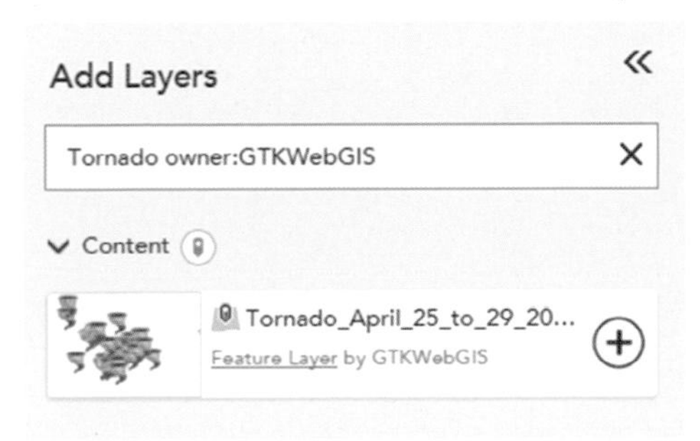

15. **Navigate the scene for a better view of the tornados.**

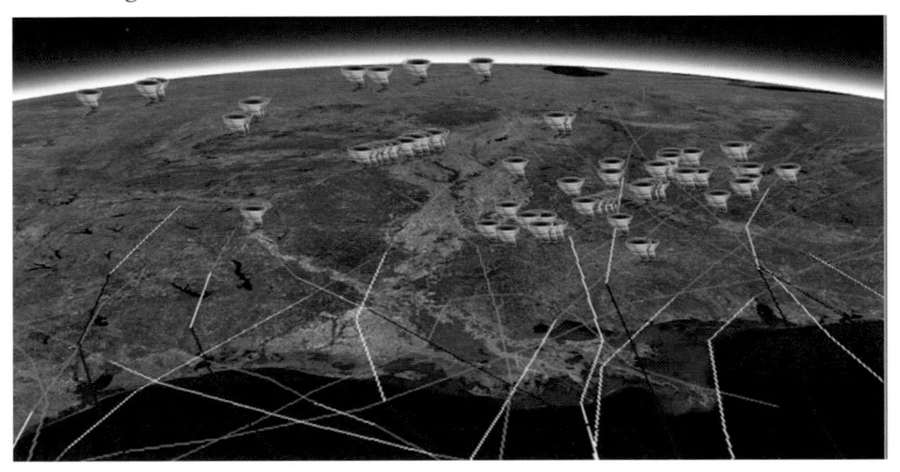

The feature layer you just added uses a 2D icon symbol, but these symbols appear as if they are standing on the surface. You will change the symbol from 2D to 3D later in this section.

Next, you will add a cartographic scene layer.

16. **In the Layers search box, type Western Pacific Typhoon, and press Enter, or click the Search button.**

In the results, you will see web scenes and perhaps feature layers, depending on the content, which may change. Next, you will add a web scene.

17. **Select a web scene, and click the Import button ⊕.**

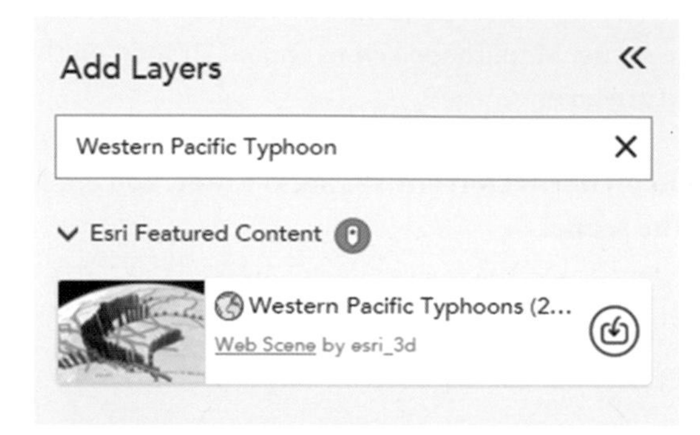

Next, you will navigate to the typhoon areas.

18. **At the bottom of the screen, click one of the slides (or bookmarks).**

You will see typhoons represented as cylinder symbols, with greater heights representing higher wind speeds and darker colors representing lower barometric pressures.

19. **Click the X button of the layer search box to clear the search result.**

You will see Esri featured content.

20. **Click More to expand the list of Esri featured content (you may see different content than the content shown in the figure).**

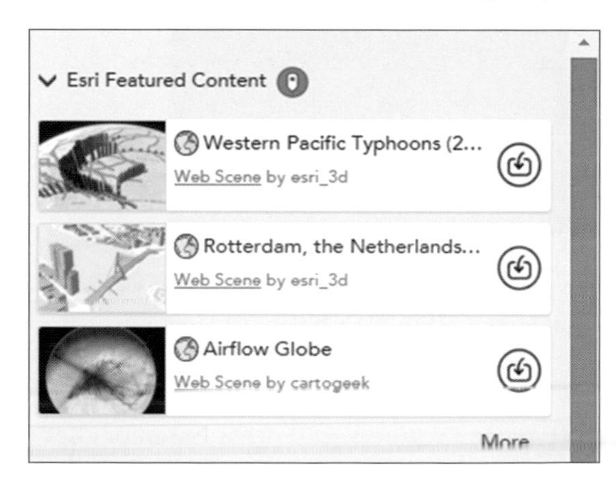

Next, you will add two web scene layers to your scene.

21. Under **Esri Featured Content**, browse to find a city without imagery textures, such as **Portland**, and click the **Import** button ⊚ to import the city's web scene to your scene.

22. Under **Esri featured Content**, browse to find a city with photorealistic texture (you may need to click the **More** button again), such as **Montreal**, and import the web scene to your scene.

23. To view the scene layers you just added, click the **Layers** button to see the layer list, point to each of the layers, and click the **Zoom to** button to see each layer.

You may need to tilt the view or zoom in to get a closer look at each layer. You will notice the difference between photorealistic scenes and cartographic scenes.

24. At the bottom of the **Contents** pane, click **Done**.

25. Click **Save Scene**.

SAVE SCENE

26. Perform the following actions in the **Save Scene** window:

- Set the **Title** as **Natural Disasters**.
- Set the **Summary** as **This web scene contains historic earthquakes, tornados, typhoons, and 3D cities.**
- Set the **Tags** as **earthquakes, tornados, typhoons, GTKWebGIS**.
- Click **Save**.

Next, you will explore the layer options.

27. In the **Contents** pane, scroll down to find the tornado layer, click the arrow ▾, and click **Configure Layer**.

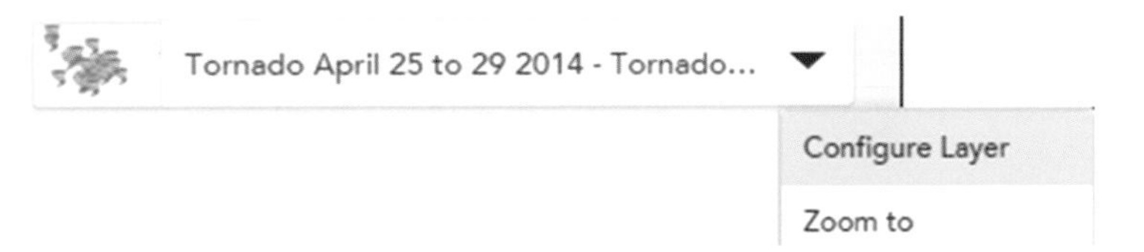

In the **Configure Layer** pane, you can change layer settings, such as **Elevation Mode**, **Transparency**, and **Symbols** for this feature layer.

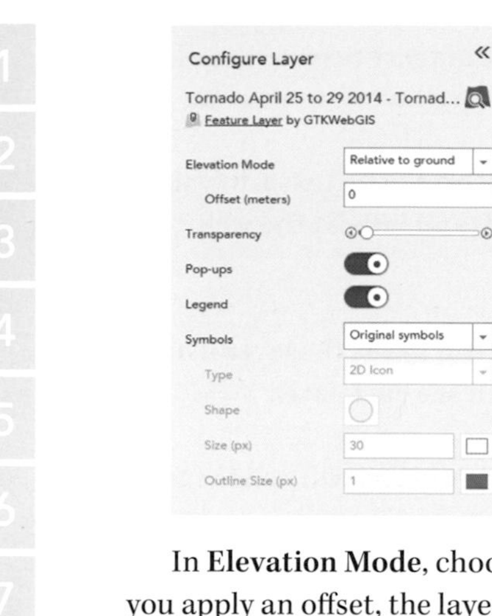

In **Elevation Mode**, choosing **Relative to ground** aligns the data to the ground elevation. If you apply an offset, the layer aligns to the ground and vertically offsets from the ground based on a height value entered in meters. Choosing **Absolute height** positions the data vertically based on the feature's geometry z-value position, with the following conditions:

- If the geometry does not contain z-values, the default elevation is sea level.
- When you apply an offset, you will see your layer displayed at the offset value in meters from the z-value position or the sea level position.

For a map image layer, such as the **Historic Disasters** layer, you do not have the option to change elevation and symbols settings.

28. **Click Symbols, and choose Change Symbols; for Type, choose 3D Object; for Shape, choose a shape; and notice that the tornado symbol has changed.**

ArcGIS scene viewer provides limited capabilities to configure 3D symbols. For advanced 3D symbology capabilities, you must use ArcGIS Pro.

29. **Click Symbols, choose Original Symbols, and then click Done.**

Next, you will learn how to work with slides in web scenes. Slides show different views of your scene. Slides are similar to bookmarks in web maps, but slides can have different basemaps, layers, sun and shadow settings, in addition to extents. You already may have imported some slides from the scenes you imported earlier. In the next steps, you will add additional slides.

30. **Click Slides in the Contents pane.**

Under **Slides**, each slide in the list has a thumbnail and a title, an **X** button to delete the slide, and a **Refresh** button to update the slide.

31. **Click the Layers button, and under Layers, turn on the tornado layer, and turn off the other layers.**

32. **Navigate the scene, click on a slide (or bookmark), choose a camera view you like, click Capture Slide, notice the new slide on the list, and set the title of that slide as Tornados.**

33. **Click a different slide, notice the change, click the Tornados slide again, and then click Done.**

You will see the camera view revert back to the view you set. Next, you will set the initial view of your scene.

34. **Turn on all the layers, and then navigate the scene to a camera position that you would like users to see at your initial view.**

35. **Click Done, click Save Scene to open the Save Scene window, and click Save.**

In this section, you learned the basic operations of ArcGIS scene viewer and created a web scene by adding several different types of layers, including a map image layer, feature layer, and two scene layers. Map image layers drape on the surface of your scene (you can change the symbols of feature layers to 3D), and scene layers stand on the surface.

9.2 Create a scene in ArcGIS Pro

ArcGIS Pro provides advanced tools and detailed controls for you to create 3D scenes. In this section, you will create a global scene to visualize a CSV file of earthquakes in 3D.

1. **On your desktop computer, start ArcGIS Pro.**

2. **If you are starting ArcGIS Pro for the first time after installation, follow these instructions for licensing unless your instructor or administrator instructs you differently:**

    - **For License Type, choose Named User License from the list.**
    - **For Licensing Portal, choose ArcGIS Online or Portal for ArcGIS if you have one.**

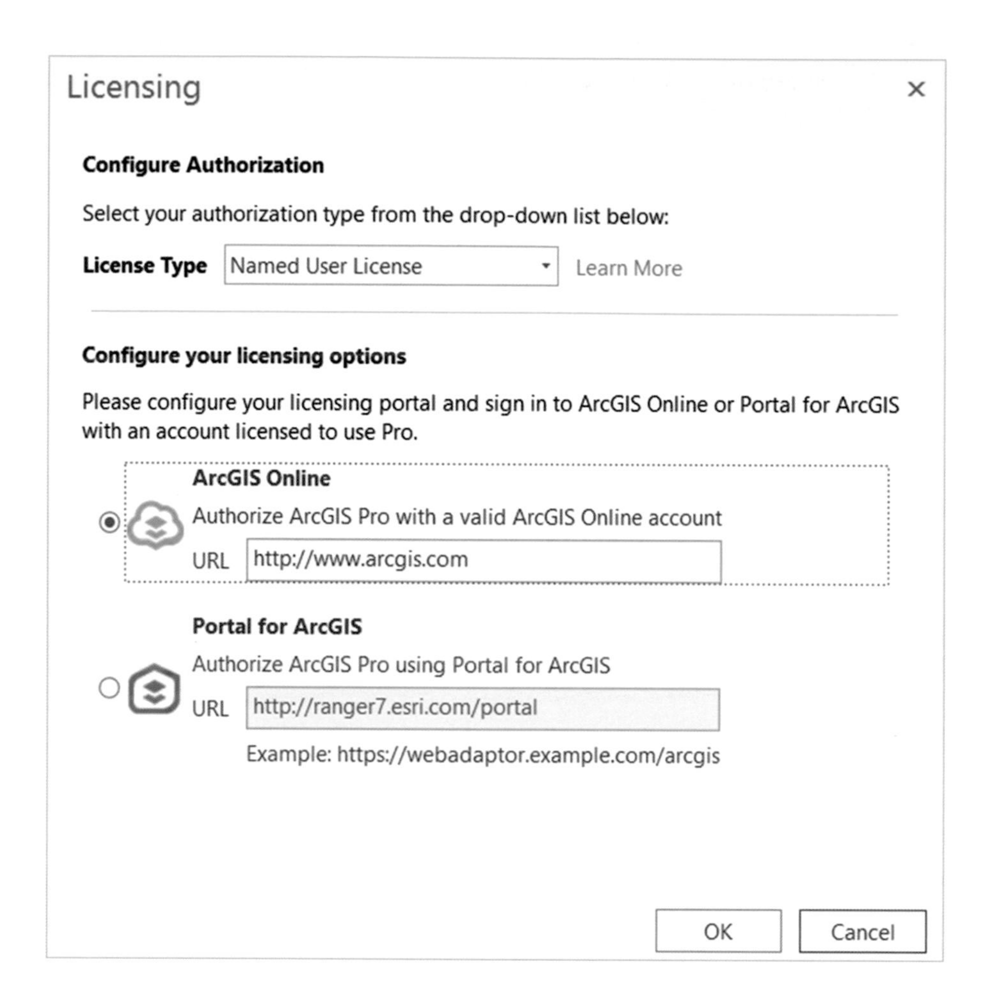

- **Click OK, and sign in with your ArcGIS Online or Portal for ArcGIS account.**

3. **Under Create a New Project, choose the Global Scene.aptx Template.**

Aptx stands for the ArcGIS Pro project template file extension. By selecting this extension, you will create a global scene.

4. **In the Create a New Project window, for Name, specify Earthquakes; for Location, select C:\EsriPress\GTKWebGIS\Chapter9; and click OK.**

ArcGIS Pro creates an empty scene and loads with a default globe view.

5. **Navigate the scene using your mouse.**

   - **Press and hold the left mouse button to pan the view.**
   - **Scroll your mouse wheel to zoom in and out. You can also press and hold the right mouse button to zoom.**
   - **Press and hold the mouse wheel to rotate and tilt the scene.**

Next, you will add an earthquake layer to the scene. A CSV file of earthquakes has been provided in the sample data for the book.

6. **Click the Map tab in the ribbon menu, click the Add Data arrow, and choose XY Event Data.**

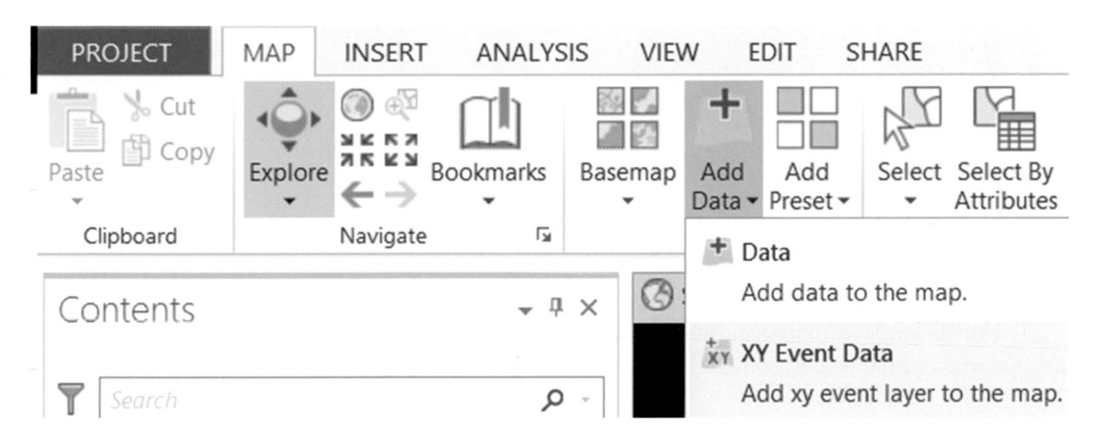

A table with X/Y fields is a kind of event layer. This CSV file is considered an event layer. The **Make XY Event Layer** geoprocessing pane appears.

7. **In the Make XY Event Layer pane, perform the following actions:**

   - **For XY Table, select C:\EsriPress\GTKWebGIS\Chapter9\data\1.0_day.csv.**
   - **Leave the X Field as longitude and the Y Field as latitude.**
   - **Leave the Z Field empty.**
   - **Set the Layer Name as Earthquakes.**
   - **Leave Spatial Reference as GCS_WGS_1984.**
   - **Click Run.**

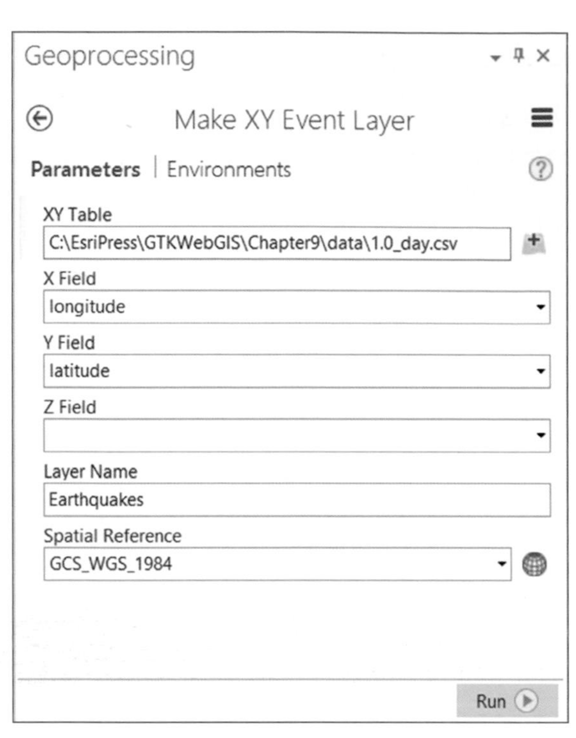

The resulting layer will be added to the map. Next, you will need to export this event layer to a new format, which is a layer in a file geodatabase.

8. **In the Contents pane, right-click the Earthquakes layer, and choose Data > Export Features in the context menu.**

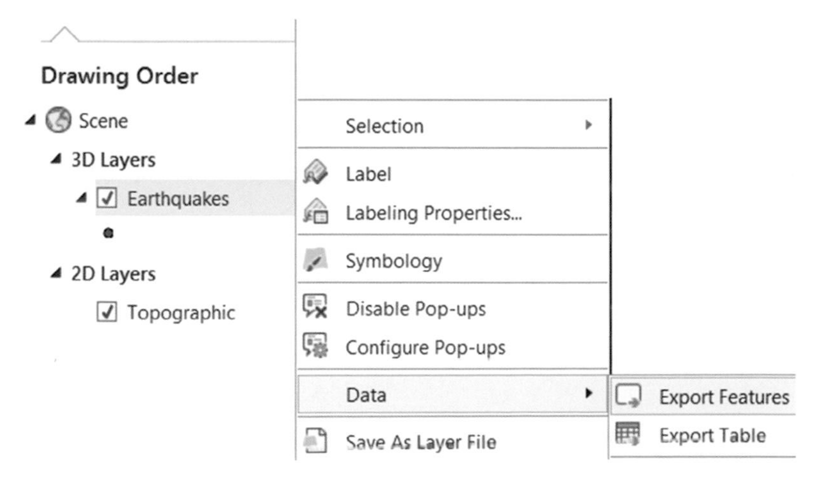

You will see **Copy Features** appear in the **Geoprocessing** pane.

9. **In the Geoprocessing pane, leave the Input Features as Earthquakes. For the Output Feature Class, browse to C:\EsriPress\GTKWebGIS\Chapter9\data\data.gdb, and set the Name as Earthquakes, and click Save.**

10. **Click Run.**

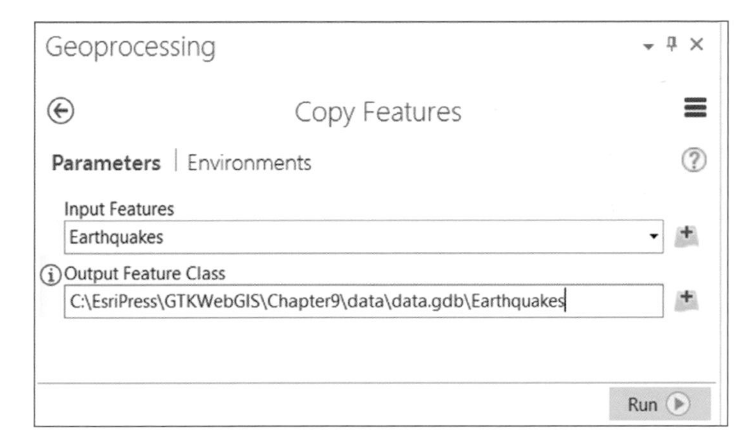

The resulting layer adds to the 2D Layers group in the **Contents** pane. Next, you will remove the two earthquakes layers before adding the layer back in a different way.

11. **In the Contents pane, right-click each of the Earthquakes layers, and choose Remove.**

Next, you will add the new earthquake layer to the scene using preset symbols. Preset symbols provide a convenient way for you to configure 3D symbols.

12. **Click the Map tab in the top ribbon menu, click Add Preset, and click Thematic Shapes.**

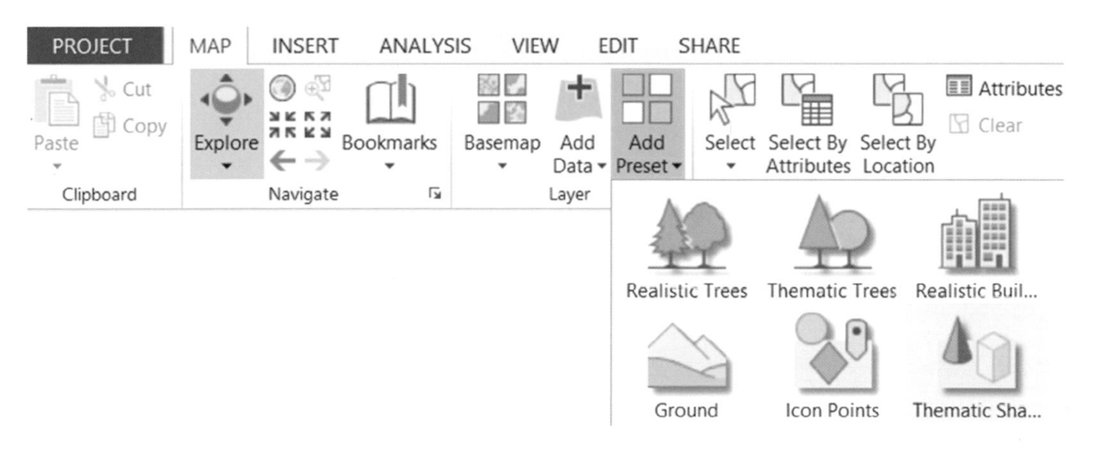

13. Browse to **C:\EsriPress\GTKWebGIS\Chapter9\data\data.gdb**, and select **Earthquakes**, and click **Select.**

The earthquakes layer adds to the scene. You may not see the layer because ArcGIS Pro automatically set a visible scale range, and your current scale may be out of the range. In this case, you can navigate the scene to the layer.

14. While the **Earthquakes** layer is highlighted in the **Contents** pane, click the **Appearance** tab in the ribbon.

15. In the **Visibility Range** group, set both **In Beyond** and **Out Beyond** to **<None>**.

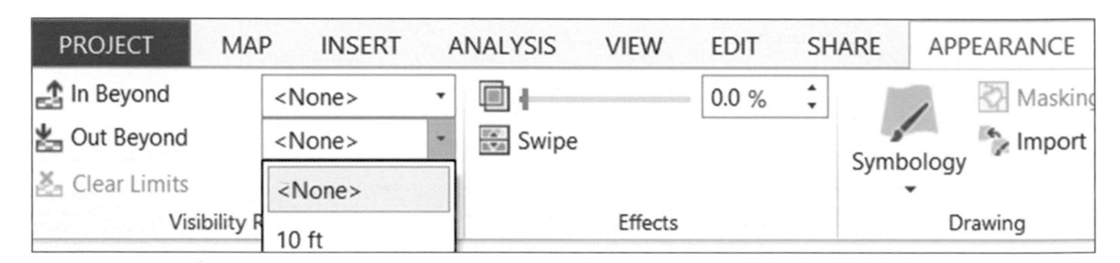

Next, you will configure the earthquakes symbol.

16. Right-click the **earthquakes** layer in the **Contents** pane, and choose **Symbology**.

The **Symbology** pane of the **Earthquakes** layer appears.

17. In the **Symbology** pane, perform the following actions:

- Select **Standing Cylinder.**
- For **Color Using**, choose **mag.**
- For **Color Ramp**, choose **Yellow to Red.**
- For **Height**, choose **mag** (the unit is meters).
- Set the **Scale** at **200000.**
- Unselect **Use aspect ratio.**
- Set the **Width** at **30000.**

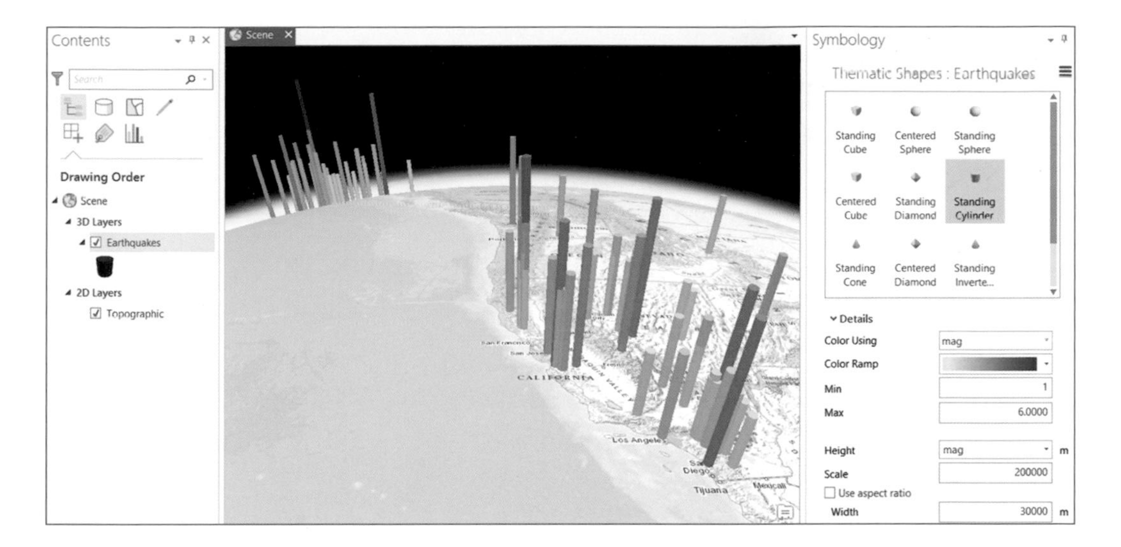

You should see the earthquakes as cylinder symbols. If not, you can navigate the scene to the United States to get a good camera view.

Next, you will configure the label for this layer.

18. **In the Contents pane, right-click the Earthquakes layer, and choose Label.**

The default label for the earthquakes layer appears in the scene view. You will change the label to use the **mag** field.

19. **In the Contents pane, right-click the Earthquakes layer, and click Labeling Properties to bring up the Label Class pane.**

20. **In the Label Class pane, perform the following actions:**

- **Delete the existing expression in the Expression box.**
- **In the Fields list, find and double-click mag.**
- **Click Apply.**

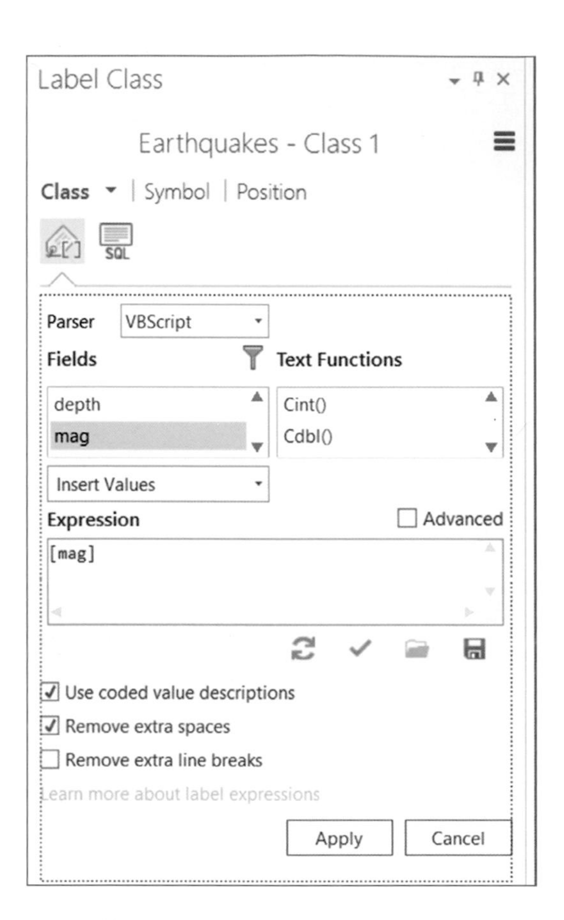

You will see the earthquakes labeled with their magnitudes. Next, you will configure the pop-up of the layer.

21. **In the Contents pane, right-click the Earthquakes layer, and choose Configure Pop-ups.**

22. **In the Pop-ups pane, click the Edit pop-up element icon in Fields.**

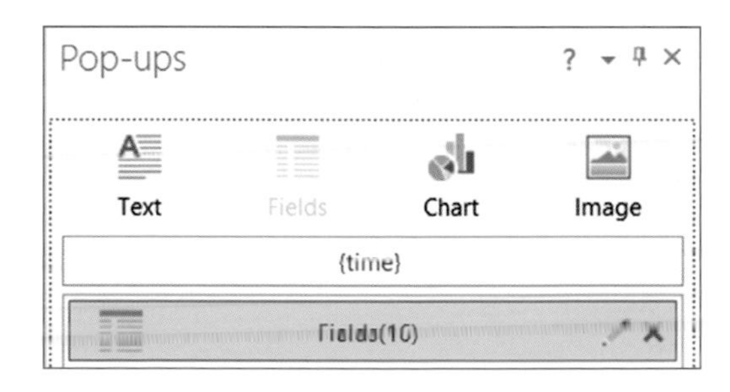

23. In the **Fields Options** list, unselect all the fields except **time**, **depth**, **mag**, and **place**.

24. Click the **Back** button ⊖.

25. In the **Pop-ups** pane, click **Image**, and then click the edit icon ✎ in **Image**.

26. In the **Image Options** pane, perform the following actions:

    - **For Title, remove the text.**
    - **For Caption, specify Click the icon to see details.**
    - **For Source URL, specify http://esrimapbook.esri.com/GTKWebGIS/chapter9/earthquake.png** ⁓**.**
    - **For Hyperlink, specify http://earthquake.usgs.gov/earthquakes/eventpage/{id} (id is a field name in the earthquake layer).**
    - **Click the back button** ⊖**.**

27. Click an earthquake on the map to see the pop-up.

28. In the pop-up, click the icon image, which should bring up a USGS page showing the details of the earthquake.

This capability allows you to link to additional information about your geospatial features. Next, you will change the mode of the scene to get a sense of a local scene.

29. Close the web browser.

30. Click the **View** tab, and click **Local** to switch the view to local scene mode.

31. As you navigate the view in this mode, notice that the terrain and layers display on a projected planar surface.

32. **Click the View tab, and click Global to change the view back to global mode.**

If necessary, navigate the scene until the globe appears.

33. **On the Quick Access Toolbar at the top left corner of the ribbon, click the Save button to save your project.**

In this section, you created a global scene. You started with a CSV file of earthquakes, converted the file into a layer in a file geodatabase, and added the new layer to your scene using preset symbols. Then you configured the 3D symbols to display earthquake magnitudes, configured the layer label and pop-ups, and compared local and global views.

9.3 Share a web scene

The scene you created in the previous section is viewable by only you, locally using ArcGIS Pro. In this section, you will publish the scene to ArcGIS Online or Portal for ArcGIS as a web scene so that you and more users can view this scene using web browsers.

1. **Continuing from the previous section, click the Project tab in ArcGIS Pro.**

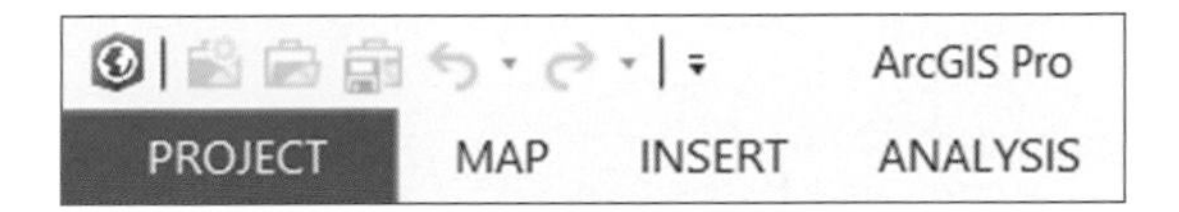

If you exited in the previous section, just start ArcGIS Pro, and open your scene project.

2. **At the far left of the screen, click the Portals tab. Make sure you are connected to ArcGIS Online or your Portal for ArcGIS (if you have multiple portal connections, right-click and set the portal you would like to use as the active portal).**

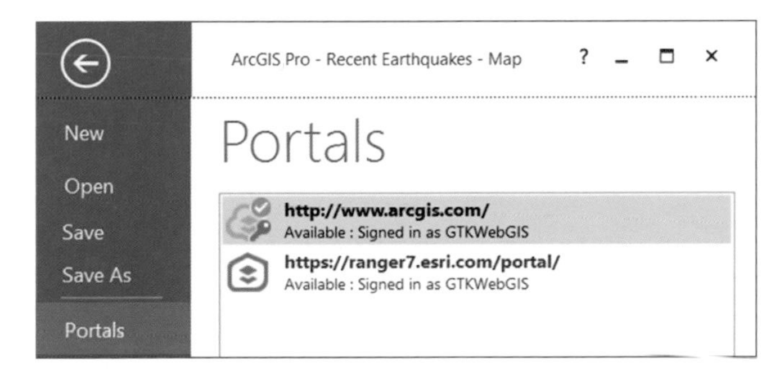

Your portal connections may be different than the ones you see in the illustration.

3. **In the upper-left corner of the screen, click the Back Arrow button.**

4. **Click the Share tab, and click Web Scene.**

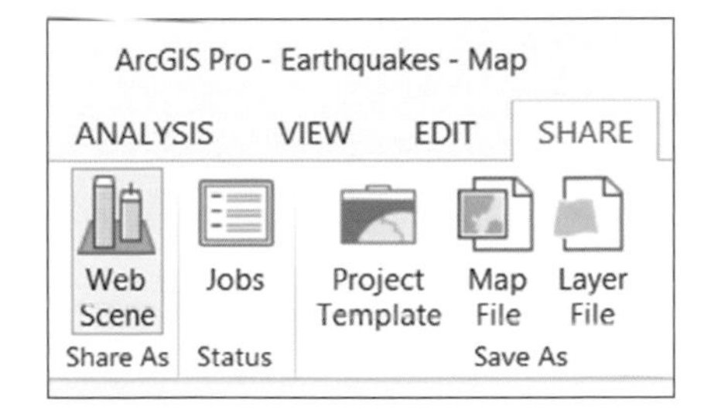

The **Share Web Scene** pane appears.

5. In the **Share Web Scene** pane, perform the following actions:

   - **For Name, specify Earthquakes.**
   - **For Summary, specify Earthquakes on 2016/1/18.**
   - **For Tags, specify Earthquakes.**
   - **For Sharing Options, select Everyone.**
   - **Click Analyze to check for any potential performance issues or errors that you may need to resolve before you can share the web scene.**

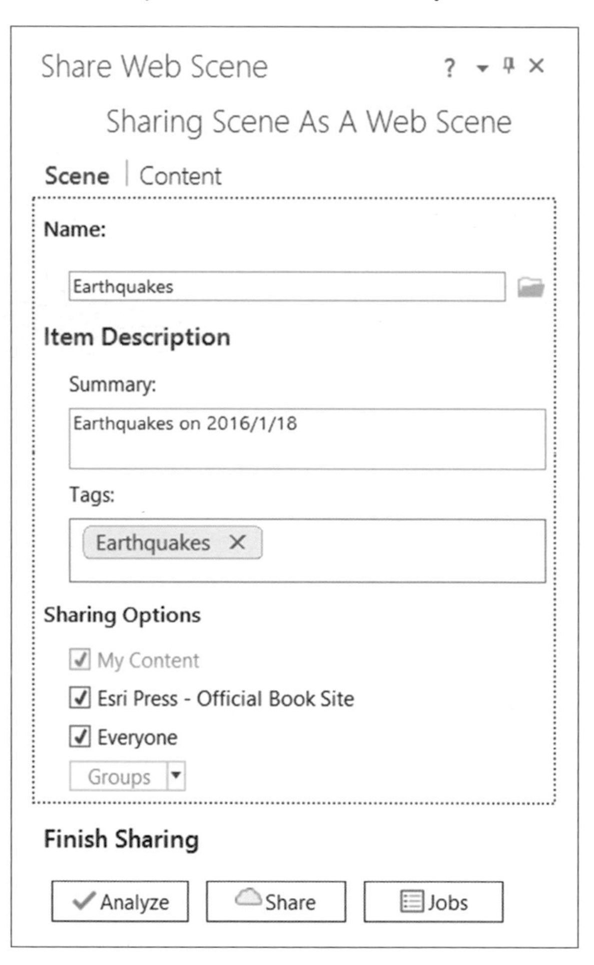

There are no errors. You will see the following warning: **Layer does not have a feature template set.**

6. **Double-click the warning to bring up the explanation page about this warning (you can ignore the warning in this case).**

7. **In the Share Web Scene pane, click Share to publish your web scene.**

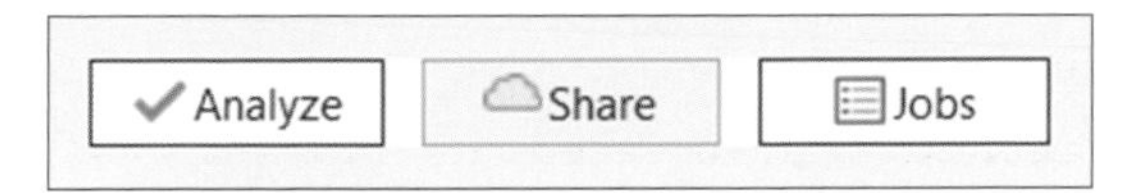

8. **Click Jobs to monitor the status of your web scene. You may need to scroll down to see the Jobs button.**

As your layer is processed and shared, a progress bar will note the completion of each stage of the process until the full scene has been shared successfully.

9. **Once you have shared the web scene successfully, click the Share tab at the bottom, and then click the Manage the web scene link to open the item details page of the web scene in ArcGIS Online.**

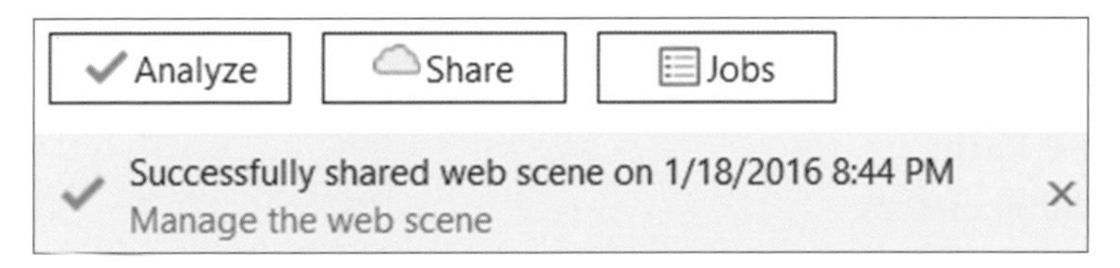

You will see the item details page of your web scene open in your web browser. Alternatively, you can find the web scene from your ArcGIS Online or Portal for ArcGIS **My Content** list.

10. **Click the item thumbnail to view your scene in ArcGIS scene viewer.**

Next, you will add a complementary layer to further enhance your scene.

11. **Sign in with your ArcGIS Online account.**

12. **In the scene viewer, click the Modify button to open the Contents pane.**

13. **Click Add Layers.**

14. **Search for plate boundaries owner:GTKWebGIS, click the Add button to add the layer to your scene, and click Done.**

15. **In the Contents pane, point to the Tectonic_Plates — Tectonic Plates layer, click the arrow ▾, click Rename, specify the name as Tectonic Plates, and press Enter.**

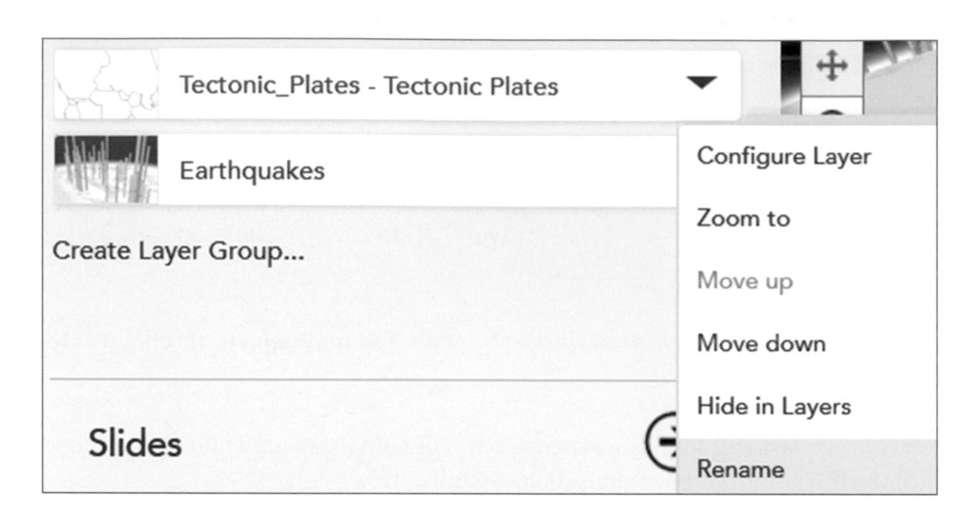

16. Point to the **Tectonic Plates** layer, click the arrow ▾ again, and click **Configure Layer.**

17. In the **Configure Layer** pane, perform the following actions:

- **Change Transparency to 30%.**
- **Change Symbols to Change symbols.**
- **Change Type to 3D Path.**
- **Change Size to 30000 meters.**
- **Click Done.**

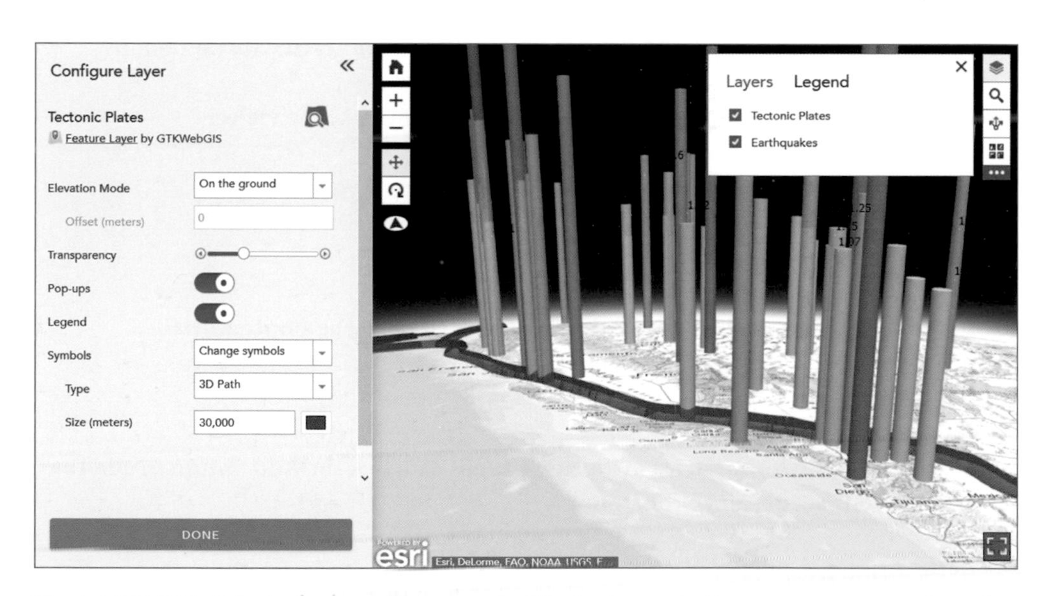

The tectonic plate boundaries will appear in 3D. As shown in the scene, most earthquakes happened near the plate boundaries.

18. **Click Save Scene.**

19. **In the Save Scene window, click Save.**

You have created a web scene. You may feel that there are some tools and menus you do not want your users to access because the tools may confuse your users. Next, you will eliminate the unneeded tools and menus.

20. **In your web browser address bar, add &ui=min to your current page URL, and press Enter.**

The scene is reloaded with fewer tools and menus.

21. **Share the current page URL (your scene UR) with your users.**

You can share your URL with users via email, text, and other ways.

QUESTIONS AND ANSWERS

1. **I created photorealistic 3D models in CityEngine or other tools. How can I bring them into a web scene in ArcGIS Online or Portal for ArcGIS?**

   Answer: You can export your 3D models to multipatch format, which is a GIS industry standard developed by Esri. It uses collections of geometries to represent 3D objects. Multipatch features can be used to construct 3D features in ArcGIS, save existing data, and exchange data with other non-GIS 3D software packages such as Collaborative Design Activity (COLLADA) and SketchUp.

   To add the multipatch layer into your scene in ArcGIS Pro, you can click the **Map** tab, **Add Preset**, and **Realistic Building**, and then select your multipatch layer. You can then publish the layer to Portal for ArcGIS 10.4 now. Soon, you will be able to publish the layer on ArcGIS Online.

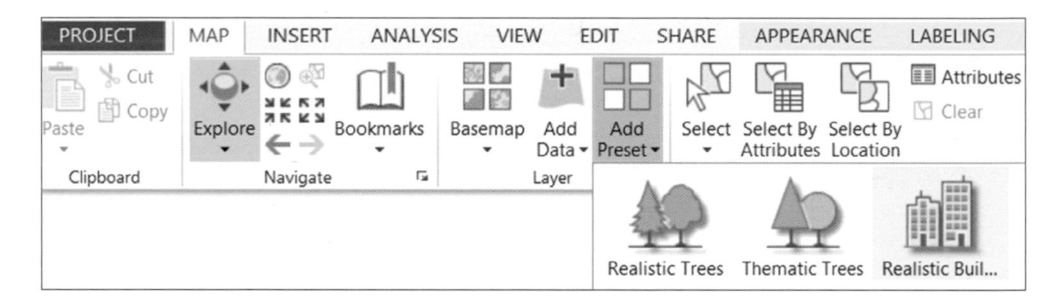

2. **In ArcGIS Pro, how can I configure 3D symbols for line and polygon layers?**

   Answer: Perform the following actions:

   - Move the layer to the **3D layers** group.
   - Click the **Appearance** tab.
   - Remove the visibility range of the layer if needed.
   - In the **Extrusion** group on the ribbon, choose an extrusion type, and specify the extrusion height; for example, set the extrusion height as population density multiplied by 100 meters.

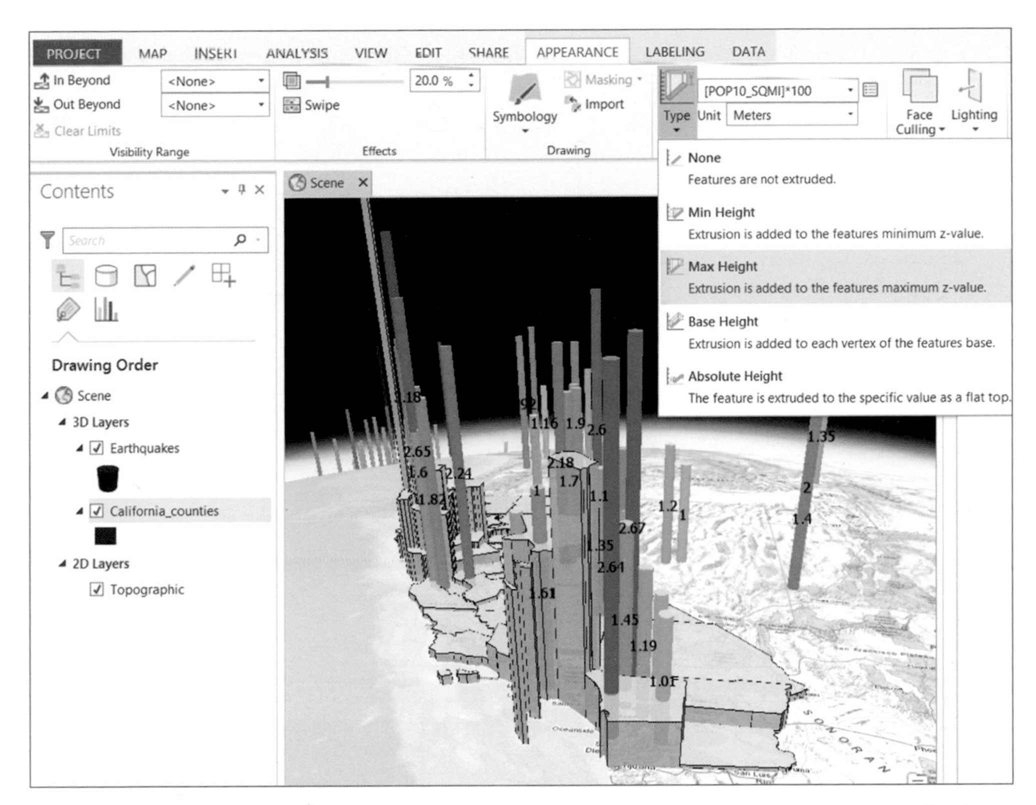

3. **I published an extruded line or polygon layer to ArcGIS Online, but the lines or polygons do not extrude in the scene viewer. They appear draped on the ground surface. Why does this issue arise, and how can I work around this issue?**

   Answer: Extruded features are not yet directly supported in ArcGIS Online and Portal for ArcGIS. Future releases of ArcGIS Online and Portal for ArcGIS will provide an enhanced workflow. The current workaround is to convert the extruded layer to a multipatch layer and then publish your scene. Perform the following steps to convert your layer to a multipatch layer:

   - Click the **Analysis** tab and then click **Tools**.

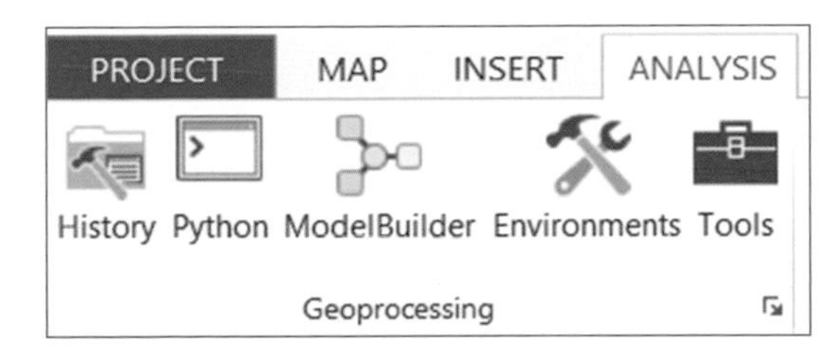

- In the **Geoprocessing** pane, search for **Layer 3D to Feature Class**. Click the first tool found, use your extrusion layer as the input feature layer, and run the tool. The resultant layer will be added to your scene.

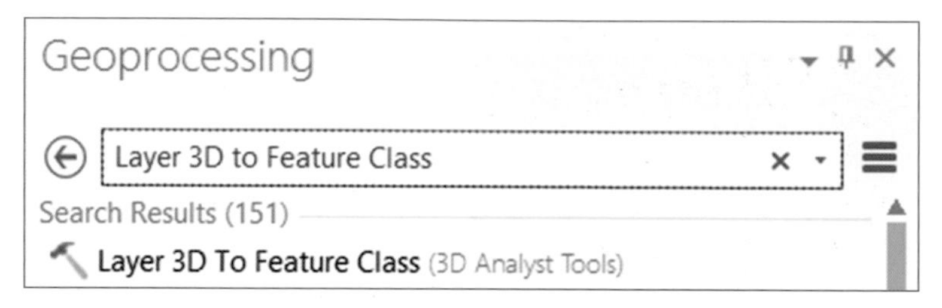

- Remove your original extrusion layer.

- Save and share your scene to Portal for ArcGIS (version 10.4 and later).

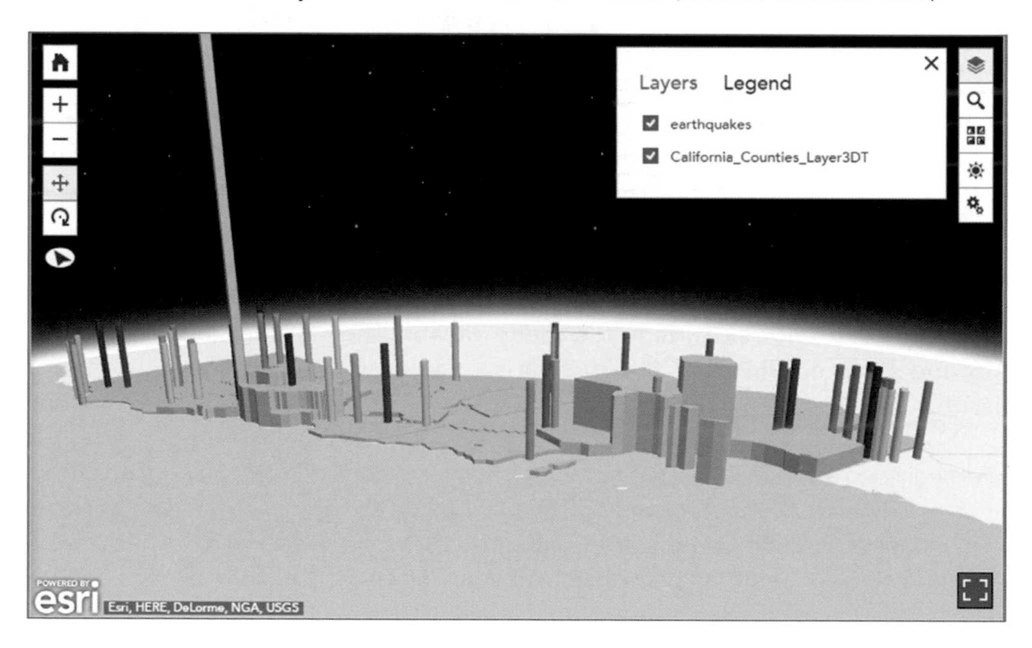

ASSIGNMENTS

Assignment 9: Create a 3D web scene to show recent earthquakes or other point data layers.

 Hints:

- You can download the latest earthquakes from the USGS website using the short URL, **http://on.doi.gov/1Jpg5og.** Note, for the purpose of a quick homework, choose a small CSV file, such as the one for earthquakes with magnitudes greater than 1 in the past one day.
- If your data is in CSV format, add the data as an event layer, and convert the data to a shapefile or a layer in a file geodatabase.
- Add the shapefile or the file geodatabase layer via preset symbols.

What to submit: Send an email to your instructor with the subject line **Web GIS Assignment 9: Your name**, and include the following information:

- **The URL to display your web scene in ArcGIS scene viewer**

Resources

ArcGIS help documents

ArcGIS Earth product information site, http://www.esri.com/software/arcgis-earth.

"ArcGIS Pro Help," http://pro.arcgis.com/en/pro-app/help/main/welcome-to-the-arcgis-pro-app-help.htm.

"Author a Web Scene," http://pro.arcgis.com/en/pro-app/help/mapping/map-authoring/author-a-web-scene.htm.

"Go 3D," http://pro.arcgis.com/en/pro-app/help/mapping/map-authoring/go-3d-with-arcgis-pro.htm.

"Make Your First Scene," https://doc.arcgis.com/en/arcgis-online/create-maps/make-your-first-scene.htm.

"Share a Web Scene," http://pro.arcgis.com/en/pro-app/help/sharing/overview/share-a-web-scene.htm.

"View Scenes in the Scene Viewer," https://doc.arcgis.com/en/arcgis-online/use-maps/view-scenes.htm.

Esri training and tutorial website

"Get Started with ArcGIS Pro," https://learn.arcgis.com/en/projects/get-started-with-arcgis-pro.

"Mapping the Third Dimension," in *The ArcGIS Book*, Christian Harder (Ed.), http://learn.arcgis.com/en/arcgis-book/chapter6.

"Sharing 3D Content with ArcGIS," by Esri, http://training.esri.com/gateway/index.cfm?fa=catalog.webCourseDetail&courseid=2959.

Esri videos

"3D Analyst: An Introduction," by Deepinder Deol and Jinwu Ma, http://video.esri.com/watch/4712/3d-analyst-an-introduction.

"3D Cartographic Techniques: An Introduction," by Nathan Shephard and Kenneth Field, http://video.esri.com/watch/4698/3d-cartographic-techniques-an-introduction.

"ArcGIS for 3D Cities: An Introduction," by Brian Sims, Dan Hedges, and Thorsten Reitz, http://video.esri.com/watch/4728/arcgis-for-3d-cities-an-introduction.

"Scene Services and the Open Indexed 3D Scene Format," by Thorsten Reitz and Chris Andrews, http://video.esri.com/watch/4445/scene-services-and-the-open-indexed-3d-scene-format.

Chapter 10
Getting started with ArcGIS API for JavaScript

You have learned how to create web apps using various ArcGIS configurable apps. These ready-to-use web apps provide tremendous functionality, but they may not meet all your project requirements. In such cases, you must program your own apps or customize some existing web apps. JavaScript is the most popular programming language on the web. Built on top of JavaScript, ArcGIS API for JavaScript provides libraries for you to develop custom web GIS apps. In this chapter, you will learn to use JavaScript API quickly and easily so that you can adapt individual JavaScript samples and combine multiple JavaScript samples.

Learning objectives

- *Understand the basics of ArcGIS API for JavaScript.*
- *Debug JavaScript.*
- *Adapt a JavaScript sample.*
- *Combine multiple JavaScript samples.*
- *Develop 2D and 3D GIS apps using JavaScript.*

This chapter in the big picture

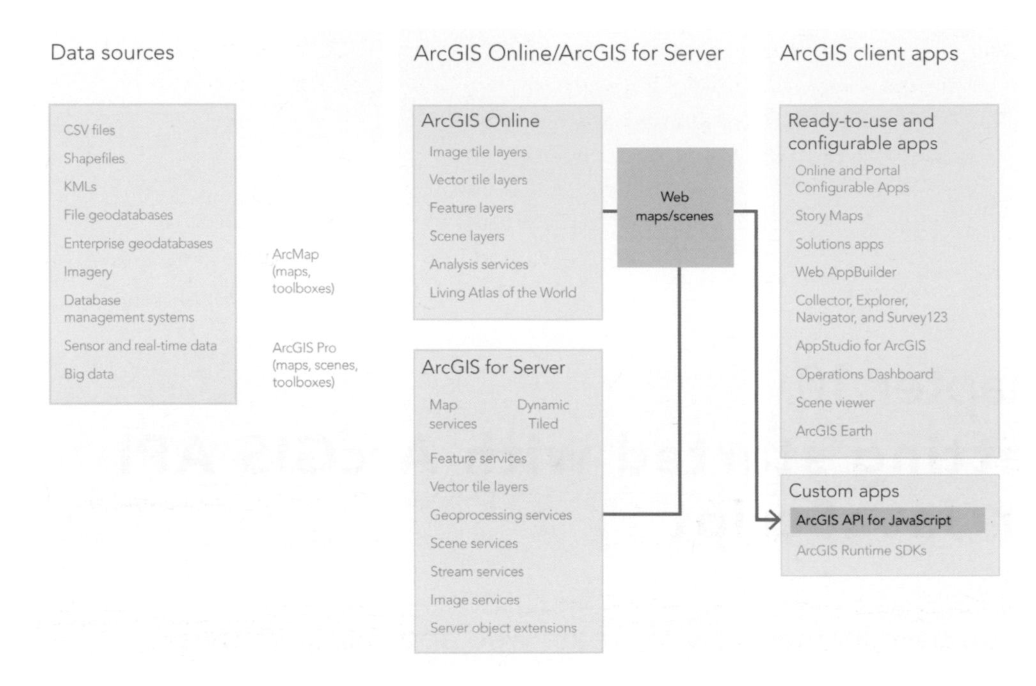

ArcGIS offers many ways to build web applications. The green line highlights the technology that this chapter teaches.

Why JavaScript?

Almost all web pages today include some JavaScript code. JavaScript is a bridge between web browsers and servers. JavaScript interacts with servers to use the servers' capabilities and works with web browsers to make web pages dynamic and interactive.

Unlike web services, which run on the server side, JavaScript typically runs on the browser side. JavaScript has more cross-platform capabilities than native desktop apps and mobile apps. All desktop and mobile web browsers support JavaScript. With responsive web design approaches, a single JavaScript web app can run on both desktop browsers and mobile browsers and across a wide range of screen sizes.

To use JavaScript, you do not need a professional integrated development environment (IDE), and you do not need to compile JavaScript programs. You can write JavaScript code using any clear text editor, and then load and run the source code in a web browser.

HTML and CSS

In general, using JavaScript is easy in the beginning. However, once you begin to program with JavaScript, you must know the basics of Hypertext Markup Language (HTML) and Cascading Style Sheet (CSS). JavaScript works in conjunction with HTML and CSS to build web apps.

- HTML is a markup language used to package your content.
- CSS is a formatting language used to style content.
- JavaScript creates dynamic and interactive features for your web pages.

You can quickly learn the basics of HTML, CSS, and JavaScript at the tutorial website **http://www.w3schools.com**.

JSON

ArcGIS API for JavaScript relies on JavaScript Object Notation (JSON), a lightweight and easy to understand format for storing and exchanging data in the following ways:

- Data is in field_name:value pairs.
- Data is separated by commas.
- Curly braces hold objects.
- Square brackets hold arrays.

For example, the following JSON code describes an array of two students. Each `student` is represented with a `name` field and `name` value, a `hobby` field and `hobby` value, and an `address` field and `address` value. Each `address` value—with a `street` field and a `street` value, a `zip` field and a `zip` value—is another JSON object.

```
{
  "students": [{
         "name": "John",
         "hobby": "Basketball",
         "address": {
                "street": "380 New York St",
                "zip": 92373
         }
  }, {
         "name": "Lisa",
         "hobby": "Movie",
         "address": {
                "street": "270 State Avc",
                "zip": 92000
         }
  }]
}
```

ArcGIS REST API

The JavaScript API relies on the ArcGIS Representational State Transfer (REST) API. In previous chapters, you learned how to publish various web services using ArcGIS Online and ArcGIS for Server. These services and web maps show their capabilities via the ArcGIS REST API. ArcGIS JavaScript APIs call the server via the REST API. In terms of server-side capabilities, the JavaScript API can deliver only what the REST API can deliver.

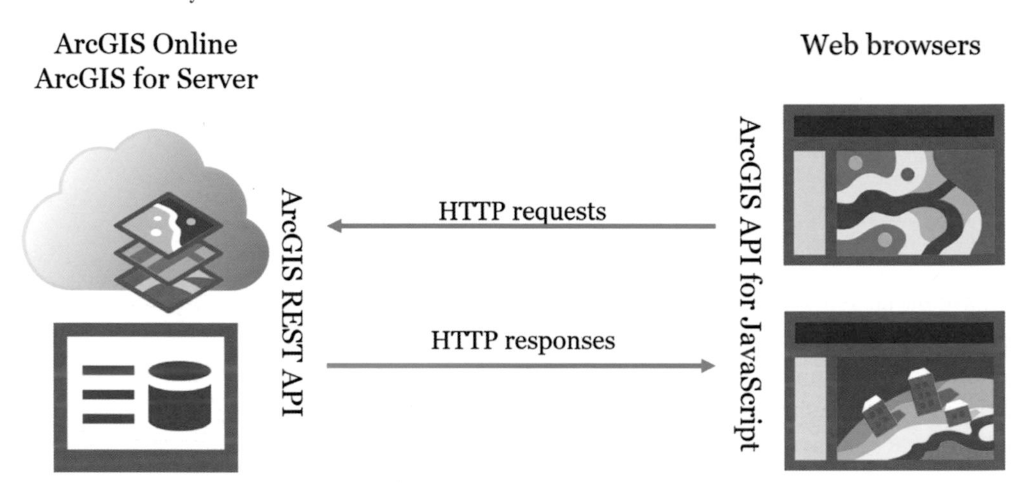

ArcGIS for JavaScript relies on the ArcGIS REST API to interact with ArcGIS for Server, ArcGIS Online, and Portal for ArcGIS.

With the ArcGIS REST API, all resources and operations are exposed through a hierarchy of endpoints, or URLs. The REST endpoint of a map service is typically in the format of http(s)://<machine_name>/arcgis/rest/services/<folder_name>/<service_name>, such as **http://sampleserver6.arcgisonline.com/arcgis/rest/services/Wildfire/MapServer**.

By default, the endpoint URL for the first layer in this map service appends a layer number, starting with 0 for the first layer, such as **http://sampleserver6.arcgisonline.com/arcgis/rest/services/RedlandsEmergencyVehicles/MapServer/0**.

These REST URLs are what you use in JavaScript to call (or run) the map service to generate a map or the layer to do a query. As you learned in the previous chapters, ArcGIS REST Services Directory is a good place to explore the ArcGIS REST API. For example, a map service supports the export operation. If you click the **Export Map** link on the Services Directory page of the map service, you get a URL such as this: **http://sampleserver6.arcgisonline.com/arcgis/rest/services/RedlandsEmergencyVehicles/MapServer/export?format=png8&bbox=117.29546425851748%2C34.036800257765584%2C-117.11795646748254%2C34.12201888281171&bboxSR=4326&size=1908%2C916&f=html**.

If you change **f=html** to **f=json**, the new URL is the URL that the JavaScript API sends to ArcGIS for Server to request the map service to generate a map. The response in JSON format would resemble the following screen capture.

sampleserver6.arcgisonline.com/arcgis/rest/services/RedlandsEmergencyVehicles/MapServer/export?format=png8&bb

{"href":"http://sampleserver6.arcgisonline.com/arcgis/rest/directories/arcgisoutput/RedlandsEmergencyVehicles MapServer/x_____x2Aoqp2xDc-wF9_eV5CU5-w..x_____x_ags_mapd39b47d3fbe542d4845ef5a77289115c.png","width":1908,"height":916,"extent":{"xmin":-117.29546425851753,"ymin":34.036800257765584,"xmax":-117.11795646748259,"ymax":34.12201888281171,"spatialReference":{"wkid":4326,"latestWkid":4326}},"scale":39098.53249475937}

Response to a map request in JSON format.

You may find the response difficult to read. The response is meant for the JavaScript to parse and present. If you want a more readable JSON response, you can use the JSON format—that is, use **f=pjson** instead of **f=json** in the request URL.

ArcGIS API for JavaScript

ArcGIS API for JavaScript is built on top of HTML5, which includes HTML, JavaScript, and CSS. The API runs on the client side. The API has two main capabilities:

- Interact with GIS servers, including ArcGIS Online, Portal for ArcGIS, and ArcGIS for Server, to provide mapping, querying, editing, analysis, and other GIS functions.
- Display the responses of GIS servers in maps, views, pop-ups, charts, and other formats, and interact with users.

The cloud hosts ArcGIS API for JavaScript library. You do not have to download and install the library manually. You simply reference the API using the <script> and <link> tags, as you will see in the tutorial section. Organizations that require more secure access (such as organizations that do not allow an external Internet connection) have the option to download and host the API on their local network.

You will generally follow six steps when you use ArcGIS API for JavaScript to write a web app:

1. Reference the ArcGIS API for JavaScript.
2. Load API modules needed for your functions.
3. Create your map or scene.
4. Create your 2D map view or 3D scene view.
5. Define the page content, especially the space(s) to hold your maps and scene views (divs).
6. Style the page.

When you adapt an existing sample or combine multiple samples, you will not use all the steps because some of the steps are already done in the samples.

As a developer, you often will use the ArcGIS API for JavaScript website, **http://developers.arcgis.com/javascript**. On this site, you can find the guide, API references, samples, and JavaScript API forum on GeoNet (**geonet.esri.com**), a place where you can share, chat, and collaborate with other ArcGIS users.

The Help document site for ArcGIS API for JavaScript. The site may look different than the image you see here.

ArcGIS API for JavaScript builds on top of Dojo, a JavaScript framework or toolkit that helps reduce development efforts and enhance cross-browser capability. If you have experience with other JavaScript frameworks, such as Bootstrap, React, and JQuery, you can use them with the API too.

ArcGIS API for JavaScript main classes

ArcGIS API for JavaScript provides many classes with which you can create objects. This list includes the most commonly used classes:

- **View:** Includes the **MapView** class and the **SceneView** class.
    - **MapView**: Displays maps with 2D renderers. **MapView** properties include extent, center, rotation, scale, and more.
    - **SceneView**: Displays 3D scenes and maps with 3D renderers. **SceneView** properties include center, rotation, scale, camera, and more.
- **Map**: Manages layers that can be added and removed from a **Map** dynamically or can be loaded from an existing web map.
- **WebScene**: Manages scene layers that can be added to and removed from a **WebScene** dynamically. A **WebScene** can also be loaded from an existing web scene.
- **Layer**: The fundamental component of maps and web scenes. This class includes subclasses representing the feature layer, graphics layer, dynamic map layer, image layer, tiled layer, stream layer, vector tile layer, web tiled layer, elevation layer, and scene layer.
- **Render**: Contains drawing information for layers. This class includes subclasses such as **SimpleRenderer**, **ClassBreakRenderer**, and **UniqueValueRenderer**.
- **Symbols**: Displays points, lines, polygons, and text in 2D. This class includes many subclasses, such as **SimpleMarkerSymbol**, **SimpleLineSymbol**, **SimpleFillSymbol**, and more.
- **3D Symbols**: Displays points, lines, polygons, and text in 3D. This class includes many subclasses, such as **IconSymbol3DLayer**, **LineSymbol3DLayer**, **ExtrudeSymbol3DLayers**, and more.
- **Task**: Includes subclasses such as **QueryTask**, **Geoprocessor**, and **RouteTask**. These subclasses allow you to perform attribute or spatial searches on individual layers or all layers in a map service, use geoprocessing tasks, find optimal routes, and more.

- **Portal:** Provides a way to build apps that work with content from ArcGIS Online and Portal for ArcGIS. For example, you can load an ArcGIS Online web map or web scene with the layers already configured with renderers and pop-ups.

Class properties, methods, and events

Each ArcGIS API for JavaScript class often includes many of the following properties, methods, and events:

- **Properties:** A **MapView**, for example, includes extent, center, rotation, scale, zoom, and other properties. You can get and set the properties of an object with the simple notation `objectName.propertyName`. To get and set the zoom level of a map view, for example, you can use `mapView.zoom`.
- **Methods:** Methods are actions or functions that a class can perform. You can call or run the method in this format: `objectName.methodName(parameters)`. For example, you can add a layer to a map using the add method of a map: `map.add(layer)`.
- **Events:** Events happen to the elements in your app, such as when an object is ready, started, changed, completed, moved, or displayed, or when an error has occurred. You may trigger some functions in response to these events.
  - To monitor an object's property change, use the `.watch(property, callback)` method. For example, you can watch and handle a map view's extent change with the following code:

```
mapView.watch("extent", function(response) {
    console.log("the response object is the new extent");
});
```

  - To handle the result of a task or a class, use **promise**. Promises play an important role in the 4.0 version of the ArcGIS API for JavaScript. In essence, a promise is a value that "promises" to be returned whenever the process completes. Promises are commonly used with the `.then(callback, errback)` method. For example, you can handle the result of a **queryTask** with the following code:

```
queryTask.execute(parameters).then(
            function getResults(queryResult) {},
            function getErrors(err) {}
);
```

Explore ArcGIS API for JavaScript samples in the sandbox

You can explore the samples to learn ArcGIS for JavaScript quickly and easily. Most samples at the API website have an **Explore in the sandbox** button. You can click the button, modify the sample source code in a text box in your web browser, and run your code in the browser without installing your own web server.

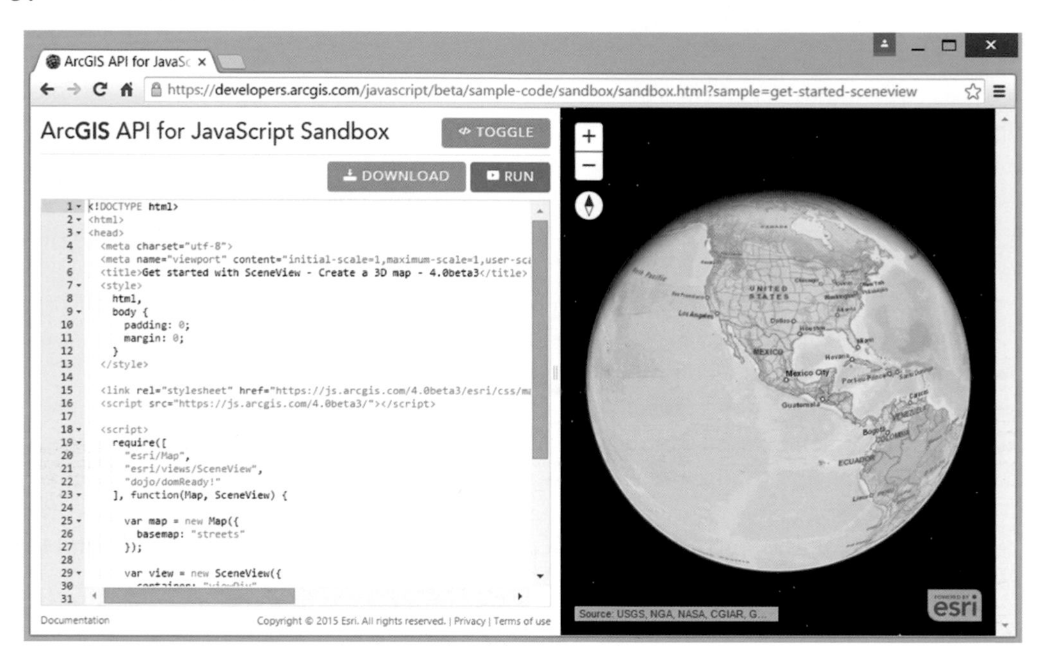

You can learn to use and explore code samples directly in your browser.

Adapting samples and combining samples

Using samples to start your web app development project is cost effective. You can use samples from the API website, ArcGIS Online or Portal for ArcGIS configurable apps, solutions apps, apps at Esri GitHub (**https://github.com/Esri**), or Web AppBuilder for ArcGIS Framework.

In general, adapting a JavaScript sample for your own requirements involves three important steps:

1. Replace the web service or layer URL(s).
2. Replace the attribute field names with names for the new service(s) or layer(s) and the related values of the fields.

3. Replace the related symbols, which often involves feature layers and graphics. For example, a sample app highlights some lines using a line symbol. If you want this sample to work with a point layer, match the feature type by changing the line symbol to a point symbol.

Combining multiple samples can be more involved, but in general, you must load the required modules that each sample needs. Normally, you should keep the source code from different samples as separate functions or classes. This practice can help you avoid code conflicts and more easily maintain your code.

Incorporating widgets

A widget is a self-contained component that you can easily incorporate into your JavaScript apps. ArcGIS API for JavaScript provides many widgets, such as the **BasemapToggle** widget. Adding a widget to your web app generally takes four steps:

1. Define a div in the HTML body to hold the widget.
2. Load the module(s) in the require function.
3. Define the widget properties, such as the associated map, layer, or fields.
4. Call the startup method of the widget.

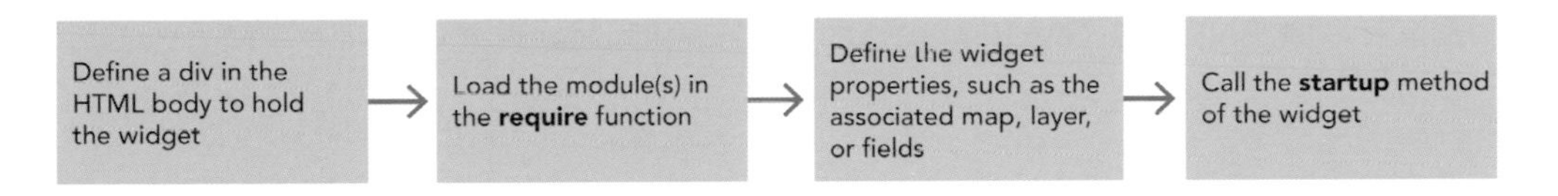

The steps to add a widget to your web app.

You will notice that the widgets for ArcGIS API for JavaScript differ from the widgets for ArcGIS Web AppBuilder.

Add layers to your apps

You can add layers to your app in two ways:

- **Web map or web scene approach**: A web map or web scene already has stylized layers included, with pop-ups configured.
- **Dynamic approach**: You must choose the type of ArcGIS API layer. Your choice of some data sources is easy. For example, you must use **FeatureLayer** for hosted feature layers, **ArcGISTiledLayer** for tiled ArcGIS map services, **VectorTileLayer** for cached vector tiles, and **StreamLayer** for data streams using HTML5 WebSockets. However, other sources can make your choice more difficult. For example, you often must choose between the **ArcGISDynamicLayer** and **FeatureLayer** approach for displaying layers of a dynamic map

service. For **ArcGISDynamicLayer**, the server generates the map and then returns the map image to the client side. For **FeatureLayer**, the server returns the vector data to the client side, and the client side draws the map.

IDE and debugging

Inevitably, you will make typos or other errors while writing programs. JavaScript IDEs such as WebStorm, Sublime, and Microsoft Visual Studio can ease your programming experience by providing syntax highlighting and IntelliSense. IntelliSense, a context-aware, code-completion feature, reduces misspellings, typos, and other common mistakes.

As a developer, you often must use debugging tools. Otherwise, even identifying a small typo can be frustrating. Chrome, Firefox with Firebug add-on, Safari, and Internet Explorer (version 10 and later) all provide developer tools. These tools have the following capabilities:

- Display JavaScript errors that arise in the consoles at runtime.
- Pause the execution of your code at the breakpoints you set so that you can examine the variable values and the status of the document object model (DOM).
- Monitor network activities so that you can analyze the network usage and improve the performance of your code.
- Inspect HTML elements and modify their styles and layout in real time.

This tutorial

In this tutorial, you will create several web apps, including a 3D map to display world populations with extruded polygons and an app to display earthquakes and tectonic boundaries in 2D and 3D views. You will also link two views so that you can pan or zoom the 2D view as the 3D view follows along.

Data: You are provided several hosted layers and a web scene as the required data.

System requirements:

- Chrome web browser for viewing your apps and learning debugging. Chrome allows you to view most HTMLs using the **file:///** URL. Your Chrome must support WebGL in order to display 3D scene views. You can find detailed system requirements by visiting the ArcGIS scene view requirements page.
- Notepad++, Notepad, a pure text editor, or a professional JavaScript development environment for editing JavaScript code. You can read your JavaScript code more easily in Notepad++ than in Notepad. You can download Notepad++ at **http://www.notepad-plus-plus.org**. By using this application, you will not need to learn a professional IDE for this chapter.

- ArcGIS Online for serving the basemaps, hosted layers, and a web scene. These contents are shared at the public level. Everyone can access the contents; therefore, you will not need an account for this tutorial.
- ArcGIS API for JavaScript 4.0 or newer.

10.1 Understand the basics of 2D views and 3D views

You will find this section online, starting at the **Samples** page of the ArcGIS API for JavaScript website. Next, you will explore two samples as an introduction to 2D and 3D views.

1. **Go to the appropriate website, and click Get started with 2D.**
2. **Read the step-by-step instructions, and become familiar with the steps for developing a JavaScript app.**

You do not need to understand every single line of the source code.

3. **Scroll up, and click Explore in the sandbox.**
4. **On the left, read the source code, and on the right, navigate the app.**
5. **Click the browser back button twice to return to the Sample page.**
6. **Click Get started with 3D, and repeat steps 2 through 4.**

10.2 Adapt a 3D scene view sample

In this section, you will display the world population by country in 3D. Your goal is to learn how to adapt a simple JavaScript sample.

Notes:

- The following steps involve some typing. To avoid typos, refer to the source code at C:\EsriPress\GTKWebGIS\Chapter10\polygon_extrusion_key.html, or simply copy the piece of source code you need as you work on the following steps.
- You will change some JavaScript code in the ArcGIS JavaScript API sandbox. Please do not refresh the sandbox page or leave the sandbox page; otherwise, you will lose your changes.
- JavaScript is case sensitive, and indentation does not matter in JavaScript. However, the best practice is to indent for user readability.

1. **Start Chrome, go to an ArcGIS API for JavaScript Sandbox sample, such as http://arcg.is/27e0PrD. Select and delete all the source code in the text area on the left, and click Run.**

Typically, you can choose a sample (that is close to your project requirements) from the ArcGIS API for JavaScript samples. For this section of the tutorial, a JavaScript sample has been provided to you at C:\EsriPress\GTKWebGIS\Chapter10\polygon_extrusion.html. You will copy and paste the source code into the sandbox text area in the next step.

2. **Start Notepad++ or another JavaScript editor, open C:\EsriPress\GTKWebGIS\Chapter10\polygon_extrusion.html, select all the source code, and copy and paste the source code to the text area on the left side of the sandbox page. Review the source code, and click Run.**

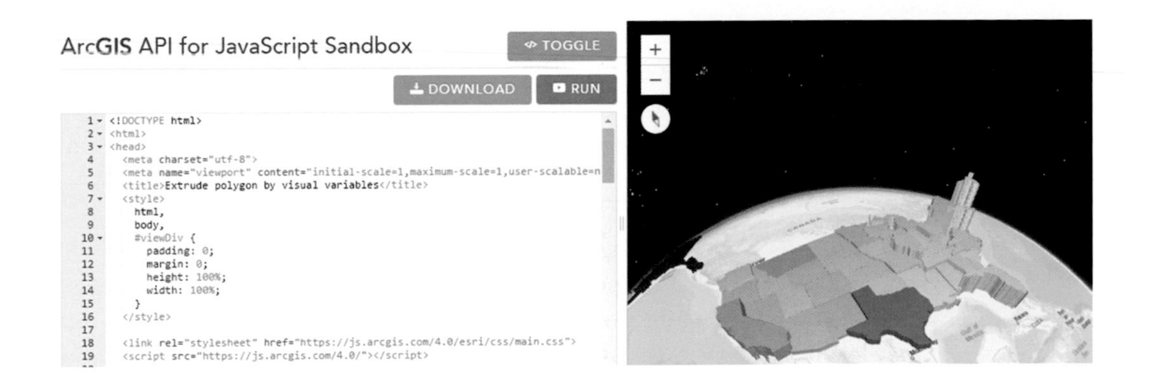

This app displays a 3D global scene view with a basemap and a feature layer of all the states in the United States. This scene extrudes the states based on their population density values and fills the states with colors, ranging (or ramping) from white to red according to the geographic size of each state.

You will navigate the scene as you would in the ArcGIS scene viewer. In other words, you will left-click and drag to pan, right-click and drag to tilt and rotate, and use the mouse wheel to zoom.

Next, you will change the scene from global view to local view. Changing the scene allows you to experiment with the local scene and see all countries of the world at a glance. You set the `viewingMode` property of the view object to be local. You can find more information on this property at the ArcGIS API for JavaScript website (click **API Reference** > **esri/views** > **SceneView** > **Properties** > **viewingMode**).

3. **In the source code, add `viewingMode: "local"`, to the view object, as illustrated, and then click Run.**

Next, you will change the state layer to a world country layer. The layer URL is **https://arcgis.storymaps.esri.com/arcgis/rest/services/SevenBillion/PopulationGrowth_1960_2010/MapServer/0.**

1 2 3 4 5 6 7 8 9 10

4. **In the source code, find the `featureLayer url`, and change its value to `https://arcgis.storymaps.esri.com/arcgis/rest/services/SevenBillion/PopulationGrowth_1960_2010/MapServer/0`.**

```
51        //Create featureLayer and add to the map
52        var featureLayer = new FeatureLayer({
53          url: "https://arcgis.storymaps.esri.com/arcgis/rest/services/SevenBillion/PopulationGrowth_1960_2010/MapServer/0"
54        });
55        map.add(featureLayer);
56
```

Note: The http/https protocol of the feature layer URL must be the same http/https protocol of your current page. If your current page URL uses https, the layer URL must use https. If your current page URL uses http, the layer URL must use http. If you mix the protocols, you may not be able to add this layer because of your web browser's security control in displaying mixed http and https contents.

This 2D layer displays with a 3D symbol. The 3D symbol extrudes the polygons according to the `POP07_SQMI` field, as indicated by the `sizeInfo` type. The `minSize` and `maxSize` define the minimal and maximum heights of extrusion, and `minDataValue` and `maxDataValue` correspond to the min and max value of the POP07_SQMI field. Next, you will change the 3D symbol to extrude the countries based on the population field.

5. **In the source code, under `size`, perform the following actions:**

   - **Change the value of `field` to `POP2007`.**
   - **Change the value of `maxSize` to `10000000` meters. This new value will extrude the countries higher.**
   - **Change the value of `minDataValue` to `11992` and the value of `maxDataValue` to `1321851888`.**

```
59        var extrudePolygonRenderer = new SimpleRenderer({
60          symbol: new PolygonSymbol3D({
61            symbolLayers: [new ExtrudeSymbol3DLayer()]
62          }),
63          visualVariables: [{
64            type: "size",
65            field: "POP2007",
66            minSize: 40000,
67            maxSize: 10000000,
68            minDataValue: 11992,
69            maxDataValue: 1321851888
70          }, {
```

Note: Refer to the **Questions and Answers** section for details about how to find the minimum and maximum field values.

6. **Click Run, and notice that the countries extrude based on their population.**

Next, you will change the layer symbology to shade the countries with colors according to their area values.

7. **In the source code, under `color`, do the following:**
    - **Change `field` to `SQKM`.**
    - **Change the value of `minDataValue` to `0.25` and the value of `maxDataValue` to `16897294`.**

```
}, {
  type: "color",
  field: "SQKM",
  minDataValue: 0.25,
  maxDataValue: 16897294,
  colors: [
    new Color("white"),
    new Color("red")
  ]
}]
```

8. **Click Run, zoom out the map, and notice that different colors shade the countries according to their size by area.**

Next, you will change the camera settings of the view so that you can see most of the countries in the world at the initial view.

9. **In the source code, find the camera settings, change the `position` value to `[-170, -89.96, 30000000]`, the `tilt` value to `60`, and the `heading` to `15`, and click Run.**

These values set the camera at longitude –170, latitude –89.96, elevation 30,000,000 meters, looking at 15 degrees east away north, and looking down 60 degrees.

In this section, you adapted a sample by changing the layer's URL, attribute names, and values. You also changed the camera settings to get a comprehensive initial view. If you want to save the code you developed, you can click **Download**.

10.3 Debug JavaScript and monitor HTTP traffic

You can easily work with the sandbox source editor in the beginning, but the editor has some limitations. If you refresh the page or leave the page, you lose all your work. In addition, the editor is not easy or convenient when you must debug your code.

In this section, you will edit the JavaScript file locally using an editor, create an intentional error, and learn how to debug JavaScript. Learning this skill can save you a lot of frustration and greatly increase your productivity when you develop apps.

Note: Typically, you should deploy your JavaScript code to a web server first, load your web apps in a web browser via http or https protocol, and then debug your code. To keep the tutorial shorter and easier, you will debug local files directly without deploying your JavaScript code. See the **Questions and Answers** section for related information.

1. **Start Chrome, navigate to C:\EsriPress\GTKWebGIS\Chapter10\3dView.html, and notice the scene view is the same as the scene view you developed in the previous section.**

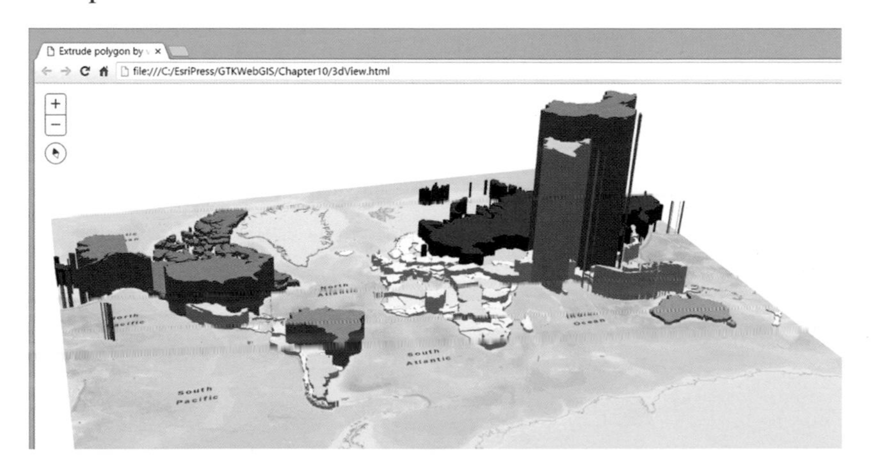

Next, you will introduce intentional errors and debug your code using the developer tools.

2. **Start Notepad++ or another JavaScript editor, and open C:\EsriPress\GTKWebGIS\Chapter10\3dView.html.**

You may recognize that the code is the same code that you developed in the previous section.

3. **In the source code, find** `new SceneView` **and change it to** `new sceneView` **(change the S to lowercase). Save your program.**

4. **In Chrome, reload C:\EsriPress\GTKWebGIS\Chapter10\3dView.html. Notice that nothing is displayed.**

Next, you will see that Chrome can tell you where the code is wrong.

5. **With the Chrome window as your active window, press the F12 key to open the Developer Tools interface.**

Alternatively, you can open the **Developer Tools** by clicking the **Options** button and clicking **More tools** > **Developer tools.**

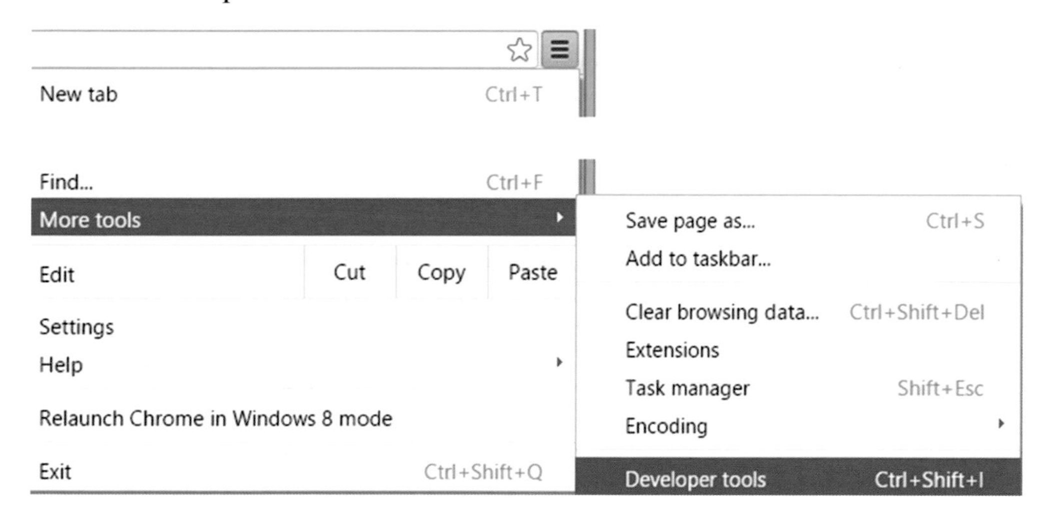

6. **In the Developer Tools window, click Console in the top menu banner. Notice the error message, along with the line that contains the error. If you do not see the error message, reload the page in Chrome.**

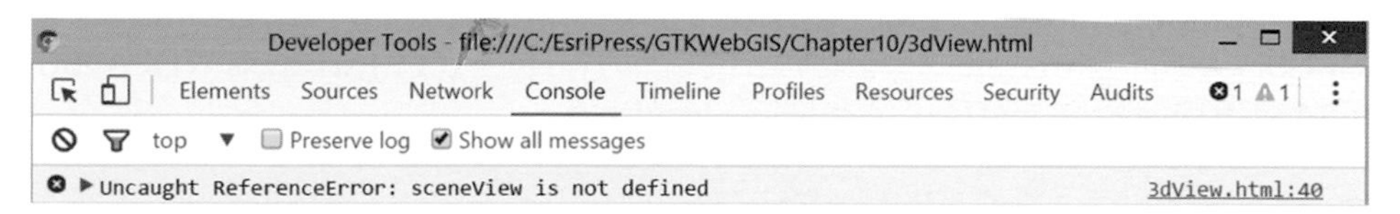

7. **Click the file name and the line number of the error.**

The source code appears with the error line annotated. This annotation shows you the location of the errors so that you can more easily find the problem.

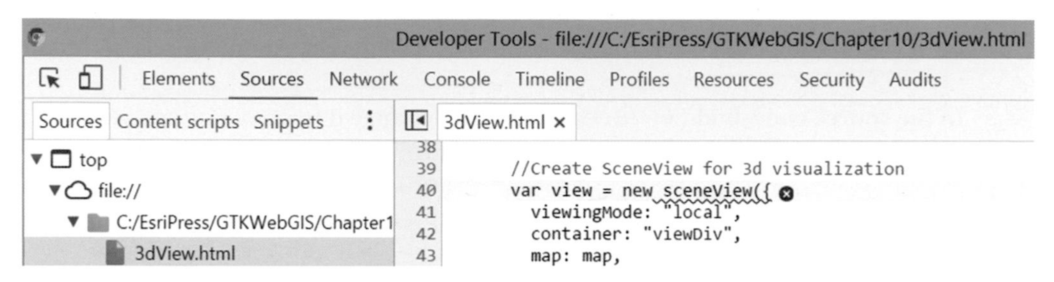

8. **In Notepad++, locate the line with the error, as indicated by the browser console. Change new sceneView back to new SceneView (change the s back to uppercase S), and save your program.**

9. **Reload your page in your Chrome browser. Notice that the scene appears in the browser, and no error messages appear in the console.**

In this section, you read the error messages in the console window and quickly corrected the error.

In addition to providing the console tool, web browser **Developer Tools** provide far more JavaScript debugging capabilities, including setting up breakpoints, stepping through your source code, examining variable values, inspecting HTML elements, changing code and styles on the fly, and tracking the HTTP tracks between clients and servers. Learning these tools can help you quickly discover the cause of issues and fix them efficiently.

10.4 Combine a 2D view and a 3D view

In this section, you will combine a 2D map view and a 3D scene view in one app. You will use two samples, **Add FeatureLayer to your Map** and **Load a basic WebScene.** The two files are available at the ArcGIS for JavaScript website. You also can find the two files at **C:\EsriPress\GTKWebGIS\Chapter10\2d3d_Views.html** and **load_webScene.html.**

Note: This section involves a lot of typing. You can avoid typing by copying the source code you need from **C:\EsriPress\GTKWebGIS\Chapter10\2d3d_Views_key.html**. If you run into errors in this section, refer to this file, or debug your code using the debugging skill you learned previously.

1. **Start Chrome, and open C:\EsriPress\GTKWebGIS\Chapter10\2d3d_Views.html. Notice the app displays only a 2D map view with an operational layer.**

You will replace the operation layer with an earthquakes layer.

2. **Open C:\EsriPress\GTKWebGIS\Chapter10\2d3d_Views.html with Notepad++ or another JavaScript editor.**

3. **In the source code, replace the feature layer URL with** `http://services2.arcgis.com/No7KRrFgpO516cMP/arcgis/rest/services/Earthquakes_2016_1_18/FeatureServer/0`, **which is the URL of a hosted earthquake layer.**

You can copy this long URL from **2d3d_Views_key.html.**

4. **In the source code, remove the line** `// Carbon storage of trees in Warren Wilson College`.

The line is a comment. The comment is no longer valid and no longer needed.
The current view has an extent specified. You will remove the extent so that the view will use the default extent, which is the whole world.

5. **In the source code, remove the extent setting of the** `view`. **Also remove the comma in the line preceding the extent line.**

```
var view = new MapView({
  container: "viewDiv",
  map: map,

  extent: { // autocasts as new Extent()
    xmin: -9177811,
    ymin: 4247000,
    xmax: -9176791,
    ymax: 4247784,
    spatialReference: 102100
  }
});
```

6. **Save the source code in Notepad++, and reload the file C:\EsriPress\GTKWebGIS\Chapter10\2d3d_Views.html in your Chrome web browser.**

You will see the page display a 2D world map and an earthquake layer.

When you combine the source code of multiple samples, the variable names may conflict with each other, and the source code can grow too long and become hard to read. To avoid such problems, you often need to move some code of the sample into a function. Moving the code will change the variables of the samples to a local scope and the code of the samples into encapsulated chunks; thus, you can avoid variable conflicts and make your code easier to read and manage.

7. **In the source code, add the following highlighted lines:**

```
    FeatureLayer, Extent
) {
    function create_2dView() {

    }
```

Adding these new lines creates an empty JavaScript function named **create_2dView**.

8. **Move the source code from the `var map` line to the `map.add` line into the brackets of the function `create_2dView`. The function name and brackets are highlighted so that you can more easily see them.**

```
    FeatureLayer
  ) {
      function create_2dView() {
          var map = new Map({
            basemap: "hybrid"
          });

          var view = new MapView({
            container: "viewDiv",
            map: map
          });

          /*******************
           * Add feature layer
           *******************/

          var featureLayer = new FeatureLayer({
            url:
"https://services2.arcgis.com/No7KRrFgpO516cMP/arcgis/rest/services/
 Earthquakes_2016_1_18/FeatureServer/0"

          });



          map.add(featureLayer);
      }
        });
```

9. **Add the highlighted line,** `create_2dView();` **to call the function you just created.**

```
    FeatureLayer
  ) {
        create_2dView();
```

10. **In Notepad++, save your file.**

11. **In your Chrome, load C:\EsriPress\GTKWebGIS\Chapter10\2d3d_Views.html. Notice that the app still works fine without any changes.**

Next, you will add the tectonic boundaries layer.

12. **In Notepad++, copy the lines from** `var featureLayer = new` **... to** `map.add(featureLayer);`**, and paste them right underneath the location where they already are. In the new lines (highlighted), do the following:**

    - **Change** `var featureLayer` **to var** `featureLayer1` **so that the new variable name does not conflict with any existing variable names.**
    - **Change the** `url` **value to the URL of the tectonic layer, which is** `http://services2.arcgis.com/No7KRrFgpO516cMP/arcgis/rest/services/Tectonic_Plates/FeatureServer/0`**.**

If the URL is too long to type, you can copy the URL from **2d3d_Views_key.html.**

- **Change the** `map.add(featureLayer)` **to** `map.add(featureLayer1)`**.**

```
map.add(featureLayer);
var featureLayer1 = new FeatureLayer({
url:
"https://services2.arcgis.com/No7KRrFgpO516cMP/arcgis/rest/services/
Tectonic_Plates/FeatureServer/0"
});
map.add(featureLayer1);
```

13. **Save your file in Notepad++. In Chrome, go to C:\EsriPress\GTKWebGIS\Chapter10\2d3d_Views.html. Notice that the layer displays both the earthquakes layer and the tectonic layer.**

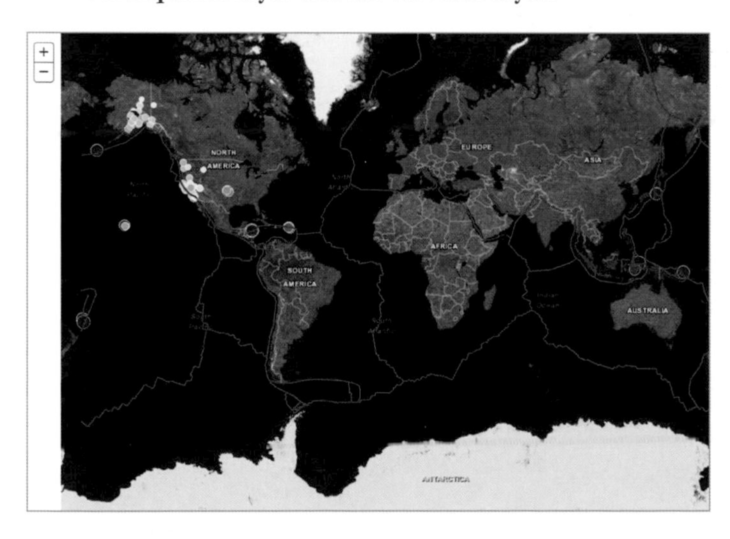

Next, you will combine your code with another sample.

14. **Open C:\EsriPress\GTKWebGIS\Chapter10\load_webScene.html with Notepad++.**

When you combine two JavaScript samples, you do not need to duplicate some lines, such as the basic HTML tags, the BODY tags, and the reference to ArcGIS API for JavaScript. However, you must combine the required modules from both samples.

The `"esri/views/SceneView"`, `"esri/portal/PortalItem"`, and `"esri/WebScene"` packages in load_webScene.html are not included in the **2d3d_Views.html**. You will add these modules in the next step.

15. **In Notepad++ in 2d3d_Views.html, add `"esri/views/SceneView"`, `"esri/portal/PortalItem"`, and `"esri/WebScene"`, to the required modules, and add `SceneView`, `PortalItem`, and `WebScene` in the argument list to refer to these packages.**

The order of the arguments must match the order of the modules, and you will keep the commas. The code you will add in this step is highlighted.

```
require([
  "esri/Map",
  "esri/views/MapView",
  "esri/layers/FeatureLayer",
  "esri/views/SceneView",
  "esri/portal/PortalItem",
  "esri/WebScene",
  "dojo/domReady!"
],
function(
  Map, MapView,
  FeatureLayer,
  SceneView, PortalItem, WebScene
) {
```

16. **In Notepad++ in 2d3d_Views.html, create an empty new function named `create_3dView` underneath the end of the `create_2dView` function.**

```
  map.add(featureLayer);
}
```

```
function create_3dView() {

}
```

17. **Copy the highlighted lines from load_webScene.html, and paste them to the** `create_3dView` **function in 2d3d_Views.html.**

You do not need to copy the comment lines, which start with /* and end with */

```
function create_3dView() {
    var scene = new WebScene({
        portalItem: new PortalItem({
           id: "affa021c51944b5694132b2d61fe1057"
        })
    });
    varview = new SceneView({
        map: scene,
        container: "viewDiv"
    });
}
```

18. **In Notepad++ in 2d3d_Views.html, change the value of the scene** `id` **to** `ae2631226f9b4883942a1d2423e29772`**, which is similar to the web scene you created in the 3D scene chapter.**

```
portalItem: new PortalItem({
  id: "ae2631226f9b4883942a1d2423e29772"
})
```

You can copy the value of the id from **2d3d_Views_key.html** to **2d3d_Views.html.**

19. **In Notepad++ in 2d3d_Views.html, add the line** `create_3dView();` **to call the** `create_3dView` **function.**

```
create_2dView();
create_3dView();
```

You can refer to the **2d3d_Views_key.html** file if you want to confirm that your code is correct. Notice that the map view and the scene view both use the `viewDiv` as the container, which is a conflict. You will change the identification (id) of the containers so they are different.

20. **In Notepad++ in 2d3d_Views.html, change the** `container` **for the map view to** `viewDiv_2d`**.**

```
var view = new MapView({
  container: "viewDiv_2d",
```

21. **In Notepad++ in 2d3d_Views.html, change the** `container` **for the scene view to** `viewDiv_3d`**.**

```
var view = new SceneView({
    container: "viewDiv_3d",
```

Next, you will create the containers in the HTML body.

22. **In Notepad++ in 2d3d_Views.html, change the** `BODY` **section to the following highlighted lines:**

```
<body>
<div id="viewDiv_2d"></div>
<div id="viewDiv_3d"></div>
</body>
```

These two lines create two containers with the identifications (ids) `viewDiv_2d` and `viewDiv_3d`, respectively.

Next, you will define the positions of the two containers using CSS.

23. **In Notepad++ in 2d3d_Views.html, scroll up to the Styles section, and add the highlighted lines in this step.**

These lines will position `viewDiv_2d` to occupy the left half of the page and `viewDiv_3d` to occupy the right half of the page. You can leave the styles for `#viewDiv` in the source code or delete them.

```
<style>
  html,
  body {
    padding: 0;
    margin: 0;
    height: 100%;
    width: 100%;
  }
```

```
#viewDiv_2d{
  float:left;
  height:100%;
  width: 50%;
}

#viewDiv_3d{
  float:right;
  height:100%;
  width: 50%;
}
</style>
```

24. Scroll to the top, find the page <title> **line, change the value of** title **to** Integrate 2D and 3D Views.

<title>**Integrate 2D and 3D Views**</title>

25. In Notepad++, click the Save button.

26. In Chrome, load C:\EsriPress\GTKWebGIS\Chapter10\2d3d_Views.html.

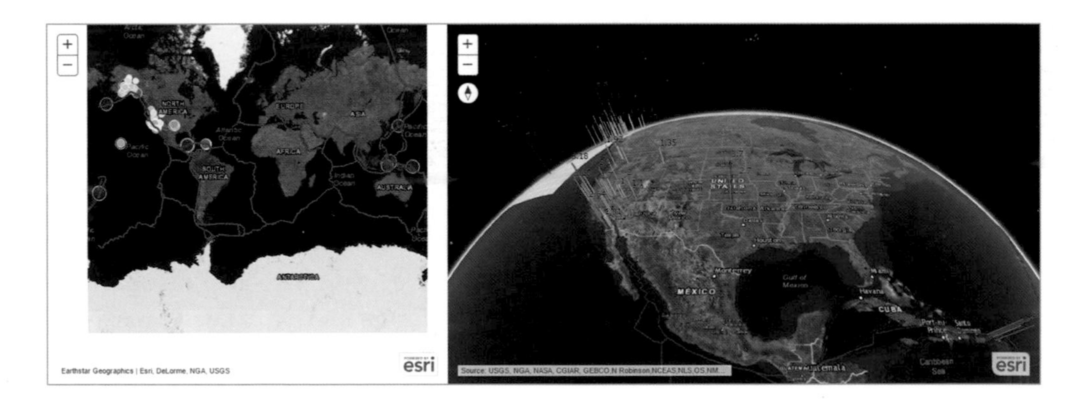

In this section, you combined two samples to create an app. This app allows you to visualize the same data on the same page in a 2D map view and a 3D scene view.

10.5 Handle JavaScript events

In the previous section, you developed two independent views. In this section, you will link the views by making the 3D scene view follow the 2D map scene.

Note: This section involves a lot of typing. You can avoid typing by copying source code from **C:\EsriPress\GTKWebGIS\Chapter10\2d3d_Views_linked_key.html**. If you encounter errors in this section, refer to this file as well, or debug your program using the debugging skill you learned earlier.

1. **Start Notepad++, open the file C:\EsriPress\GTKWebGIS\Chapter10\2d3d_Views linked.html, which is the same as the file you created at the end of the previous section.**

2. **Add the following highlighted lines and highlighted comma. These changes set the** `center` **of the map view to longitude** `-100` **and latitude** `40` **and the** `zoom` **level to** `4`**.**

```
var view = new MapView({
  container: "viewDiv_2d",
  map: map,
  zoom: 4,
  center: [-100, 40]
});
```

3. **In Notepad++, click the Save button. In Chrome, load C:\EsriPress\GTKWebGIS\Chapter10\2d3d_Views_linked.html. Notice that the initial view of the map centers on the United States.**

To link the 2D view and the 3D view, you will access 3D view in the `create_2dView` function. Currently, the 3D scene view is a local variable, and you cannot access this variable from the `create_2dView` function. You will change the 3D view to a global variable in the following steps.

4. **Add the following highlighted line above the** `create_2dView()` **line:**

   var view_2d, view_3d;

This line declares one global variable for the 2D map view and one global variable for the 3D scene view.

1 2 3 4 5 6 7 8 9 10

5. **Change**

```
var view = new MapView({
to
view_2d = new MapView({
```

By removing var at the beginning of the line and changing the variable name from `view` to `view_2d`, you will initialize the global variable `view_2d` instead of creating a local variable.

6. **Similarly, change**

```
var view = new SceneView({
to
view_3d = new SceneView({
```

7. **Add the following highlighted lines to the `create_2dView` function:**

```
map: map,
zoom: 4,
center: [-100, 40]
});
view_2d.then(function(){
      view_2d.watch("extent", function(response){
            if (response){
                        view_3d.center = response.center;
            }
      });
});
```

The code `view_2d.then` means that when the map of the 2D view has loaded, the event will trigger the function that includes. `view_2d.watch("extent", function(response)`. The code `view_2d.watch` monitors the extent's change of the 2D view. When you change the extent, the code triggers the second function. The new extent is stored in the `response` argument. The code `view_3d.center = response.center;` means that the code uses the center of the 2D view to update the center of the 3D view.

8. **Click the Save button in Notepad++.**

9. **In Chrome, refresh the page C:\EsriPress\GTKWebGIS\Chapter10\2d3d_views_linked.html.**

10. **Pan the 2D map view, and notice that the 3D scene view follows.**

Next, you will program the 3D view to follow the scale change of the 2D view.

11. **Add the following highlighted lines:**

```
view_2d.then(function() {
    view_2d.watch("extent", function(response){
        if (response){
                view_3d.center = response.center;
        }
    });
    view_2d.watch("scale", function(response){
        if (response){
                view_3d.scale = response;
        }
    });
});
```

The new lines monitor the change of the 2D view's scale. When the scale changes, the code sets the scale of the 2D view to the 3D view.

12. **Save the file in Notepad++.**

13. **In Chrome, refresh the page C:\EsriPress\GTKWebGIS\Chapter10\2d3d_views.html.**

14. **In the 2D view, zoom in and out, and observe that the 3D view follows.**

Notice that 3D views do not have a consistent scale across the view area. The actual scale and extent of the 3D view vary with the camera position and tilt angle. You can change the camera position and tilt angle of the 3D view manually (using right-click and drag).

Next, you will change the code so that when you rotate the 2D view, the 3D view will follow.

15. **In Notepad++ in 2d3d_Views_linked.html, add the following highlighted lines:**

```
view_2d.watch("scale", function(response){
        if (response){
                view_3d.scale = response;
        }
});
```

```
        view_2d.watch("rotation", function(response){
            if (response){
                view_3d.camera.heading = 0 - response;
                view_3d.animateTo({
                        target: view_3d.camera
                });
            }
        });
    });
```

The new lines watch, or monitor, the change of the 2D view's rotation. If the rotation changes, the code you just added sets the camera of the 3D view to look in the same direction (heading) as the 2D view. The direction of the 3D view is a property of the view's camera. The rotation of the 2D view and the camera of the 3D view measure in opposite directions. The code `0 - response` corrects the values so that the 2D view and the 3D view rotate in the same direction.

16. **Save the file in Notepad++.**

17. **In Chrome, refresh the page C:\EsriPress\GTKWebGIS\Chapter10\2d3d_views.html.**

18. **In the 2D view, right-click and drag to rotate the map, and notice that the 3D view follows.**

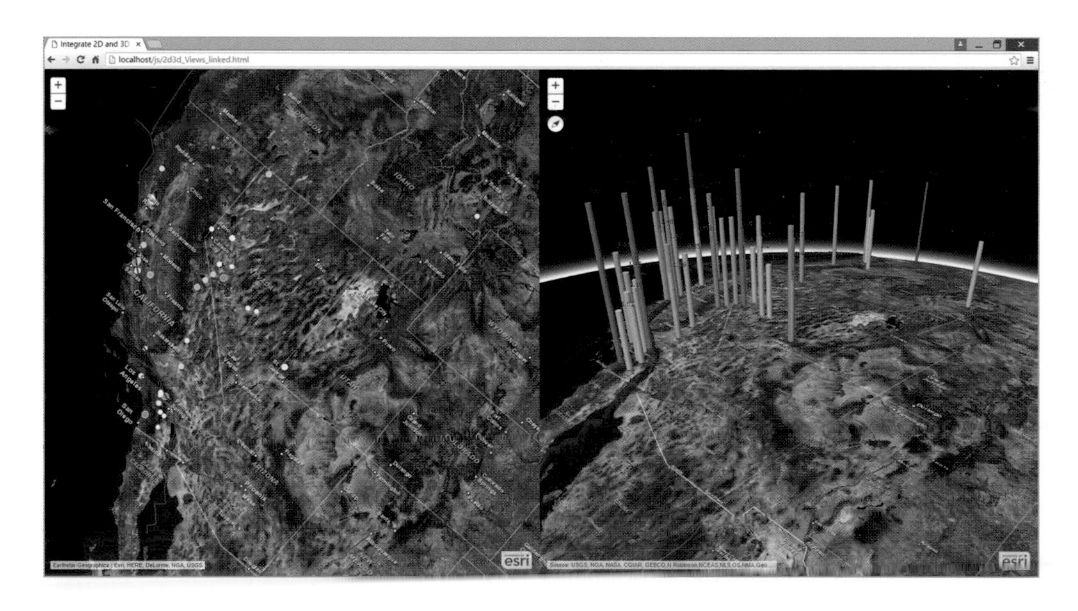

In this section, you linked the 2D view and the 3D view by monitoring the 2D view's extent, scale, and rotation properties. The code automatically updated the center, scale, and camera heading properties of the 3D view.

In this tutorial, you learned the general steps needed to create a JavaScript app using ArcGIS API for JavaScript, as well as the steps to adapt and combine samples. You learned how to check JavaScript error messages, how to add layers to your app, and how to handle events. You created simple apps, and once you learn additional programming skills, you can enhance the apps further. These apps demonstrated the flexibility and fun of ArcGIS API for JavaScript.

QUESTIONS AND ANSWERS

1. **When I specified extrusion symbols in the tutorial, I needed to get the minimum and maximum values of an attribute field. How can I get those values?**

   Answer: You can get those values using the ArcGIS REST Services Directory. In a web browser, go to the URL of a layer, click **Query** at the bottom of the page, or directly go to the layer URL appended with **/query**, for example:

   https://arcgis.storymaps.esri.com/arcgis/rest/services/SevenBillion/PopulationGrowth_1960_2010/MapServer/0/query. Assuming you need to get the minimum and maximum of the POP2007 field, on the query page, perform the following actions:

   - For **Where**, specify **1=1** (or other values if you want to filter your data).
   - For **Out Fields**, specify **POP2007**.
   - For **Return Geometry**, select **False**.
   - For **Output Statistics**, specify the following:

```
[{
  "statisticType": "min",
  "onStatisticField": "POP2007",
  "outStatisticFieldName": "Min"
},{
```

```
  "statisticType": "max",
  "onStatisticField": "POP2007",
  "outStatisticFieldName": "Max"
}]
```

- Click **Query (GET)** or **Query (POST)** to run the query.
- See the minimum and maximum values in the return record.

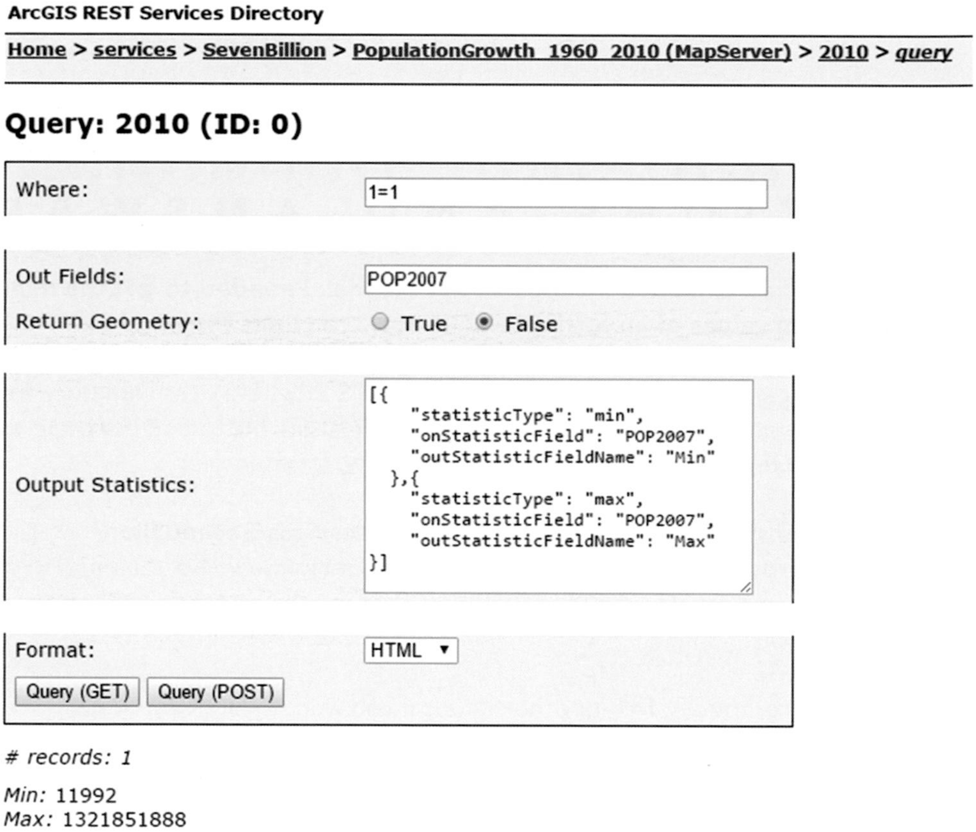

ArcGIS REST Services Directory

Home > services > SevenBillion > PopulationGrowth_1960_2010 (MapServer) > 2010 > *query*

Query: 2010 (ID: 0)

Where: 1=1

Out Fields: POP2007

Return Geometry: True False

Output Statistics:

```
[{
    "statisticType": "min",
    "onStatisticField": "POP2007",
    "outStatisticFieldName": "Min"
  },{
    "statisticType": "max",
    "onStatisticField": "POP2007",
    "outStatisticFieldName": "Max"
}]
```

Format: HTML

Query (GET) Query (POST)

 # records: 1

Min: 11992
Max: 1321851888

2. **I changed my JavaScript code and refreshed the page in my web browser. However, I cannot see the latest changes in the browser. What is wrong? How do I fix it?**

   Answer: Browser caching is causing the problem.

   Browsers often cache a version of the web pages you visited and use this cached version to speed up performance. However, this cache can cause browsers to use an outdated version of an updated page.

   If you have this problem, search online for proper browser settings.

3. **In my web browser, instead of using http://... URL to load my page, I use the following URL: file:///C:/EsriPress/GTKWebGIS/Chapter10/2d3d_Views_key.html. Is this okay?**

   Answer: Yes, you can view and develop simple JavaScript apps using the **file** URL.

   You should deploy official web apps and complex web apps to a web server and access these apps via **http** URLs.

4. **I shared my app URL (http://localhost/WebGIS/app.html) with my users. However, my users said they could not see my app. Why?**

   Answer: **Localhost** is relative and points to the computer that the users are currently using. Therefore, whenever you go to this URL on your computer, this URL points to your computer. If users go to this URL on their computers, then the URL points to their computers, which do not have your app deployed. So the users will not have access to your app.

   Although **localhost** is convenient for running your app on your own computer, when you share your app URL with your audience, use the correct host name instead of **localhost**.

5. **Microsoft IIS (Internet Information Services) is a commonly used web server. How do I know whether I have IIS installed on my Windows computer? How do I install IIS?**

   Answer: To verify whether your Windows computer has IIS installed, simply start a web browser and type **http://localhost or http://your_computer_name**. Typically, a web page appears, which means that IIS has been installed. Otherwise, search online for help documents that explain how to install IIS.

6. **How do I deploy a JavaScript web app to IIS?**

   Answer: You simply copy your web app files to a folder under **c:\inetpub\wwwroot** and give everyone read access to the folder. Search for online help documents that explain the details.

7. **A layer in my map service has more than 1,000 hurricanes. In my JavaScript, I added this layer as a feature layer to my app. However, when I view this web app, not all the hurricanes display. How can I fix this problem?**

   Answer: As discussed in previous chapters, browsers draw the feature layers. Web browsers cannot draw an unlimited number of features. To avoid overloading web browsers, ArcGIS for Server sets a default limit, which returns up to 1,000 features per request. This limit explains why you do not see all the hurricanes.

   You can fix this problem in several ways:

   - Use the dynamic map service layer approach. In this approach, the server draws the map so that it is not subject to the limit.

   - Increase the feature return limit, for example, to 2,000 or more. But increasing the limit may overload the web browsers of your end users.

   - Use the on-demand mode of the feature layer, and couple the layer with the correct scale dependency. When you take these steps, ArcGIS for Server will return only those hurricanes within the current map extent. To ensure that there are no more than 1,000 hurricanes in any map view, set the correct scale dependency.

ASSIGNMENTS

Assignment 10: Adapt a JavaScript sample or combine multiple JavaScript samples to develop a web app using ArcGIS API for JavaScript.

Data (you have several options):

- Use the hosted layers or map services you published before.
- Find layers at ArcGIS Online or ArcGIS Open Data (**http://opendata.arcgis.com**).
- Find layers at **https://sampleserver6.arcgisonline.com/arcgis/rest/services**.
- Use any other data (but different from the data layers in the original samples).

What to submit: Send an email to your instructor with the subject line **Web** GIS Assignment 10: **Your name**, and include the following files:

- The source code in a ZIP file
- Screen captures showing that your web app works

1 2 3 4 5 6 7 8 9 10

Resources

Note: This chapter was written just as ArcGIS API for JavaScript 4.0 was released. As a result, few public resources existed for this chapter at the time. You can search for ArcGIS API for JavaScript at the Esri training website and Esri video website as more resources become available.

ArcGIS API for JavaScript website

"ArcGIS API for JavaScript," https://developers.arcgis.com/javascript.

"GeoNet Forum for ArcGIS API for JavaScript," https://geonet.esri.com/community/developers/web-developers/arcgis-api-for-javascript.

Esri Training website

"Training Catalog," http://www.esri.com/training/main/training-catalog (search for ArcGIS API for JavaScript).

Esri videos

Esri videos, http://video.esri.com (search for ArcGIS API for JavaScript).

Appendix A

Image credits

Chapter 1

Images courtesy of Esri; Esri; Esri; Esri; Esri; Esri; Esri; Esri; Left side: ArcGIS Online National Geographic basemap (DC GIS, DDOT, DRES, GSA, OCTO, NHD, VITA, METI/NASA, Esri, DeLorme, HERE, USGS, USDA, EPA, EarthSat, Intermap, IPC, TomTom); Photos by Allen Carroll, Esri; Right side: City of Redlands, City of Riverside, County of Riverside, Esri, HERE, DeLorme, Intermap, iPC, USGS, USDA, EPA. Photo by Pinde; Esri; Esri; Esri; Esri; City of Redlands, City of Riverside, County of Riverside, Esri, HERE, DeLorme, Intermap, iPC, USGS, USDA, EPA; City of Redlands, City of Riverside, County of Riverside, Esri, HERE, DeLorme, Intermap, iPC, USGS, USDA, EPA; Esri; Esri; Esri; Esri; Esri; Esri; Esri; City of Redlands, City of Riverside, County of Riverside, Esri, HERE, DeLorme, Intermap, iPC, USGS, USDA, EPA; Esri; Esri; Esri; Esri; Esri; Esri; City of Redlands, City of Riverside, County of Riverside, Esri, HERE, DeLorme, Intermap, iPC, USGS, USDA, EPA; City of Redlands, City of Riverside, County of Riverside, Esri, HERE, DeLorme, Intermap, iPC, USGS, USDA, EPA; Esri; Esri; Esri; Esri; Esri; Esri; City of Redlands, City of Riverside, County of Riverside, Esri, HERE, DeLorme, iPC, NGA, USGS | Esri, HERE.

Chapter 2

Images courtesy of Esri; elevation service: USGS; Esri; Esri; Esri; Esri; Esri; Esri; Esri; Esri, HERE, DeLorme, USGS, NGA, USDA, EPA, NPS; Esri; Esri, HERE, DeLorme, USGS, NGA, USDA, EPA, NPS; Esri, HERE, DeLorme, FAO, USGS, NGA, EPA, NPS; Esri; Esri; Census, Esri, DeLorme, FAO, NOAA, EP; Esri; US Census, Esri, DeLorme, FAO, NOAA, EP; US Census, Esri, DeLorme, FAO, NOAA, EP; Esri; Esri; Esri; Esri; US Census, Esri, DeLorme, NGA, USGS | Esri, HERE, DeLorme; Esri; Esri; Esri; Esri; Esri; Esri; US Census; Esri; Esri; US Census; Wikipedia; US Census; US Census; US Census; Esri; Esri; Esri; Esri; Esri; Esri; Esri; Esri; Esri, HERE, DeLorme, NGA, USGS | Esri, US Census Bureau | Esri, HERE, DeLorme; Esri; Census, Esri, DeLorme, NGA, USGS | Esri, HERE, DeLorme; NOAA National Climatic Data Center, Esri, DeLorme, NGA, USGS | Esri, HERE, DeLorme.

Chapter 3

Images courtesy of Esri; Esri; ArcGIS Online Light Grey basemap (Esri, HERE, DeLorme, MapmyIndia, © OpenStreetMap contributors, and the GIS user community); Esri; Esri; Esri; Esri; Esri; Esri, DeLorme, NAVTEQ, TomTom, Intermap, AND, USGS, NRCAN, Kadaster NL, and the GIS User Community; Esri; Esri; Esri; Esri; Esri; Esri; Esri; Esri; Esri; Esri; Esri; Esri; Esri; Esri; Esri, DeLorme, NAVTEQ, TomTom, Intermap, AND, USGS, NRCAN, Kadaster NL, and the GIS User Community; Esri; Esri; Esri; Esri; Esri; Esri; Esri; Esri; Esri, DeLorme, NAVTEQ, TomTom, Intermap, AND, USGS, NRCAN, Kadaster NL, and the GIS User Community; ArcGIS Online topo basemap (County of Los Angeles, Esri, HERE, DeLorme, Intermap, TomTom, USGS, USDA, EPA); Esri.

Chapter 4

Images courtesy of Esri; Department of Education National Center for Education Statistics, basemap: Esri, HERE, DeLorme, NGA, USGS | Esri, HERE, DeLorme; US Census Bureau; Esri; Esri, DeLorme, FAO, NOAA, USGS, EPA | Esri, US Census Bureau; Esri; City of Lawrence, Esri, HERE, DeLorme, INCREMENT P, Intermap, USGS, METI/NASA, NGA, EPA, USDA | Esri, US Census Bureau; Esri; Esri; Esri; Esri; US.Census Bureau; Esri; Esri, DeLorme, NGA, USGS | Esri, US Census Bureau | Esri, HERE, DeLorme; Esri; Esri, DeLorme, NGA, USGS | Esri, US Census Bureau | Esri, HERE, DeLorme; Esri; Esri, DeLorme, FAO, NOAA, USGS; Esri; Department of Education National Center for Education Statistics, basemap: Esri, HERE, DeLorme, NGA, USGS | Esri, HERE, DeLorme; Esri; Esri; Esri.

Chapter 5

Images courtesy of Esri; Esri; Esri; Esri; Esri; Esri; Esri; Esri; Esri; Esri; Esri, HERE, DeLorme, USGS, NGA, USDA, EPA, NPS; Esri; Esri, DeLorme, FAO, USGS, NOAA, EPA; Esri; Esri; Esri, DeLorme, NAVTEQ, TomTom, Intermap, AND, USGS, NRCAN, Kadaster NL, and the GIS User Community; Esri; Esri; Esri; Esri; Esri, HERE, DeLorme, USGS, METI/NASA, NGA, USDA, EPA; Esri; Esri; Esri; Esri.

Chapter 6

Images courtesy of Esri; Esri; Esri; Esri; Esri; ©2013 Esri, DeLorme, HERE, iPC, USGS, METI/NASA, EPA, USDA; Esri; Esri; Esri; Esri; Esri; Esri; Esri; Esri; Esri; Esri; Esri; Esri; Esri; By Esri. Data from USGS, National Atlas, NOAA National Climatic Data Center; Esri.

Chapter 7

Images courtesy of Esri; Esri; Esri; Esri; Esri; Esri; USGS; Esri; Esri; Esri; Esri; Esri; Esri; Esri; Esri; Esri; City of Long Beach, Bureau of Land Management, Esri, HERE, DeLorme, NGA, USGS; Esri; Esri; Esri; Esri; Esri; Esri; Esri; Esri; Bureau of Land Management, Esri, HERE, DeLorme, NGA, USGS; Esri; Railroads: Courtesy of US Bureau Transportation Statistics; Rivers: Courtesy of ArcWorld; Esri; Esri; Esri; Esri; Esri; Esri; Esri; Esri; Esri; Esri; Esri; Esri; Esri; Esri; Railroads: Courtesy of US Bureau Transportation Statistics; Rivers: Courtesy of ArcWorld; Basemap: Esri, HERE, DeLorme, MapmyIndia, © OpenStreetMap contributors, and the GIS user community; Railroads: Courtesy of US Bureau Transportation Statistics; Rivers: Courtesy of ArcWorld; Basemap: Esri, HERE, DeLorme, MapmyIndia, © OpenStreetMap contributors, and the GIS user community; Esri.

Chapter 8

Images courtesy of Esri; Esri; Esri; Esri; Esri; Esri; Courtesy of Mozilla Foundation, Instagram, and Twitter, Inc; Esri; Esri; Basemap data courtesy of Basemap; data courtesy of City of Riverside, County of Riverside, Bureau of Land Management, Esri, HERE, DeLorme, INCREMENT P, NGA, USGS; Esri; Esri; Esri; Esri; Esri; Esri; Esri; Basemap data courtesy of City of Redlands, City of Riverside, County of Riverside, Cal-Atlas, Esri, DeLorme, HERE, Intermap, iPC, TomTom, USGS, USDA, EPA; Esri; Esri; Basemap data courtesy of City of Redlands, City of Riverside, County of Riverside, Cal-Atlas, Esri, DeLorme, HERE, Intermap, iPC, TomTom, USGS, USDA, EPA; Esri; Basemap data courtesy of City of Redlands, City of Riverside, County of Riverside, Cal-Atlas, Esri, DeLorme, HERE, Intermap, iPC, TomTom, USGS, USDA, EPA; Esri; Esri; Basemap data courtesy of Esri, FAO, NOAA; Esri; Esri; Esri; Esri; Esri; Esri; Esri; Esri; Esri; Esri; The Android robot is reproduced or modified from work created and shared by Google and used according to terms described in the Creative Commons 3.0 Attribution License; Esri; Esri; Esri.

Chapter 9

Images courtesy of Esri; Esri; Esri, DeLorme, FAO, NOAA, USGS, EPA, and US Census; USDA FSA, DigitalGlobe, GeoEye, Microsoft, City of Montreal, Canada, Esri Canada, Esri; USDA FSA, DigitalGlobe, GeoEye, Microsoft, City of Montreal, Canada, Esri Canada, Esri; Esri; Esri; Esri; Esri; USDA FSA, DigitalGlobe, GeoEye, Microsoft; Esri; Esri; Esri; Earthstar Geographics, and NOAA Storm Prediction Center; Esri; Esri; Esri; Esri; Esri; Esri; Esri; Esri; Esri; Esri; Esri; Esri; Esri; Esri, DeLorme, FAO, NOAA, USGS, EPA, NPS; Esri; Esri; Esri; Esri; Esri; Esri; Esri; Esri; Esri; Esri; Esri; Esri, DeLorme, FAO, NOAA, USGS, EPA, NPS; Esri; Esri, DeLorme, FAO, NOAA, USGS, EPA, NPS, US Census; Esri; Esri; Esri, DeLorme, FAO, NOAA, USGS, EPA, NPS, US Census.

Chapter 10

Images courtesy of Esri; Esri; Esri; Esri; USGS, NGA, NASA, CGIAR, GEBCO, N Robinson, NCEAS, NLS, OS, NMA, Geodatastyrelsen and the GIS User Community | Esri, HERE, DeLorme, NGA, USGS; Esri; Census, USGS, NGA, NASA, CGIAR, GEBCO, N Robinson, NCEAS, NLS, OS, NMA, Geodatastyrelsen and the GIS User Community | Esri, HERE, DeLorme, NGA, USGS; Census, USGS, NGA, NASA, CGIAR, GEBCO, N Robinson, NCEAS, NLS, OS, NMA, Geodatastyrelsen and the GIS User Community | Esri, HERE, DeLorme, NGA, USGS; Census, USGS, NGA, NASA, CGIAR, GEBCO,N Robinson, NCEAS, NLS, OS, NMA, Geodatastyrelsen and the GIS User Community | Esri, HERE, DeLorme, NGA, USGS; Esri; Esri; Esri; World Bank, Census, USGS, NGA, NASA, CGIAR, GEBCO,N Robinson, NCEAS, NLS, OS, NMA, Geodatastyrelsen and the GIS User Community | Esri, HERE, DeLorme, NGA, USGS; World Bank, Census, USGS, NGA, NASA, CGIAR, GEBCO, N Robinson, NCEAS, NLS, OS, NMA, Geodatastyrelsen and the GIS User Community | Esri, HERE, DeLorme, NGA, USGS; Esri; Esri; Esri; Esri; Earthstar Geographics | Esri, DeLorme, NGA, USGS; Earthstar Geographics | Esri, DeLorme, NGA, USGS; USGS, NGA, NASA, CGIAR, GEBCO, N Robinson, NCEAS, NLS, OS, NMA, Geodatastyrelsen and the GIS User Community | Earthstar Geographics | Esri, HERE, DeLorme, NGA, USGS; USGS, NGA, NASA, CGIAR, GEBCO, N Robinson, NCEAS, NLS, OS, NMA, Geodatastyrelsen and the GIS User Community | Earthstar Geographics | Esri, HERE, DeLorme, NGA, USGS; Esri.

Appendix B

Data credits

\\EsriPress\GTKWebGIS\Chapter1\Locations.csv, courtesy of Esri.

\\EsriPress\GTKWebGIS\Chapter1\map_tour_thumbnail.jpg, courtesy of Esri, DeLorme, HERE, USGS, iPC, METI/NASA.

\\EsriPress\GTKWebGIS\Chapter2\images\thumbnail.jpg, courtesy of Esri, DeLorme, HERE, Census.

\\EsriPress\GTKWebGIS\Chapter2\Top_50_US_Cities.csv, courtesy of US Census.

\\EsriPress\GTKWebGIS\Chapter2\New_Orleans.txt, courtesy of US Census.

\\EsriPress\GTKWebGIS\Data\Chapter3\311Incidents.csv, courtesy of Esri.

\\EsriPress\GTKWebGIS\Chapter3\Assignments_data\Incidents.zip, courtesy of Esri.

\\EsriPress\GTKWebGIS\Chapter3\Assignments_data\Incidents_gdb.zip, courtesy of Esri.

\\EsriPress\GTKWebGIS\Chapter6\Data.gdb\Earthquakes, courtesy of USGS, downloaded from National Atlas website.

\\EsriPress\GTKWebGIS\Chapter6\Data.gdb\Hurricanes, courtesy of NOAA National Climatic Data Center.

\\EsriPress\GTKWebGIS\Data\Chapter6\US_Cities.gdb\US_Cities, courtesy of US Census Bureau.

\\EsriPress\GTKWebGIS\Data\Chapter7\Planning.tbx, courtesy of Esri.

\\EsriPress\GTKWebGIS\Data\Chapter7\Planning.tbx\Select_Sites, courtesy of Esri.

\\EsriPress\GTKWebGIS\Data\Chapter7\Site_Selection.mxd, courtesy of Esri.

\\EsriPress\GTKWebGIS\Data\Chapter7\Assignment_Data\Scratch, courtesy of Esri.

\\EsriPress\GTKWebGIS\Data\Chapter7\Assignment_Data\natural_disasters.mxd, courtesy of Esri.

\\EsriPress\GTKWebGIS\Data\Chapter7\Assignment_Data\ExtractData.tbx, courtesy of Esri.

\\EsriPress\GTKWebGIS\Data\Chapter7\Assignment_Data\ExtractData.tbx\ExtractData, courtesy of Esri.

http://esripressbooks.maps.arcgis.com/home/item.html?id=9a447f0a6eaf4b48a12faaf5d575fa66, courtesy of Esri.

http://esripressbooks.maps.arcgis.com/home/item.html?id=fa13134327664764b406654ef91fd2e5, courtesy of Esri.

http://esripressbooks.maps.arcgis.com/home/item.html?id=6689c681877c4ec6a74d34e17f02544f, courtesy of Esri.

http://sampleserver5a.arcgisonline.com/arcgis/rest/services/LocalGovernment/Recreation/FeatureServer, courtesy of Esri.

http://esripressbooks.maps.arcgis.com/home/item.html?id=b3d20be9269e40679d39f9e9f523e66e, courtesy of Esri.

http://esripressbooks.maps.arcgis.com/home/item.html?id=65b878583009427283b1c406a0b19d9e, courtesy of Esri.

\\EsriPress\GTKWebGIS\Chapter9\Data\1.0_day.csv, courtesy of USGS.